CONCRETE SHEAR IN EARTHQUAKE

This volume consists of papers presented at the International Workshop on Concrete Shear in Earthquake, held at the University of Houston Texas, USA, 13-16 January 1991.

Sponsored by
National Science Foundation

CONCRETE SHEAR IN EARTHQUAKE

Edited by

T. C. C. HSU

and

S. T. MAU

Department of Civil and Environmental Engineering,
University of Houston, Texas, USA

CRC Press
Taylor & Francis Group
Boca Raton London New York

CRC Press is an imprint of the
Taylor & Francis Group, an **informa** business
A TAYLOR & FRANCIS BOOK

CRC Press
Taylor & Francis Group
6000 Broken Sound Parkway NW, Suite 300
Boca Raton, FL 33487-2742

First issued in paperback 2019

ISBN-13: 978-1-85166-729-1 (hbk)
ISBN-13: 978-0-367-86400-2 (pbk)

Visit the Taylor & Francis Web site at
http://www.taylorandfrancis.com

and the CRC Press Web site at
http://www.crcpress.com

British Library Cataloguing in Publication Data

Concrete shear in earthquake.
I. Hsu, T.C.C. II. Mau, S.T.
624.1762

Library of Congress CIP data applied for

PREFACE

Wall-type and shell-type reinforced concrete structures subjected to membrane shear have traditionally been utilized to dissipate seismic energy in earthquake regions in Japan, China, and other Southeast Asian countries. Such design practice has evoked considerable controversy over the years for two reasons. First, this practice is opposite to the American earthquake design philosophy in which the ductility of a structure is provided through flexural deformation, while suppressing shear failure. Second, the lack of a theory to predict the shear ductility of reinforced concrete structures makes it difficult to evaluate the utilizable ductility.

A considerable number of experiments involving membrane shear have been carried out during the last two decades around the world, particularly in Canada, Germany, Japan, New Zealand, Switzerland, and the United States. It was found that the ductility of a shear panel depends largely on the amount and arrangement of reinforcement. With proper design, it appears possible to utilize the shear ductility of wall-type and shell-type structures to resist earthquake.

In the past 20 years, major theoretical advances have also been made in predicting the behavior of reinforced concrete subjected to membrane shear. These theories are capable of predicting not only the shear strength, but also the shear deformations of the entire post-cracking history. In other words, it is now possible to predict the shear ductility and the shear hysteresis response of structures.

The recent advances in the experimental and theoretical studies of membrane shear in reinforced concrete have opened new possibilities in earthquake engineering. To explore these new opportunities, a workshop was held on January 14 – 16, 1991, at the Hilton Hotel and Conference Center on the campus of the University of Houston, Houston, Texas, USA. This workshop brought together a panel of international experts in concrete shear and in earthquake design for the following objectives: to evaluate the state-of-the-art of shear research in earthquake engineering, to examine the prospects and limits of utilizing the membrane shear ductility, and to identify the areas of future research and development needs.

All of the 46 papers presented in the workshop are included in this proceedings volume. These papers deal with various types of shearwalls, nuclear containment structures, high-rise buildings, box bridges, I-girders, columns, and shells. They represent the most up-to-date knowledge of concrete shear in earthquake. Included in this volume are also the related discussions and comments on the future research and development needs.

T. C. C. Hsu
S. T. Mau

ACKNOWLEDGMENT

The International Workshop on Concrete Shear in Earthquake, held at the University of Houston on January 14 – 16, 1991, was sponsored by National Science Foundation Grant No. BCS-8902176. Dr. S. C. Liu, program director, was most generous in his time and effort to help organize the workshop.

The workshop was strongly encouraged and supported by Dr. Roger Eichhorn, Dean of the College of Engineering, and by Dr. Michael O'Neill, Chairman of the Department of Civil and Environmental Engineering. Paula Davenport, departmental administrator, handled all the financial affairs. Their cooperation is acknowledged.

The success of the workshop could not have been achieved without the help of the workshop assistants: Abdeldjelil Belarbi, Xiao-Bo Pang, Li-Xin Zhang, Wei-Su Yin, and Jason Godfrey. Special thanks are due Suzie Shead, who compiled and edited this proceedings volume and transcribed the oral discussions.

Last, but not least, the advice and the assistance of the session chairmen and the session reporters were most valuable in guiding the workshop. The enthusiastic efforts of the 58 participants, who came from all around the world, were especially appreciated.

CONTENTS

Theme II: Theoretical Studies of Membrane Shear Behavior

OPENING REMARKS

by Thomas T. C. Hsu, Workshop Organizer

Ladies and Gentlemen,

It is indeed my pleasure to welcome many old friends to Houston. I am also looking forward to making new acquaintanceships and establishing new friendships.

We are gathered here for the next three days to study the topics of concrete shear in earthquake. This problem of shear in reinforced concrete had been studied for nearly a hundred years. For the first 65 years of this century, studies in this area were primarily done in an empirical manner. I recall a famous civil engineer, Professor Hubert Rusch of Germany, saying in a lecture at Cornell University in 1962 that he had studied the shear problem all his life, but he still did not understand shear. The past 25 years, however, have witnessed major advances, primarily through the development of the various types of truss model theories. These theories can satisfy not only the equilibrium condition, but also the compatibility condition and the constitutive laws. As a result of these achievements, we can now predict not only the shear strength, but also the entire load-deformation history of some simple structures. Studies of shear behavior under cyclic loading have led to the formulation of hysteretic rules for shear, particularly those in Japan. In all, the level of research has reached the stage where a rational treatment of shear in structures subjected to earthquake is now possible. Indeed, many of you in this distinguished audience are the persons who have done the research and prepared the stage for the new advances that we hope to achieve in this workshop.

The goals of this workshop are three-fold. One is to focus and define our present knowledge on shear in reinforced concrete and its application to earthquake. The state-of-the-art knowledge is compiled in the workshop proceeding volume, which includes 46 papers. The second goal is to explore new ways of utilizing shear ductility of reinforced concrete structures to help dissipate earthquake energy, and to define its limitations.

Thirdly, we will seek out directions for future studies, identify new research areas, and make recommendations to the National Science Foundation.

My colleague, Professor S. T. Mau, joins me in welcoming you all to this workshop at the University of Houston. We feel most fortunate to have the enthusiastic response and participation of 58 distinguished engineers representing 13 nations. S. T. and I would like to acknowledge the generous support of the National Science Foundation of the United States, without which this meeting of international minds would not be possible.

Now, to get the ball rolling, I will go ahead and define the scope of this workshop. There are generally two types of shear in reinforced concrete. One is the punching shear, or out-of plane shear. The other is the membrane shear, or in-plane shear. Punching shear in reinforced concrete is resisted mainly by concrete and therefore induces brittle failure. This type of shear failure should be suppressed in earthquake design, and therefore is not a topic for this workshop. In contrast, membrane shear is resisted by both steel reinforcement and concrete struts forming a truss action. With proper design, ductile failure can be achieved by ensuring the yielding of steel to precede the crushing of concrete. This ductile shear deformation can be utilized to help dissipate earthquake energy. In this workshop, the term "shear" should refer only to membrane shear, not punching shear.

The term "beam shear" includes a whole spectrum of situations. For a shallow beam without web reinforcement, the action of shear is similar to the punching shear. Shear in such a beam is excluded from this workshop. On the other end of the spectrum, we have deep beam with web reinforcement. As long as the web is subjected to membrane shear, the beam is a valid subject for this workshop.

Two types of structures involving membrane shear are particularly pertinent to this workshop. The first is the shear walls, including cantilever shear walls and framed shear walls. I am particularly confused about the behavior of framed shear walls and hope to learn something from this workshop. The second type is nuclear containment structures. These are massive structures with walls four feet deep.

This two-and-a-half-day workshop will be divided into five sessions, each conducted by a chairperson and a reporter. The responsibilities of the chairperson are to introduce the speakers, to perform the unenviable job of keeping the time, and to lead the discussion after each group of presentations. Each speaker is allotted ten minutes, except the state-of-the-art papers which are allotted 20 minutes each. The task of the reporters is to provide a summary of the discussions, which will be included in the final proceeding volume. They will also help to draft the recommendations for NSF.

I will now turn over the meeting to the first session chairman, Professor Thürlimann, and reporter, Professor Balaguru.

I hope you will enjoy the workshop.

THEME I:

EXPERIMENTAL STUDIES OF MEMBRANE SHEAR BEHAVIOR

STATE OF THE ART OF EXPERIMENTAL STUDY RELATED TO IN-PLANE SHEAR IN REINFORCED CONCRETE SHELL ELEMENT IN JAPAN

YUKIO AOYAGI
Earthquake Engineering Department,
Central Research Institute of Electric Power Industry
1646 Abiko Abiko-shi Chiba, JAPAN

ABSTRACT

Significant features of experimental researches related to in-plane shear,which have been conducted in Japan since 1975, are described. The review contains three categories of tests, that is, reinforced concrete plate elements, small-sized hollow cylinders and large-sized prestressed concrete containment safety verification models.

INTRODUCTION

In Japan, experimental studies related to the shear walls in building structures have been conducted since the early part of the 1950s to enhance the seismic resistance. Based on the results, rationalization of seismic design of building has also been achieved remarkably. On the other hand, a need to construct concrete containment structures in nuclear power stations led to activation of experimental research concerning in-plane shear in reinforced concrete shear elements.

The experimental studies, which have been conducted during the past fifteen years, include the shell element tests using plate-type specimens and hollow cylindrical type specimens. The former can be divided into two categories, that is, the case in which yielding in reinforcements determines the ultimate behavior and the case in which shear compression crushing of concrete strut governs the ultimate strength. The cylindrical models with or without internal pressures are subjected to reversals of horizontal or torsional forces at the top level of the models.

The paper mainly describes the procedures and test results concerning the above mentioned in-plane shear experimental researches, which have been performed since around 1975 in Japan.

PLATE ELEMENT TESTS

Two main experimental researches have been performed using plate elements, which are subjected to in-plane shear.

One is the experiment which was conducted by the author et al. and was intended to investigate the mechanical behaviors when yielding of reinforcements is a governing factor. Another is an enlarged Collin's type experiment, which is being conducted by Dr. Ohmori et al.

The main purpose of using the plate element specimens was to investigate typical critical stress conditions occurring in reinforced or prestressed concrete containments subjected to earthquake forces combined with or without accident internal pressures.(Figure-1).

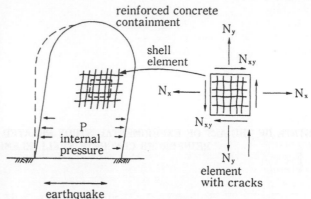

Figure 1. Stress conditions in concrete containment subjected to earthquake loading

Author's experiments [1],[2]
Plate specimens with plan dimensions of test area 150 × 150 cm and a thickness of 10 cm(Figure-2) were used for the test of orthogonally as well as triangularly reinforced concrete shell elements. Around the periphery of the specimen, slitted thickened load introductory zones were attached to anchor the reinforcement and to facilitate uniform introduction of in-plane forces. The loads were applied by what is called a tournament scheme. 30 specimens were tested. Two percentages of steel were used and deviation angles of the principal forces with respect to the orthogonal reinforcement directions were varied from 0 ° to 45 °. The disadvantages of this test method are the limited

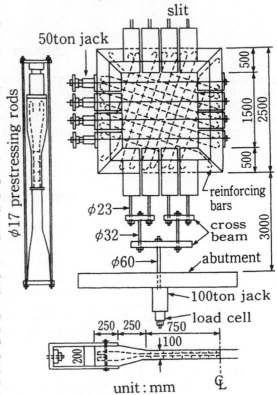

Figure 2. Method of applying in-plane forces to reinforced concrete plate specimens employed by the author

magnitude of shear forces applied to the specimens and the inability to introduce reversible shear stresses in the specimens.

Based on the experimental results governing crack angles, interface shear rigidities across a crack, average reinforcement stress etc. were derived.

Dr. S. Ohmori's experiments [3]

A series of loading tests on reinforced concrete panels subjected to reversed and cyclic in-plane stresses has been planned and is now under way at Kajima Institute of Construction Technology. The main object of the experimental program is to obtain useful information for developing a new hysteresis loop model of the concrete and the reinforcement.

Several reinforced concrete panels have already been tested. The test specimens were 2500 mm square and 140 mm thick and deformed rebars were arranged in two layers in orthogonal directions. In the test panels, the percentage of reinforcing steel was varied from 0.8% to 2.0%, but always with the same reinforcement ratio of ρ_{sx} and ρ_{sy} in orthogonal directions.

The test panels were loaded by a newly developed testing facility as shown in Figure-3. This facility can apply any combination of in-plane shear and normal forces to test panels by automatically controlled 24 closed loop hydraulic actuators and a network of links. The maximum capacity of load and stroke of each actuator are ± 125 ton and ± 125 mm respectively. All tests were conducted in reversed cyclic pure in-plane shear.

A typical shear stress-strain hysteresis curve is shown in Figure-4.

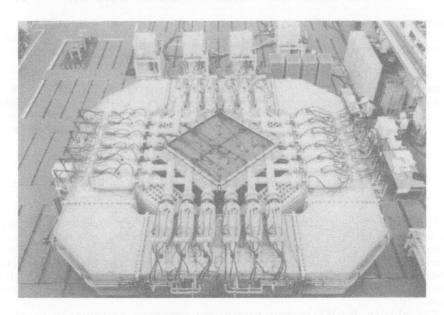

Figure 3. Overall view of testing facility employed
by Dr. Ohmori et al.

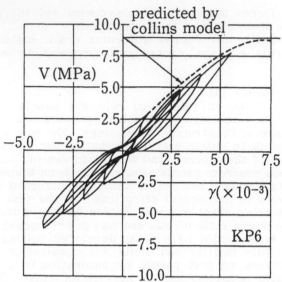

Figure 4. An example of shear stress-strain relationships
obtained by Dr. Ohmoris experiment
($\rho_x=\rho_y$=2.0%, reversed loading)

TESTS OF SMALL SIZED HOLLOW CYLINDRICAL SPECIMENS

In plane shear tests using hollow cylindrical models are divided into two types ; one is horizontal reversed loading test, and another is torsional reversed loading test.

Horizontal loading tests [4],[5],[6]

The main purpose of the experiments were to investigate the effects of the presence of internal pressure and the difference of reinforcement arrangements on strength and hysteretic behavior of reinforced concrete containments when subjected to earthquake forces. A BWR (Boiling Water Reactor) Mark III Type reinforced concrete containment was adopted as a prototype. The scale of the models was 1/25. The dimensions and the loading schemes are illustrated in Figure-5.

An example of the envelopes for hysteretic load-top slab displacements is shown in Figure 6, which gave the following conclusions;

(1) The strength capacity of three-way reinforced concrete containments against lateral forces combined with internal pressure is somewhat inferior to that of orthogonally reinforced one if compared on the condition that the volumetric reinforcement ratios are the same for the two cases of rein-forcement arrangements. However, three-way reinforcement improves initial shear rigidity as well as ultimate horizontal deformability for lateral forces.

(2) The ability for three-way reinforced concrete containment to absorb strain energy in the range of large deformations is superior to that of orthogonally reinforced one. The equivalent viscous damping coefficient for the former is markedly larger than that for the latter, especially at the increased deformational stages.

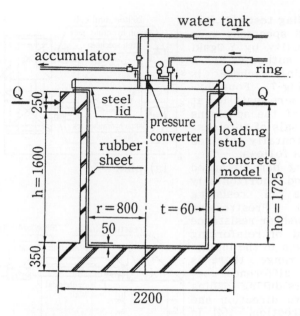

Figure 5. Configuration and loading set-up for cylindrical
reinforced concrete model subjected to horizontal
forces with presence of internal pressure
(1/25 scale)

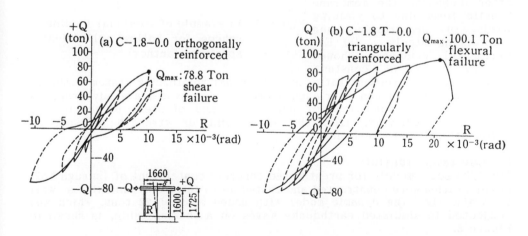

Figure 6. Hysteretic curves for horizontal displacements at the top
stub of 1/25 scale models without internal pressure.

Torsional loading tests [7],[8]
A total of 18 specimens were tested torsionally by Y.Ogaki et al. and the author et al. A typical loading scheme is depicted in Fig-7. Four hydraulic jacks were installed at the top slab of the specimens to exert reversals of torsions.

The experimentally obtained results were as follows:
(1) Reinforcing bars arranged vertically and horizontally are effective in resisting shear. (2) Prestressing steels possess shear resistance similar to that of reinforcing bars. (3) Even in the inelastic response range, there is scarcely any difference between behaviors during loading in the positive direction and the reverse direction. (4) In the case of prestressed concrete cylindrical specimens the tangential shear strength may be approximately evaluated by a restraining force on concrete elements remaining after deducting the membrane tensile force due to yielding loads of prestressing and reinforcing steels. (5) Experimental evidences showed that for the case of a constant volumetric reinforcement ratio, an orthogonal reinforcing system combined with diagonal bars and a triangular reinforcing system are advantageous for the loading conditions of pure shear and of shear plus membrane tension, respectively, from the view point or strength as well as deformational characteristics.

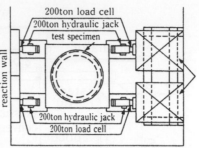

(a) plan

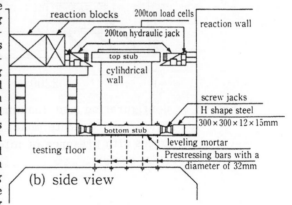

(b) side view

Figure 7. An example of torsional loading scheme for hollow cylindrical specimens (author)

Dynamic tests [9],[10]
Two 1/30 scale models for prestressed concrete containment of Tsuruga No2. power station were constructed and tested up to failure dynamically as well as statically. The dynamic model with added mass of 22 tons, which was subjected to simulated earthquake waves on a shaking table, is shown in Figure-8.

The ultimate shear stress attained, which caused horizontal shear slip at the mid-height region, was around 60kgf/cm²),for both dynamic and static models. Equivalent viscous damping coefficients were measured approximately as 2% at S_1 level and 6% at S_2 level, respectively, which almost corresponded to the values obtained from the static test. (As to the definition of S_1 and S_2 earthquake, see foot note∗).

Another dynamical test was conducted by Mr. T.Endo et al.

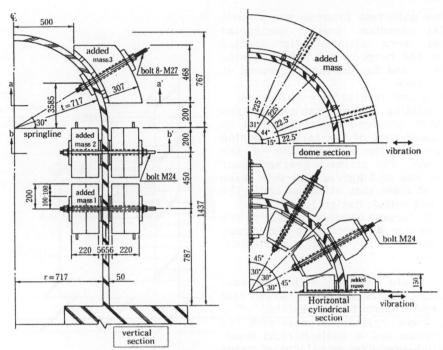

Figure 8. Cross sections of dynamically loaded 1/30 scale PCCV
model with added mass.

LARGE-SIZED PRESTRESSED CONCRETE CONTAINMENT MODELS

As a last stage of experimental works related to the verification of
Japanese drafted containment code [11] two large scale prestressed
concrete models were tested up to failure by reversed static horizontal
loading. The test conducted by Y. Ogaki et al. was intended to investigate
the ultimate behaviors at S_2 earthquake loading without internal pressure,
while the main purpose of the test performed by the author et al. was to
make clear the ultimate behaviors at S_1 earthquake with presence of
internal pressure.

1/8 scale model without internal pressure[12].
The test model consisted of a hemispherical top dome, a cylinder with
several openings and a base mat slab, modelled at a scale of one eighth
(1/8) of the designed prototype. The dimensions were diameter of 5.38 m,
height from base of 6.45 m, and cylinder-wall thickness of 19 cm (Figure-9).

) According to " Evaluation Guide on Seismic Design of Nuclear Reactor
Facilities for Electric Power Generation " S_1 and S_2 are defined as follows:
  S_1 is mainly determined from historical data for destructive earthquakes
and the most notable active fault. S_2 is defined as one of which intensity
exceeds the historical maximum but not be infinite and determined from
seismotectonics and active faults dislocated within 50,000 years. Right
down earthquake having magnitude 6.5 is also required to assume its
occurrence as to every site in Japan for S_2.

The main test program was a horizontal loading test. Horizontal forces were alternatively applied within the range of the seismic design loads S_1 and S_2, and then increased up to collapse.

The test results were as follows:
(1) Sliding shear explosive type collapse occurred horizontally at at the lower part of the largest opening located at the web portion of the specimen. The ultimate average shear stress was 60.3 kgf/cm² and this value was 1.84 times that of stress under the severest seismic design load (S_2).
(2) The largest opening located at the web portion did not exert significant influence on total mechanical behavior.

1/15 scale model with internal pressure and temperature gradient [13]
The model structure consisted of a rigid base, cylindrical wall with two buttresses and a hemispherical dome. Internal pressures were applied using hot water and leakage was prevented using a flexible rubber liner inside the model. To simulate the LOCA (loss of coolant accident) condition, the internal pressure and inner temperature were kept at 4.0 kgf/cm²) and 80 ℃ respectively. Alternate lateral forces were applied to the LOCA conditioned model up to the ultimate state of the structure (Figure-10)

The lateral force was reached to 4.1 times the design earthquake load (S_1), and tangential shear stress of the wall was 5.87 $\sqrt{f_o'}$ or 0.31 $\sqrt{f_o'}$ at the ultimate state. (Where f_o' is concrete compressive strength in kgf/cm²).

Summary of ultimate shear stresses obtained by cylindrical model tests.
Ultimate shear stresses (τ_u) obtained by cylindrical model tests are plotted against web reinforcement capacity ($\rho_w\sigma_y$) in Figure-10. τ_u and $\rho_w\sigma_y$ are normalized by $\sqrt{f_o'}$. (ρ_w : web reinforcement ratio, σ_y : yield point of reinforcing steel including prestressing steel(kgf/cm²) f_o' : concrete compressive strength(kgf/cm²).

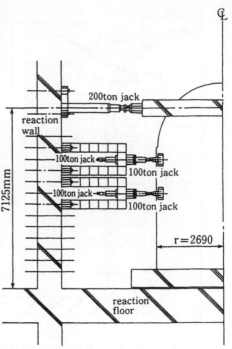

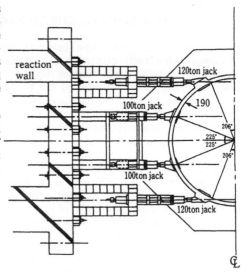

Figure 9. Loading scheme for 1/8 scale PCC model subjected to lateral forces.(℄:symmetry)

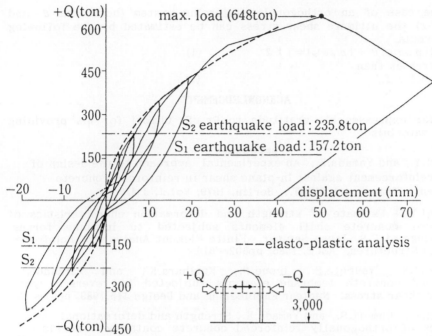

Figure 10. Measured and computed displacement at spring line, 3.0m
above base, thermal and internal pressure combined (1/15
scale PCCV model)

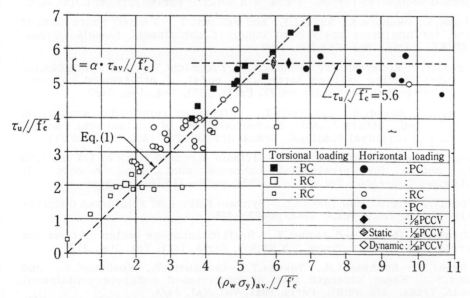

Figure 11. Relationships between measured ultimate shear stresses
and web reinforcement capacity ratios normalized by
$\sqrt{f_c'}$ (unit:kgf/ cm²)

In the case of an orthogonal reinforcing system (horizontal θ and vertical z) the ultimate shear stress can be estimated by the following simple formula.

$$\tau_u = [(\rho_w \sigma_y) \theta + (\rho_w \sigma_y) =] / 2 \qquad (1)$$

but not greater than

$$\tau_u = 5.6 \sqrt{f_o'} \qquad (2)$$

ACKNOWLEDGEMENT

The author expresses his gratitude to Dr. S. Ohmori for his providing valuable materials.

REFERENCES

1. Aoyagi,Y., and Yamada,K., An experimental approach to the design of network reinforcement against in-plane shear in reinforced concrete containment. Trans. 4th SMiRT., Berlin, 1979, Vol.J, J 4/7

2. Aoyagi,Y., Estimate of strength and deformation characteristics of reinforced concrete shell elements subjected to in-plane forces. Proceedings of U.S. Japan Seminar, Finite Element Analysis of Reinforced Concrete Structures, ASCE, 1985, pp602~619

3. Ohmori,N., Tsubota,H., Inoue,N., Kurihara,K., and Watanabe,S., Reinforced concrete membrane elements subjected to reversed cyclic in-plane shear stress. Nuclear Engineering and Design 115,1989, pp61~72

4. Aoyagi,Y., Ohmori,S., and Yamada,K., Strength and deformational characteristics of orthogonally reinforced concrete containment models subjected to lateral forces. Trans. 6th SMiRT., Paris,1981, Vol.J(a), J 4/5.

5. Aoyagi,Y., Yamada,K., and Takahashi,T., Strength and deformational characteristics of three-way reinforced concrete containment models subjected to lateral forces. Trans. 6th SMiRT., Paris, 1981, Vol.J(a),J4/6

6. Mikame,A., Okamura,K., Suzuki,T., and Fukushi,T., A mixed finite element method for nonlinear analysis of concrete containment vessels. Trans. 6th SMiRT., Paris, 1981, Vol.J(a), J3/6

7. Ogaki,Y., Kobayashi,M., Takeda,T., Yamaguchi,T., and Yoshioka,K., Horizontal loading test on large-scale model of prestressed concrete containment vessel. Trans. 6th SMiRT, Paris, 1981, Vol.,J(a), J4/2

8. Aoyagi,Y., Aoyagi,S., and Endo,T., Effects of reinforcing systems on strength and deformational characteristics of cylindrical models subjected to torsional loadings. Trans. 8th SMiRT., 1985, Vol.H, H6/11

9. Ogaki,Y., Kato,S., and Takeda,T., Ultimate strength tests of 1/8 and 1/30 scale models for prestressed concrete containment vessels. J. Prestressed Concrete., Vol,23, No.1, 1981, pp79~115,(in Japanese)

10. Endo,T., Kato,O., and Tanabe,T., Dynamic failure of reinforced concrete shells. Trans. J.C.I. Vol.4. 1982, pp423~428

11. Ohsaki,Y., Ibe,Y., and Aoyagi,Y., Drafted Japanese design criteria for concrete containment. Trnas. 6th SMiRT., Paris, 1981, Vol. J(a), J1/2

12. Ogaki,Y., Kobayashi,M., Takeda,T., Yamaguchi,T., Yoshizaki,S., and Sugano,S., Shear strength test of prestressed concrete containment vessels. Trans. 6th SMiRT., Paris, 1981, Vol.J(a), J4/3.

13. Aoyagi,Y., Ohnuma,H., Ichikawa,K., and Isobata, O., Test of a PCCV under load combination of LOCA and earthquake. Trans. 6th SMiRT., Paris, 1981, Vol.J(a), J4/9.

THE RESPONSE OF REINFORCED CONCRETE ELEMENTS SUBJECTED TO SHEAR

MICHAEL P. COLLINS
Professor, Department of Civil Engineering
University of Toronto, Canada

ABSTRACT

This paper summarizes some of the progress that has been made in developing rational models for shear behavior. To emphasize the characteristics of a rational model, the paper first briefly reviews the plane sections theory for flexure. Then the equilibrium and compatibility conditions for elements in shear are given. Experiments to investigate the stress-strain characteristics of reinforced concrete in pure shear are reviewed. Finally, some comments are made on developing models for reversed cyclic shear.

INTRODUCTION

The behavior of reinforced concrete in shear has been studied since 1899 with a particularly intensive research effort being conducted in the 15 years after the 1955 Air Force warehouse shear failures [1] (see Fig. 1). While this large effort did produce empirical equations for shear strength [2, 3], it did not result in a behavioral theory for shear comparable in rationality and generality to the plane sections theory for flexure.

A rational model for shear should make it possible to predict not only the shear strength but also the complete load-deformation response of elements subjected to shear. Further, such a theory should be capable of being extended to the seismic situation of reversed, cyclic loading.

PLANE SECTIONS THEORY FOR FLEXURE

In predicting the response of reinforced concrete beams subjected to flexure, we can restrict our attention to only longitudinal strains and their corresponding longitudinal stresses. The heart of the plane sections theory is the assumption that the longitudinal strains vary linearly over the depth of the beam [see Fig. 2(a)]. The longitudinal concrete stresses are found from the longitudinal concrete strains by using assumed stress-strain characteristics for the concrete. Usually it is assumed that in compression, the stress-strain curve obtained from a test cylinder can be used and that in tension, the concrete is not capable of resisting stress. To determine the stress in the reinforcement, it is assumed that the strain in the reinforcement is equal to the strain in the surrounding concrete and that the stress-strain characteristics obtained from a tension test of a reinforcing bar can be used. The moment-

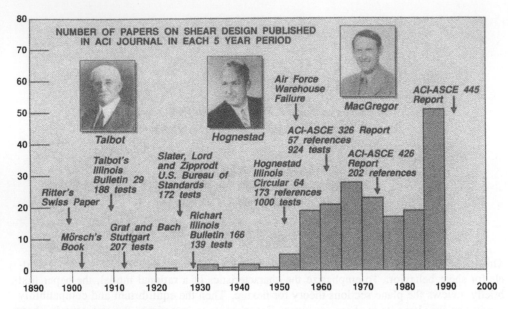

Figure 1. Research on shear.

curvature response is predicted by finding the strain distributions that result in zero axial load (i.e., $C = T$) and calculating the associated moments and curvatures.

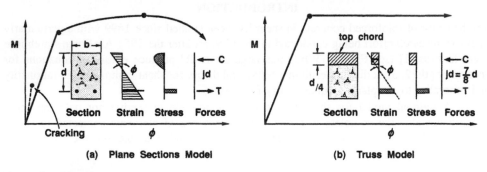

Figure 2. Moment-curvature response of reinforced concrete beam.

It is important to stress that the plane sections theory can predict the complete moment-curvature response of a reinforced concrete section and not just the value of the failure moment. This enables estimates of the energy dissipating capacity of the section to be made.

Figure 2(b) illustrates a simple truss model for a beam subjected to flexure. In this model the compression zone of the beam has been represented by a "chord" of uniformly stressed concrete. By a careful choice of the size of the top chord, this simple model could be adjusted to give an accurate estimate of the failure moment of the section. However, this model fails to give any information on the curvature capacity of the section. Once the reinforcement yields the compression force, and hence the top chord strain, stays constant while the bottom chord strain can increase indefinitely. By comparing Fig. 2(a) and (b) it can be appreciated that it is the compatibility requirements imposed by the linear strain distribution that limit the curvature capacity of the section.

Over 25 years ago Aoyama [4] demonstrated how the plane sections theory could be extended to predict the moment-curvature response of a reinforced concrete section subjected to reversed, cyclic loading. Since that time extensive work has been done to refine and verify this model and to account for parameters such as concrete confinement, spalling of concrete cover, influence of axial load, and influence of biaxial loading.

ELEMENTS SUBJECTED TO SHEAR

Figure 3 shows an element of reinforced concrete subjected to pure shear. The element contains an orthogonal grid of reinforcement with the x- and y-axes coinciding with the reinforcement directions. It is desired to predict the relationship between the applied shear stress and the resulting shear strain.

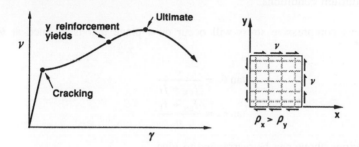

Figure 3.   Shear stress – shear strain response of reinforced concrete element.

Because there is no applied normal stress in the x-direction, the tensile stresses in the x-reinforcement, f_{sx}, must be balanced by compressive stresses, f_{cx}, in the concrete [see Fig. 4(a)]. Thus

$$f_{cx} = \rho_x f_{sx} \tag{1}$$

where ρ_x is the ratio of reinforcement area to concrete area. Similarly, in the y-direction,

$$f_{cy} = \rho_y f_{sy} \tag{2}$$

If the reinforcing bars are assumed to be carrying only axial stresses, then the shear stress on the concrete on planes x and y must equal the applied shear v.

If we know the values of the normal stress and the shear stress on the x-plane and the y-plane, we can find the normal stress and the shear stress on any other plane [see Fig. 4(b)]. The highest compressive stress in the concrete, f_2, will be

$$f_2 = \frac{f_{cx} + f_{cy}}{2} + \sqrt{\left(\frac{f_{cx} - f_{cy}}{2}\right)^2 + v^2} \tag{3}$$

while the highest tensile stress in the concrete, f_1, will be

$$f_1 = \sqrt{\left(\frac{f_{cx} - f_{cy}}{2}\right)^2 + v^2} - \frac{f_{cx} + f_{cy}}{2} \tag{4}$$

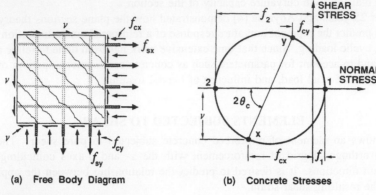

(a) Free Body Diagram   (b) Concrete Stresses

Figure 4. Equilibrium conditions.

The highest compressive stress will occur on a plane that is inclined at θ_c to the x-plane where

$$\tan \theta_c = \frac{v}{f_{cx} + f_1} \tag{5}$$

$$\tan \theta_c = \frac{f_{cy} + f_1}{v} \tag{6}$$

The equations above can be rearranged to give

$$\rho_x f_{sx} = v \cot \theta_c - f_1 \tag{7}$$

$$\rho_y f_{sy} = v \tan \theta_c - f_1 \tag{8}$$

$$f_2 = v(\tan \theta_c + \cot \theta_c) - f_1 \tag{9}$$

The strain conditions for the element subjected to pure shear can be described in terms of the Mohr's circle shown in Fig. 5. Useful relationships that can be derived from the geometry of this circle include

$$\tan^2 \theta = \frac{\epsilon_x + \epsilon_2}{\epsilon_y + \epsilon_2} \tag{10}$$

$$\epsilon_1 = \epsilon_x + \epsilon_y + \epsilon_2 \tag{11}$$

$$\gamma_{xy} = 2(\epsilon_x + \epsilon_2) \cot \theta \tag{12}$$

Before the equilibrium and compatibility equations above can be used to determine the load-deformation response we need to know the relationships between the stresses and strains for both the reinforcement and the concrete.

Following an approach pioneered by Wagner [5], the compression field theory [6, 7] assumes that the directions of the principal stresses in the concrete coincide with the directions of the principal strains (i.e., $\theta_c = \theta$). Following the procedure used in flexure, the stress in the reinforcement (f_{sx} and f_{sy}) is related to the strain in the surrounding concrete (ϵ_x and ϵ_y) by the stress-strain characteristics obtained from the tension test of a reinforcing bar.

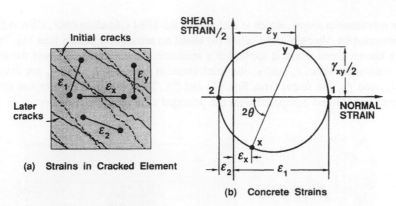

(a) Strains in Cracked Element

(b) Concrete Strains

Figure 5. Compatibility conditions.

Early versions of the compression field theory neglected the tensile stresses in the concrete ($f_1 = 0$) and assumed that the compressive stress in the concrete, f_2, could be related to the compressive strain, ϵ_2, by the stress-strain curve obtained by testing a cylinder in compression. This second assumption resulted in unconservative predictions of shear capacities. Research was then conducted to improve this assumption.

STRESS-STRAIN RELATIONSHIPS FOR DIAGONALLY CRACKED CONCRETE

It has been found that the principal compressive stress in the concrete, f_2, is a function not only of the principal compressive strain, ϵ_2, but also of the coexisting principal tensile strain, ϵ_1 [see Fig. 6(a)]. An appropriate relationship is

$$f_2 = f_{2max} \left[2 \frac{\epsilon_2}{\epsilon_c'} - \left(\frac{\epsilon_2}{\epsilon_c'} \right)^2 \right] \tag{13}$$

where

$$\frac{f_{2max}}{f_c'} = \frac{1}{0.8 + 170\epsilon_1} \leq 1.0 \tag{14}$$

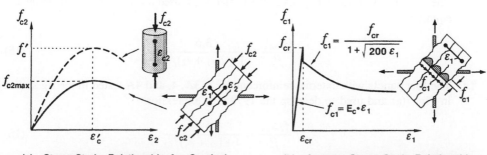

(a) Stress-Strain Relationship for Cracked Concrete in Compression

(b) Average Stress-Strain Relationship for Cracked Concrete in Tension

Figure 6. Stress-strain relationship for cracked concrete.

The relationship above, which is included in the 1984 Canadian code, CSA A23.3 [8], was recommended by Vecchio and Collins [9] based on tests of 30 panels [see Fig. 7(b)]. In such tests a known shear stress is applied to a reinforced concrete element and the resulting strains are measured. From ϵ_x and ϵ_y the reinforcement stresses f_{sx} and f_{sy} are determined and then, f_{cx} and f_{cy} are found from Eqs. (1) and (2). The principal compressive stress, f_2, is found from Eq. (3) and compared with the principal compressive strain, ϵ_2.

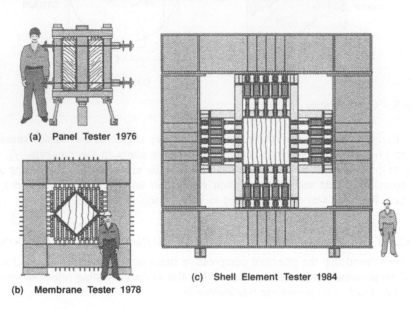

(a)　Panel Tester 1976

(b)　Membrane Tester 1978

(c)　Shell Element Tester 1984

Figure 7.　Testing reinforced concrete elements in pure shear.

Figure 8 compares the maximum compressive stresses, f_{2max}, for 55 specimens with the associated principal tensile strains, ϵ_1. Specimens plotted in this figure all exhibited compressive failures of the concrete with ϵ_2/ϵ_c' at failure exceeding 0.5. The specimens include:

(i) The four beam tests, CF1, SA2, SA3, and SA4 used in 1976 [7] to develop the equation

$$\frac{f_{2max}}{f_c'} = \frac{3.6}{1 + 2(\epsilon_1 + \epsilon_2)/\epsilon_c'} \tag{15}$$

(ii) The four panels with external reinforcement, 3, 3A, 4, and 4A tested in the rig shown in Fig. 7(a) and used to develop the 1979 [10] expression

$$\frac{f_{2max}}{f_c'} = \frac{5.5}{4 + (\epsilon_1 + \epsilon_2)/\epsilon_2} \tag{16}$$

(iii) Twelve of Vecchio's original panels [11] PV10, PV12, PV17, PV19, PV20, PV21, PV22, PV23, PV25, PV26, PV28, and PV29.

(iv) Five additional shear tests, TP1A, TP4A, PC1A, PC4, and PC7 conducted by André [12] and Chan [13] in the rig shown in Fig. 7(b).

(v) Five panels, PB13A, PB24, PB25, PB26, and PB27, first subjected to tension in the x-direction and then subjected to compression in the y-direction, by Bhide [14] using the rig shown in Fig. 7(b).

(vi) Six panels, PK02, PK03, PK04, PK06, PK07, and PK08 subjected to various loading histories by Kollegger [15] using the rig shown in Fig. 7(b).

(vii) Three shell elements, SE1, SE5, and SE6 tested in membrane shear by Kirschner and Khalifa [16] using the rig shown in Fig. 7(c).

(viii) Three partially prestressed shell elements. PP1, PP2, and PP3 tested in membrane shear by Meyboom [17].

(ix) Three shell elements, SE8, SE9, and SE10, subjected to reversed cyclic membrane shear by Stevens [18].

(x) Four high-strength concrete ($f'_c \approx 70$ MPa) shell elements, SE11, SE12, SE13, and SE14 subjected to membrane shear by Biedermann [19] and Porasz [20].

(xi) One shell element, SP8, containing confinement reinforcement tested in membrane shear by Adebar [21].

(xii) Five shell elements, EZ1, EZ2, EZ3, EZ4, and EZ5, reinforced with high-strength reinforcement ($f_y \approx 860$ MPa) subjected first to tension and then to compression by Kuchma.

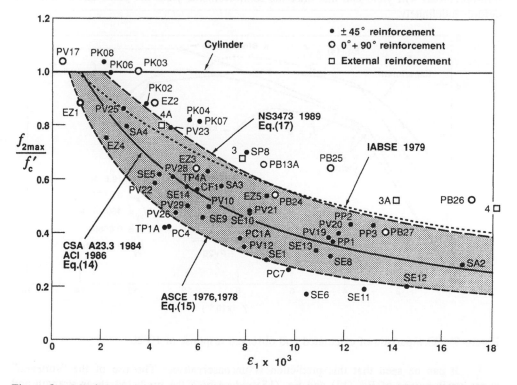

Figure 8. Maximum concrete compressive stress as a function of principal tensile strain.

Also shown in Fig. 8 are the plots of Eqs. (14), (15), and (16) as well as the expression recommended by the 1989 Norwegian concrete code [22] which is

$$\frac{f_{2max}}{f'_c} = \frac{1}{0.8 + 100\epsilon_1} \leq 1.0 \qquad (17)$$

In plotting Eqs. (15) and (16) it was assumed that ϵ'_c and ϵ_2 equalled 0.002.

It can be seen that Eq. (15) and Eq. (17) approximately bound the experimental data while Eq. (14) lies in about the middle of the scatter band.

It is of interest that specimens in which the reinforcement is parallel and perpendicular to the principal compressive stress seem to fail at higher stresses than those specimens in which the reinforcement is at about 45° to the principal stresses. As the concrete typically fails by sliding on planes at about 45° to the principal compressive stress it is possible that reinforcing bars at this inclination act as failure initiators.

The sensitivity of the calculated load-deformation response to the different assumptions about f_{2max} is illustrated in Fig. 9 which shows the shear stress – shear strain response of specimen PV20. Four predictions for the response are shown, three of which ignore tensile stresses in the concrete (i.e., $f_1 = 0$). One of the predictions ignores the influence of ϵ_1 on the concrete stress-strain response (i.e., $f_{2max} = f'_c$). This method predicts that both reinforcements will yield and that after the reinforcements yield the panel can continue to deform indefinitely.

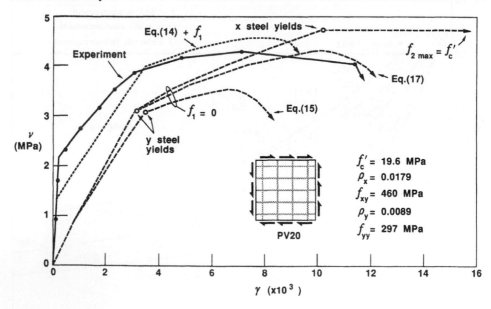

Figure 9. Comparison of calculated and observed response of PV20.

It can be seen that this prediction is unconservative. The use of the "softened" stress-strain curves of Eq. (17) and Eq. (15) reduce both the predicted shear strength and the deformation capacity of the panel. However, all three predictions based on setting f_1 equal to zero fail to capture the actual shape of the load-deformation response. The fourth prediction shown in Fig. 9 does account for tension in the concrete and as a result it gives the most accurate estimate of the actual response. The collection of assumptions used in making this fourth prediction is known as the modified compression field theory [9, 23].

In the modified compression field theory the tensile stress in the concrete is given by

$$f_1 = \frac{f_{cr}}{1 + \sqrt{200\epsilon_1}} \tag{18}$$

but

$$f_1 \leq \frac{0.18\sqrt{f'_c}\tan\theta}{0.31 + \frac{24w}{a + 16}} \quad \text{MPa units} \tag{19}$$

where w is the crack width in mm and a is the maximum aggregate size in mm.

CONCLUDING REMARKS

By considering equilibrium conditions, compatibility conditions and the stress-strain characteristics of cracked concrete it has proved possible to develop models capable of predicting, with reasonable accuracy, the load-deformation response of reinforced concrete elements loaded in shear. Further, it has been possible to extend these techniques to treat elements subjected to reversed, cyclic shear [18] (see Fig. 10). This work has shown that repeated cycles of shear stress at any level above that causing first yielding of the reinforcement will eventually cause a reinforced concrete element to fail by concrete crushing.

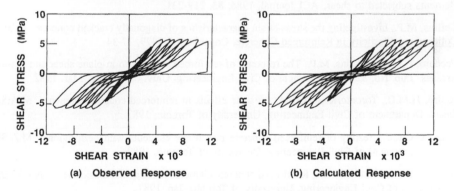

Figure 10. Comparison of calculated and observed response of SE8.

Considerable work is still required to extend our knowledge of the stress-strain characteristics of cracked concrete so that more precise predictions of the load-deformation response in shear can be made. Issues that need clarification include: the influence of reinforcing bar orientation, the influence of load history, the influence of confining reinforcement and the influence of the relative strength of the concrete paste and the aggregate.

ACKNOWLEDGEMENT

The long-term research project at the University of Toronto on reinforced concrete in shear has been made possible by the consistent funding policies of the Natural Sciences and Engineering Research Council of Canada. This continuing support is gratefully acknowledged.

REFERENCES

1. Anderson, B.G., Rigid frame failures. ACI Journal, 1957, **53**, 625-636.

2. ACI-ASCE Committee 326, Shear and diagonal tension. ACI Journal, 1962, **59**, 1-30, 277-344, and 352-396.

3. ACI-ASCE Committee 426, Shear strength of reinforced concrete members. Journal of the Structural Division, ASCE, 1973, **99**, 1091-1187.

4. Aoyama, H., Moment-curvature characteristics of reinforced concrete members subjected to axial load and reversal of bending. Proceedings of International Symposium on the Flexural Mechanics of Reinforced Concrete, ASCE-ACI, Miami, 1964, 183-212.

5. Wagner, H., Ebene Blechwandträger mit sehr dünnem Stegblech. Zeitschrift für Flugtechnik und Motorluftschiftahr. 1929, **20**, Nos. 8 to 12.

6. Mitchell D., and Collins, M.P., Diagonal compression field theory – a rational model for structural concrete in pure torsion. ACI Journal, 1974, **71**, 396-408.

7. Collins, M.P., Towards a rational theory for RC members in shear. Journal of the Structural Division, ASCE, 1978, **104**, 649-666.

8. CSA Committee A23.3, Design of Concrete Structures for Buildings, Canadian Standards Assoc., Rexdale, 1984, 281 pp.

9. Vecchio, F.J., and Collins, M.P., The modified compression field theory for reinforced concrete elements subjected to shear. ACI Journal, 1986, **83**, 219-231.

10. Collins, M.P., Investigating the stress-strain characteristics of diagonally cracked concrete. IABSE Colloquium: Plasticity in Reinforced Concrete, Copenhagen, 1979, 27-34.

11. Vecchio, F.J., and Collins, M.P., The response of reinforced concrete to in-plane shear and normal stresses. Pub. No. 82-03, Department of Civil Engineering, Univ. of Toronto, Mar. 1982.

12. André, H.M.O., Toronto/Kajima study on scale effects in reinforced concrete elements. M.A.Sc. thesis, Department of Civil Engineering, University of Toronto, 1987.

13. Chan, C.C.L.C., Testing of reinforced concrete membrane elements with perforations. M.A.Sc. thesis, Department of Civil Engineering, University of Toronto, 1989.

14. Bhide, S.B., and Collins, M.P., Reinforced concrete elements in shear and tension. Pub. No. 87-02, Department of Civil Engineering, University of Toronto, Jan. 1987.

15. Kollegger, J., Collins, M.P., and Mehlhorn, G., Tension and tension-compression tests on reinforced concrete elements., Research Report, Gesamthochschule, Kassel - Universität, Jan. 1987.

16. Kirschner, U., and Collins, M.P., Investigating the behaviour of reinforced concrete shell elements. Pub. No. 86-09, Department of Civil Engineering, University of Toronto, Sept. 1986.

17. Meyboom, J., An experimental investigation of partially prestressed, orthogonally reinforced concrete elements subjected to membrane shear. M.A.Sc. thesis, Department of Civil Engineering, University of Toronto, 1987.

18. Stevens, N.J., Uzumeri, S.M., and Collins, M.P., Analytical modelling of reinforced concrete subjected to monotonic and reversed loadings. Pub. No. 87-1, Department of Civil Engineering, University of Toronto, Jan. 1987.

19. Biedermann, J.D., The design of reinforced concrete shell elements: an analytical and experimental study. M.A.Sc. thesis, Department of Civil Engineering, University of Toronto, 1987.

20. Porasz, A., An investigation of the stress-strain characteristics of high strength concrete in shear. M.A.Sc. thesis, Department of Civil Engineering, University of Toronto, 1989.

21. Adebar, P.E., Shear design of concrete offshore structures. Ph.D. thesis, Department of Civil Engineering, University of Toronto, 1990.

22. Norges Byggstandardiseringsrad (NBR), Prosjektering av betongkonstruksjoner. Beregnings - og konstruksjonsregler (Concrete structures. Design rules. NS3473), Norges Standardiseringsforbund (NSF), Oslo, 1989, 103 pp.

23. Bhide, S.B., and Collins, M.P., Influence of axial tension on the shear capacity of reinforced concrete members. ACI Structural Journal, 1989, 86, 570-581.

EXPERIMENTAL STUDIES IN NEW ZEALAND
OF SEISMIC SHEAR EFFECTS

THOMAS PAULAY
Department of Civil Engineering
University of Canterbury
Christchurch, New Zealand

ABSTRACT

The highlights of selected studies, undertaken in recent years in New Zealand, to aid further developments in the planning and execution of seismic design strategies, are presented. Concepts of a deterministic design philosophy, in which the role of shear resistance in the hierarchy of energy-dissipating mechanisms within reinforced concrete ductile structures is identified, have been presented in a companion paper.

Emphasis is on the influence of reversed cyclic displacements, corresponding with inelastic structural response to large earthquakes, on established mechanisms of shear resistance and the identification of relevant features from experimental studies.

The issues reviewed here are: the shear strength of plastic hinges in beams, squat walls designed for significant ductility by means of flexure, squat walls with restricted ductility relying on limited inelastic shear deformations, and squat shear dominated circular bridge piers under multidirectional seismic attacks.

INTRODUCTION

In a companion paper the features of a deterministic seismic design strategy with respect to shear strength in reinforced concrete structures was outlined. It was postulated that in structures, designed to respond in a ductile manner to anticipated earthquakes, mechanisms involving inelastic response in shear should not be used. In comparison with plastic mechanisms based primarily on flexural actions, inelastic shear mechanisms, encountered in common reinforced concrete members, exhibit rapid deterioration of both stiffness and strength during imposed cyclic displacements and hence have inferior ability to dissipate seismic energy. As a consequence, the design strategy, based on a hierarchy of plastic mechanisms, aims to suppress, if possible, shear failures even under the most severe seismic shaking. Corresponding design procedures, relevant to different situations in which the control of inelastic shear deformations may be problematical, have been proposed and used in New Zealand over the last 15 years. These were supported by evidence obtained from experimental work. The aim of this paper is to highlight features of some of these experimental studies.

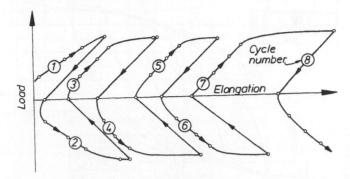

Figure 1 **Typical beam elongation-load history for a beam with reversed cycle loading**

SHEAR IN PLASTIC HINGES IN BEAMS

Because under reversed cyclic ductility demands the concrete in the flexural compression zones of plastic hinges in beams is inevitably damaged, precaution with respect to its contribution to diagonal compression strength is required. A particular feature, affecting energy dissipation in plastic hinges, is possible sliding along interconnecting wide flexural cracks under the action of reversed cyclic shear forces. Because of misfit at crack boundaries, a perfect closure of flexural cracks is no longer possible [1]. Accumulated residual tensile strain will load to lengthening of beams as shown in Fig. 1, sometimes by significant amounts, i.e., 1 to 3% of the span length of short beams.

The control of sliding shear displacement is best achieved with the use of diagonal reinforcement. Fig. 2 shows the excellent response of a test unit in which the plastic hinge was relocated from the face of a column in order to improve anchorage conditions for beam bars in the beam-column joint. In this test unit approximately 100% of the shear force, associated with plastic hinge formation, was resisted by diagonal reinforcement in tension and compression respectively. This much diagonal reinforcement will never be required. Design recommendations have been made [2] as to when and how much of the resistance of shear force in the plastic hinge region of beams should be assigned to diagonal reinforcement.

SQUAT WALLS

Shear dominated squat walls, prone to shear failures, can be readily designed to develop ductility primarily in the flexural mode. While a diagonal tension failure under seismic forces can be readily prevented, sliding at the base after yielding of the vertical reinforcement can seriously reduce the energy dissipation capacity of squat walls. In such cases again diagonal reinforcement may be used to resist a fraction of the total shear developed simultaneously with the maximum flexural strength of the wall. Fig. 3 compares the response of two half scale walls with height/length ratio of 0.57, having small flanges. Sliding displacement at moderate displacement ductilities lead to the web

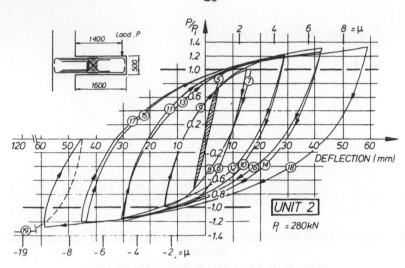

Figure 2 **Hysteretic response of a beam with diagonal reinforcement across the plastic hinge**

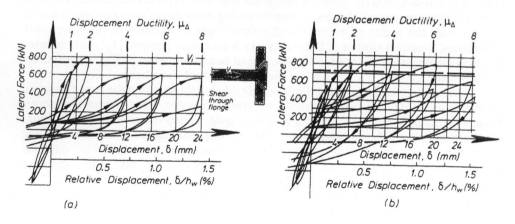

(a)   (b)

Figure 3 **Hysteretic response of squat walls with (a) conventional and (b) diagonal reinforcement**

without any diagonal reinforcement slicing through the flange (Fig. 3(a)). Poor energy dissipation combined with significant reduction of strength in the second cycle to the same displacement ductility, well below the flexural strength of the wall, are evident in Fig. 3(a). In comparison Fig. 3(b) shows improvements of simulated seismic response when approximately 30% of the shear associated with flexural strength could be resisted by diagonal reinforcement. The simple principles of design of such walls are examined in a companion paper.

SQUAT WALLS WITH RESTRICTED DUCTILITY

Walls with flexural strength in excess of that required for code-specified earthquake resistance are often encountered in practice. Such walls are usually squat with height/length ratios of less than 2, containing close to minimum mesh reinforcement. The nominal shear strength of such walls [3] may also exceed the code-specified lateral force intensity. However, because the design for strength was based on some perceived ductility capacity and consequently on reduced seismic forces, rather than on elastic response, the relatively large flexural strength of the wall could be developed during an earthquake. The correspondingly increased shear force may then result in an unexpected shear failure.

Many such walls, built and designed before the introduction of capacity design concepts [2] in New Zealand, belong to this category. Even though eventual shear failure by diagonal tension is expected, such walls exhibit some restricted ductility. Because of excess strength, their reduced displacement ductility capacity may well be adequate to meet ductility demands of the design earthquake appropriate to the site.

In a relevant experimental study [4] the response of squat cantilever walls was studied to identify hysteretic response and ability to dissipate energy when the nominal shear strength was only 60 to 78% of that developed in flexure. Wall behaviour under simulated moderate earthquakes was of particular interest. Eventual failure of three walls with aspect ratios of 1.0, 1.5 and 2.0 was, as expected, due to diagonal tension. Nominal shear strength $(V_i)_{SHEAR}$ was based on the tradition additive principle, i.e. $V_u = V_c + V_s$, where V_c and V_s are the contributions of the "concrete" and the shear reinforcement respectively. Depending on the type of code or postulated theory which a designer adopts, there will be large variations in the relative proportions of V_c and V_s. Moreover, these proportions continually change during inelastic seismic response because of the inevitable damage of concrete and consequent reduction of V_c. Some increase in the contribution of shear reinforcement also occurs because of the change in the angle of potential diagonal failure plane. It should be appreciated that it is *not* the maximum strength developed in the wall by either flexural or shear mechanisms that is of interest, but rather the strength that can be maintained after several complete reversed displacement cycles, corresponding to an anticipated ductility demands, have been experienced by the structure.

With one minor exception all three walls developed flexural strengths, based on measured material properties, in each of the two directions of the imposed lateral force. The maximum strengths thus developed during testing were 128 to 166% of the design shear strength, indicating that existing design procedures [2,3] for estimating maximum shear strength are rather conservative.

Maximum strengths were developed in these walls with aspect ratios of 1.0, 1.5 and 2.0 at displacement ductilities of $\mu_\Delta = 2.5$, 3.0 and 4.2 respectively. However, an important feature, relevant to the seismic scenario, was the significant reduction of strength encountered during *subsequent* application of cyclic displacements to the same or larger ductilities.

Figure 4 shows the response of Unit 1.5 (height/length ratio of 1.5). A very satisfactory performance is evident, provided that imposed displacement ductility does not exceed $\mu_\Delta = 3$. It is of interest to note that although serious deterioration occurred in both stiffness and strength at $\mu_\Delta = 4$, resistance did not reduce below the level predicted for shear strength. The evidence of eventual shear failure is seen in Fig 5 where the numbers indicate the sequence of the formation of major diagonal cracks.

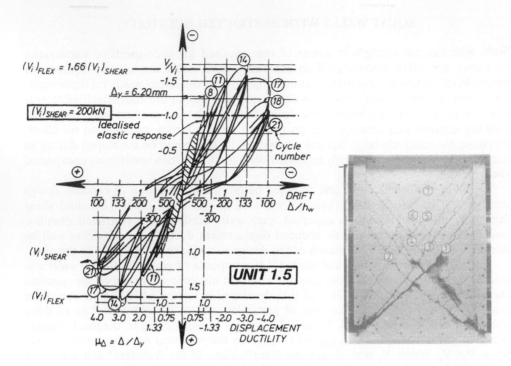

Figure 4   **Lateral force-displacement relationship for a wall with height/length ratio of 1.5**

Figure 5   **Crack pattern in test Unit 1.5**

The role of shear deformations in the development of restricted ductility in squat walls with excess flexural strength emerges from Fig. 6. Attempts were made during the test to identify separately sources of wall deflections at the loaded end at the top. These deflections, labelled in Fig. 6, were due to flexure Δ_f, shear Δ_v, bar elongations within anchorages in the foundation beam Δ_a, and sliding of the wall base along the top of the foundation beam Δ_s. It is seen that the latter two deformations remained relatively small proportions of the total. As imposed ductility was increased, expressed in Fig. 6 in terms of drift, i.e. deflection/wall height ratio Δ/h_w, the relative contribution of flexural deformations steadily decreased while those due to shear increased correspondingly.

It was found that, irrespective of the displacement ductility envisaged in the design of squat walls with restricted ductility, the usable maximum drift should be limited to

$$\Delta/h_w \ (\%) - \left(h_w/\ell_w + 0.5\right)/2 \tag{1}$$

where h_w and ℓ_w are the height and the length of the wall respectively.

Diagonal cracking in walls of restricted ductility is to be expected while they are subjected to moderate earthquakes associated with lateral forces less than 50% of that corresponding to computed shear strength. Cracking was found to start when the

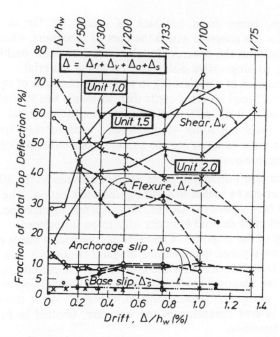

Figure 6   Components of wall deflections at different imposed drifts.

imposed drift was as small as $\Delta/h_w = h_w/(800\ell_w)$. When a small amount of web reinforcement, corresponding to typical code-specified minima, such as $\rho = 0.7/f_y$ (MPa), $(0.1/f_y$ (Ksi)) is used - a common practice for such structures - diagonal cracks become instantaneously wide, and the shear reinforcement yields. Therefore cracks will not close upon the removal of the lateral force. This is because the tensile (cracking) strength of concrete is larger than the tensile capacity of the web reinforcement provided. In terms of serviceability and damage control this phenomenon should be viewed as a serious limitation on the use of shear dominated lightly reinforced walls with restricted displacement ductility capacity.

SQUAT CIRCULAR BRIDGE PIERS
UNDER MULTIDIRECTIONAL SEISMIC ATTACK

One of the more challenging research tasks, which worldwide consumed several hundred test specimens, was the identification of all aspects of the inelastic behaviour of short columns subjected to reversed cyclic moments and shear under simultaneous application of axial compression. It is well established that, because of squatness, shear has a dominant role in determining the seismic performance of such elements, used as columns in multistorey buildings or as bridge piers. While there is an abundance of experimental data available, widely endorsed design recommendations have not emerged as yet. Because of the complexity of the inelastic seismic response of such units, this, however, is not surprising. One of the major difficulties in developing rational models, suitable for routine design, stems from our lack of success so far to establish promising constitutive laws for the concrete within the core of such piers, where, as a consequence of large seismic motions, the concrete is subjected to variable multiaxial strain patterns eventually leading to progressive disintegration of the material.

As a continuation of a project [5], in which the response of squat circular columns subjected to unidirectional reversed cyclic displacements was studied, the work reported here [6] concentrated on the effects of simulated multidirectional earthquake motions. In both projects, involving some 40 generously instrumented specimens with spiral reinforcement, the main concern was the simultaneous influence on ductility capacity of shear and axial compression.

Fig. 7 shows how horizontal forces were introduced to the 400 mm diameter and 800 mm high specimens by means of a self equilibrated test rig. The entire assembly was placed within a 10 MN capacity DARTEC universal testing machine, which then enabled the axial compression of desired intensity also to be introduced.

In the study of the response of spiral reinforcement a major difficulty arises from the inability to trace separately the causes of spiral strains, i.e. due to shear and due to the need to provide confinement for the concrete in the column core. In the presence of axial compression, diagonal cracks, caused by the shear force, eventually lead to the development of a wedge mechanism, shown in Fig. 8. After the onset of yielding of spirals, a sudden lateral expansion of the concrete, implied in Fig. 8, may lead to a dramatic loss of resistance.

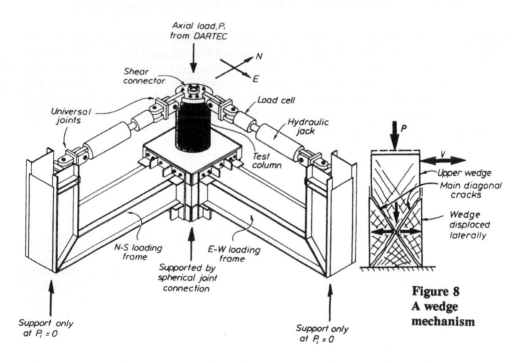

Figure 8
A wedge
mechanism

Figure 7   **Circular test columns subjected to multidirectional lateral displacements**

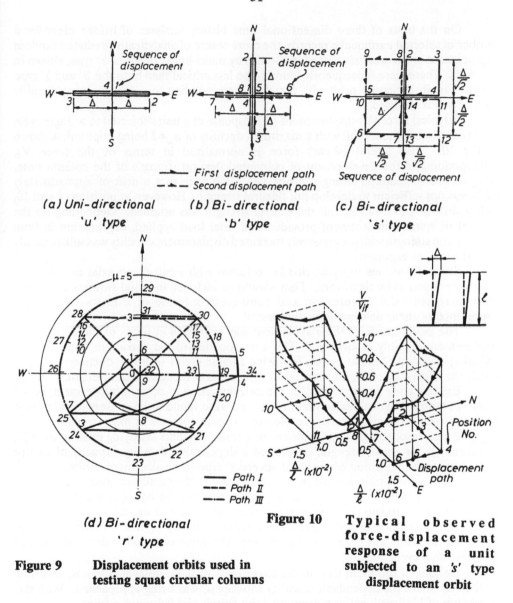

(a) Uni-directional 'u' type

(b) Bi-directional 'b' type

(c) Bi-directional 's' type

— First displacement path
– ▲ – Second displacement path

(d) Bi-directional 'r' type

— Path I
--- Path II
—·— Path III

Figure 9 Displacement orbits used in testing squat circular columns

Figure 10 Typical observed force-displacement response of a unit subjected to an 's' type displacement orbit

To enable the effect of displacement paths and history to be studied, different displacement orbits, as shown in Fig. 9, were selected. The traditional unidirection 'u' type of displacement path with progressively increasing displacement amplitude Δ was used only to allow comparisons to be made between the response of identically detailed specimens. Bi-directional displacement paths were introduced either in the form of 'b' or 's' type orbits, seen in Figs. 9(b) and (c). Sequence numbers recorded in these figures enable the paths to be followed. The bi-directional 's' type orbit was considered to impose the most severe conditions for squat columns.

On the basis of three dimensional time history analyses of bridge piers for a number of selected earthquake records, the more severe of analytically predicted random displacement orbits were simulated in the test by multi-linear orbits of 'r' type, shown in Fig. 9(d). These were subsequently found to be less critical than both the 'b' and 's' type orbits. Finally a circular orbit with a radius corresponding to a displacement ductility ratio of 4 was imposed.

A typical force (V)-displacement (Δ) response of a unit subjected to a single cycle of 's' type displacement orbit with a maximum ductility of $\mu_\Delta = 2$ being imposed, is shown in Fig. 10. The lateral (shear) force is normalized in terms of the force V_{if}, corresponding to the development of computed flexural strength of the column base, using measured material strength properties. It is seen that a drift of approximately 1.2% was not sufficient to develop full flexural strength. However, at positions 4 and 10, with a drift of $1.2\sqrt{2} = 1.7\%$, full theoretical strength was attained. Depending on the amount of spiral reinforcement provided and axial load applied, degradation in both stiffness and strength with progressively increased displacement ductility was subsequently encountered, as expected.

Shear distortions in squat circular columns with small or no axial compression load were found to be significant. They should be included in initial stiffness estimates. With increased axial compression and consequent increase of stiffness the relative importance of shear deformations diminished.

The performance of these specimens was expressed in terms of the dependable displacement ductility capacity μ_0. This corresponded to the maximum displacement, involving two displacement cycles, following similar cycles to lower ductility levels, at which the resistance of a unit reduced to 80% of its computed ideal strength.

Existing code requirements for the determination of shear reinforcement, based on 45° potential diagonal tension failure planes, were found to be conservative because observed critical diagonal cracks, developing in squat columns at much smaller angles, engaged a correspondingly larger number of spirals. All units satisfying existing code [2] requirements for shear strength, developed a dependable ductility capacity of five or more even with imposition of the most severe 's' type of displacement orbits.

Yielding of spirals, at least in the lower parts of the cantilever columns, occurred in all cases where the dependable ductility capacity μ_0 was being approached. With the imposition of bi-directional displacement orbit spirals yielded much earlier.

The magnitude of the dependable ductility capacity μ_0 under multi-directional displacement orbits was generally one less than the value obtained under conditions of unidirectional cyclic loading.

Yielding of spirals, at least in the lower parts of the cantilever columns, occurred in all cases where the dependable ductility capacity μ_0 was being approached. With the imposition of bi-directional displacement orbit spirals yielded much earlier.

The magnitude of the dependable ductility capacity μ_0 under multi-directional displacement orbits was generally one less than the value obtained under conditions of unidirectional cyclic loading.

Strength degradation at and beyond the attainment of the dependable ductility capacity μ_0 can be expected to be gradual when small axial load is being carried by a bridge column. However, under larger axial compression, strength loss beyond μ_0 is sudden. This is due to the development of the wedge mechanism shown in Fig. 8. Because predictions of ductility demands are crude, special care should be taken when relating dependable ductility capacity to estimated demand. Squat columns, when subjected to significant axial compression, are not tolerant in accommodating excess ductility demands.

In order to study the effects of shear on the ductility capacity of squat columns, shear forces corresponding to relatively large average shear stresses, of the order of 3 to 4.5 MPa (400 to 650 psi), had to be applied. This necessitated the use of relatively large vertical reinforcement content (3.2%). In most bridge piers considerably less reinforcement, and hence flexural capacity, would be encountered. In such cases the spiral reinforcement, required for purposes of confinement of the compressed concrete in the plastic hinge region, may be adequate to ensure that the effect of shear on ductility capacity will be negligible.

SUMMARY

The aim in highlighting some features of recent experimental work in New Zealand was to emphasise that conditions in potential plastic regions of beams, columns or walls represent environments in which shear mechanisms may function differently from those encountered in similar members subjected to monotonic action only. If good energy dissipation capacity is to be maintained in reinforced concrete ductile systems, the influence of shear on inelastic seismic response must be taken into account. This may necessitate the use of unconventional detailing of the reinforcement in these clearly defined plastic regions.

REFERENCES

1. Megget, L.M. and Fenwick, R.C., "Seismic Behaviour of a Reinforced Concrete Portal Frame Sustaining Gravity Loads", Proceedings of the Pacific Concrete Conference, Auckland, New Zealand, 1988, V.4, pp.41-52.

2. NZS 3101:1982, Code of Practice for the Design of Concrete Structures, Standards Association of New Zealand, 127 p.

3. ACI 318-89, Building Code Requirements for Reinforced Concrete, American Concrete Institute, Detroit, 1989, 353 p.

4. Paulay, T. and Mestyanek, J.M., "Structural Walls of Limited Ductility", Proceedings of the Pacific Concrete Conference, Auckland, New Zealand, 1988, pp.207-218.

5. Ang Beng Ghee, Priestley, M.J.N. and Paulay, T., "Shear Strength of Circular Reinforced Concrete Columns", ACI Structural Journal, V.86, No. 1, Jan-Feb. 1988, pp.45-59.

6. Wong Yuk Lung, Paulay, T. and Priestley, M.J.N., "Squat Circular Bridge Piers under Multi-Directional Seismic Attack", Research Report 90-4, Department of Civil Engineering, University of Canterbury, Christchurch, New Zealand, 276 p.

34

NEED FOR SHEAR WALLS IN CONCRETE BUILDINGS FOR SEISMIC RESISTANCE.

OBSERVATIONS ON THE PERFORMANCE OF BUILDINGS WITH SHEAR WALLS

IN EARTHQUAKES OF THE LAST THIRTY YEARS.

by

MARK FINTEL
Consulting Engineer
Boca Raton, Florida

ABSTRACT

Concrete buildings containing shear walls, whether reinforced or unreinforced have exhibited superior seismic performance in earthquakes of the last 30 years, particularly as compared with rigid frame-type reinforced concrete buildings.

While hundreds of frame-type structures collapsed or were severely damaged in past earthquakes, not a single structure containing shear walls collapsed.

Experience from past earthquakes clearly indicates that a higher overall rigidity in concrete buildings to reduce their seismic distortions and thus to insure their superior performance, can be accomplished economically by using shear walls.

This paper describes observations of shear wall performance in severe earthquakes, in which modern reinforced concrete buildings stood the test of violent shaking, starting from the Chilean earthquake of 1960, through most of the subsequent strong earthquakes, up until the most recent disastrous Armenian earthquake of December 1988.

Despite the undeniable observed superiority of shear wall-type concrete structures over concrete frame-type structures, the Codes continued, up until the last decade, to give preference to concrete ductile frame structures (which are subject to higher distortions), while placing a substantial penalty on shear walls. This code approach was due to the lack of experimental and analytical background information on shear wall behavior.

While a large body of information on shear walls has been accumulated during the 1980's, more experimental and analytical studies are needed to create a solid basis for a rational seismic design approach. The availability of such information would then encourage a wider utilization of shear walls for earthquake resistance.

INTRODUCTION

The evolution of the modern approach to earthquake engineering of buildings started in the 1950's, at a time of intense construction activity following the conclusion of the second world war.

Early attempts to provide earthquake resistance in buildings were based on rather crude assumptions about structural behavior and were handicapped by a lack of proper analytical tools as well as earthquake records.

Observations of the behavior of structures subjected to actual earthquakes, analytical studies, laboratory testing of structural elements and subassemblies, and accumulation of earthquake records over the last four decades have all contributed towards putting the subject of earthquake-resistant structural design on a rational basis.

Initially the ductile moment resistant frame evolved in the 1950's out of the moment resistant frame which, at that time, was the only system for multistory buildings for both steel and concrete. By adding ductility to the then available system, a convenient solution was created to the problem of earthquake resistance. This concept of the ductile moment resistant frame remained frozen until the late 1970's.

However, in the meantime, better and more efficient structural systems for multistory structures (both in steel and in concrete) were developed for wind resistance, incorporating shear walls or trusses for concrete or steel structures, respectively. Pure frames for highrise buildings have almost disappeared, being technically less efficient and economically not viable.

During the 1960's, 1970's and 1980's a large amount of very significant analytical and experimental research, carried out throughout the world, accumulated a wealth of sophisticated information on the earthquake response of structural systems including those containing shear walls. Also, starting from the mid 1950's a substantial body of information was assembled on performance of buildings in actual earthquakes.

Most of the analytical research in the 1950's and 1960's on the response of structures to earthquakes has emphasized the importance of a ductile moment resistant frame to reduce the seismic forces. Presupposing higher seismic forces in more rigid structures, and assuming brittle response of shear walls to in-plane lateral forces, it has been concluded that severe damage could be expected in shear wall buildings.

Based on this thinking, shear walls were considered undesirable for earthquake resistance and buildings were built primarily with moment resistant frames. While in some countries a degree of ductility was built into those frames as required by codes, in the majority of countries, and particularly in those economically less advanced, the frames were brittle and incapable of withstanding severe earthquake shaking without severe damage. Consequently, many people in seismic regions of the world live in death traps, as has been seen in many of the earthquakes of the last four decades.

This paper highlights some of the observations by the author of the behavior of buildings containing shear walls in the earthquakes of the last thirty years.

CHILE, May, 1960.

Among the first reported observations concerning shear walls were those from the Chilean earthquake of 1960, as contained in Advanced Engineering Bulletin #6, issued by the Portland Cement Association. The report states:"....the Chilean experience confirms the efficiency of concrete shear walls in controlling structural and nonstructural damage in severe earthquakes. There were instances of cracking of shear walls, but this did not affect the overall performance of the buildings. In all cases observed, the reinforcement held the walls together in good alignment, even though the amounts of steel exposed after spalling were, as a rule, less than specified by code. In essence, the walls continued to function after damage had occurred...."

SKOPJE, YUGOSLAVIA, July 1963.

In the earthquake of Skopje, Yugoslavia of July 1963, a number of buildings containing unreinforced concrete walls across the building or in the core exhibited no damage, due to inhibited interstory distortions (Fig. 1), except for some cracking at the underside of the spandrel beams (Fig. 2). This was despite the documented severe shaking undergone by these buildings. Some frame buildings collapsed in this quake.

CARACAS, VENEZUELA, July 1967.

A noteworthy case of exemplary shear wall building behavior was provided by the 17-story Plaza One Building in Caracas (Fig. 3). It was the only shear wall-type structure in Caracas that had a load carrying system consisting of walls in both directions (Fig. 4). Of the surrounding buildings several collapsed and all others suffered severe, sometimes irreparable, damage. The Plaza One Building went through the earthquake without any damage whatsoever.

The typical multistory building in Caracas, as well as in most of South and Central America, contained a relatively flexible reinforced concrete frame (Fig. 5) with brittle clay tile infill walls.

There were a number of collapsed multistory buildings in Caracas and in most of the other multistory frame buildings, the hollow clay tile infill walls suffered very severe damage.

SAN FERNANDO, CALIFORNIA, February 1971.

The six-story Indian Hill Medical Center with a shear wall-frame interactive system survived the earthquake needing only moderate repairs (Fig. 6), while the neighboring eight-story Holly Cross Hospital (Fig. 7) with a frame structure was severely damaged and subsequently demolished.

There was extensive damage to many buildings and bridges; the Olive View hospital being the most widely described in the engineering literature.

MANAGUA, NICARAGUA, December 1972.

The severe earthquake of Managua provided a particularly telling example of the difference in earthquake resistance between buildings that contained shear walls and those that did not.

The eighteen-story Banco de America, a shear wall-frame interactive system with substantial core walls (Fig. 9), and the sixteen-story Banco Central, a reinforced concrete frame structure (Fig. 10), stood across the corner from each other (Fig. 8). While the Banco de America suffered little damage, the Banco Central was very severely damaged and had to be demolished.

Another pair for comparison from Managua, although separated by a distance (Fig. 11), was the five-story concrete frame Insurance Building, which suffered severe damage, and the five-story Enaluf Building containing a relatively large core in addition to the frame. The Enaluf went through the earthquake exceptionally well.

Also, the National Theatre of Managua (Fig's 12 and 13) did not suffer any damage whatsoever, thanks to the concrete wall enclosing the auditorium.

BUCAREST, ROMANIA, March 1977.

In Bucarest, where 35 multistory buildings collapsed, hundreds of highrise apartment buildings containing concrete walls, some along the corridors, others across the buildings, remained intact and mostly without damage.

MEXICO CITY, October 1985.

The extreme extent of destruction in Mexico City provided further evidence of the consequences of not including shear walls to stiffen concrete frames of multistory buildings. About two hundred and eighty multistory frame buildings (six to fifteen stories) collapsed in this earthquake (Fig. 14); none of those contained shear walls. As in most South and Central American countries reinforced concrete frames without stiffening walls were the predominant structural system for multistory buildings.

CHILE, 1985.

The 1985 Chilean earthquake received relatively little attention in the profession, despite the fact that its magnitude was similar to that of the Mexican earthquake of the same year. This earthquake went almost unnoticed by the profession, since there were no dramatic collapses, notwithstanding the severity of the event.

The primary reason for the minimal damage was the widely used engineering practice in Chile of incorporating concrete walls into their buildings to control drift. It should be noted, that the detailing practice for shear walls in Chile generally does not follow the ductile detailing requirements of seismic regions in the US, but rather follows conventional detailing as required in previous ACI-318 Codes.

The exceptionally good performance of Chilean buildings during the 1960 and particularly the 1985 earthquake bears testimony that drift control provided by shear walls can protect relatively nonductile framing elements.

ARMENIA, December 1988.

A further evidence of the benefit of incorporating concrete walls into multistory structures, or, conversely, the negative consequence of omitting shear walls, was reaffirmed in the Armenian earthquake of 1988.

In that earthquake a total of seventy two frame-type buildings collapsed, and one hundred and forty nine were severely damaged in the four affected cities of Leninakan, Spitak, Kirovakan and Stepanavan, while at the same time all of the twenty one large panel-type buildings in those cities went through the earthquake without damage (Fig. 15).

Particularly, in the city of Spitak, which was almost totally destroyed, the only structure remaining standing and undamaged was a five-story building constructed as a large panel structure, with concrete wall panels in both directions (Fig. 16).

DISCUSSION

In previous decades significant attention was devoted to ductility details of structural systems, some of which proved inappropriate for seismic resistance of concrete structures. Ductility details incorporated into the wrong structural system are wasted, while creating a false sense of security.

During the early days of earthquake engineering many professionals confused ductility with flexibility. As a result a large number of flexible buildings were built in many seismic areas of the world. Although some of these buildings may have a reasonable degree of built-in ductility, their responses in future earthquakes have the potential to cause large economic losses, due to large interstory distortions.

In our modern buildings the cost of the structure may be as low as only 20% of the total cost, while the remaining 80% is for the architectural, mechanical and electrical components. Thus, it is of primary importance to select a structural system with the best chance of providing both life safety and property protection in future earthquakes. For concrete structures, shear walls have demonstrated the ability to fulfill these requirements at the least cost.

Considering the suitability of structural systems as related to the functional requirements of buildings, we can divide the universe of multistory buildings into residential and commercial occupancies. There is no question that for residential buildings shear walls can be used as the primary, or even the only vertical load carrying elements, thus serving the double function of carrying the loads and dividing the space. In commercial buildings where large unobstructed space is a functional requirement, a shear wall-frame interactive system provides both rigidity and space flexibility.

Ductility details for shear walls, which were developed as a result of recent laboratory tests and analytical investigations, and incorporated into some codes, have not yet been tested in actual earthquakes. The inclusion of ductility details in shear walls will unquestionably improve the ductile properties of the walls. However, to extent to which shear wall ductility is actually utilized during earthquakes, and how such ductility affects the performance of the connected frames, remains to be determined, using sophisticated dynamic response studies, or in actual earthquakes.

In order to design a shear wall to behave in a ductile manner, which requires that its strength be governed by flexure rather than by shear, its shear capacity must be known, and be larger than the shear corresponding to its moment capacity. We need to know not only the ultimate shear capacity, but also what happens between the onset of shear cracking and shear failure.

Whether and to what extent the grinding within shear cracks, caused by reversible cycles of lateral movement, can serve as an energy dissipation mechanism needs to be determined and has not yet been sufficiently investigated.

If we look at the task ahead of us, I believe that the research community can provide a great service by further developing experimental and analytical information on shear walls, so that their design and proportioning can be brought to the same level of confidence as presently available for seismic beams and columns.

The following is needed:
1. Experimentally developed force-deformation characteristics of various configurations of shear walls throughout their elastic and inelastic ranges of seismic response; knowledge of moment and shear capacities will enable us to proportion and reinforce the sections to assure ductility.
2. Analytically derived inelastic dynamic responses providing both strength and ductility demands on shear walls contained in shear wall structures as well as in shear wall-frame interactive systems.

These two types of information will give us a realistic picture of both sides of the demand-capacity inequality: the demands for strength and deformability and the available capacities. Only then are we going to be able to create better concrete structures by using ductility details judiciously, instead of wasting them on structural elements that do not enter the inelastic range during earthquake response.

CONCLUDING REMARKS

During the earthquakes of the last three decades, buildings containing shear walls exhibited extremely good earthquake performance. In most cases the shear walls were reinforced in the traditional manner for gravity and overturning, without consideration given to special details for ductility as required in recent United States Codes.

To the best knowledge of the author, who investigated and reported on the behavior of modern structures in a dozen earthquakes throughout the world, starting in 1963, not a single concrete building containing shear walls has ever collapsed. While there were cases of cracking of various degrees of severity; no lives were lost in these buildings. Of the hundreds of concrete structures that collapsed, most suffered excessive interstory distortions that in turn caused shear failures of columns. Even where collapse of frame structures did not occur and no lives were lost, the large interstory distortions of frames caused significant property losses.

The above should not be taken to imply that frame structures built by the present (1991) advanced codes would also collapse in severe earthquakes. It has been demonstrated, however, that buildings continuing shear walls, even if only conventionally reinforced, do withstand severe earthquakes, mostly without damage.

After observing the devastations and the resulting staggering loss of life in the many earthquakes, particularly those in Managua, 1972, Mexico City, 1985 and Armenia, 1988, the author believes it to be the responsibility of the engineering profession to make sure that residential buildings in particular be constructed with shear walls. Whether such walls are made of plain concrete, traditionally reinforced or reinforced for ductility will depend upon the economic capacity of a given society and on engineering judgment; however, they all protect life and in most cases also provide good protection of property.

We cannot afford to build concrete buildings meant to resist severe earthquakes without shear walls.

Figure 1. Skopje, Yugoslavia. Buildings at the River with unreinforced shear walls across the buildings.

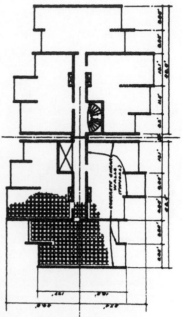

Figure 2. Same buildings - cracking of plaster at underside of spandrel beams.

Figure 4. Plaza I Building - Structural layout.

Figure 3. Caracas, Venezuela. Plaza I building. Remains of a 10-story collapsed building in front.

Figure 6. San Fernando, California. Indian Hill Medical Center.

Figure 8. Managua, Nicaragua. Banco Central and Banco de America.

Figure 5. Typical flexible concrete frame.

Figure 7. San Fernando, California. Holy Cross Hospital.

41

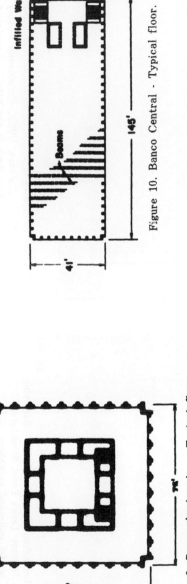

Figure 9. Banco de America - Typical floor.

Figure 10. Banco Central - Typical floor.

Figure 12. Managua, Nicaragua, National Theater.

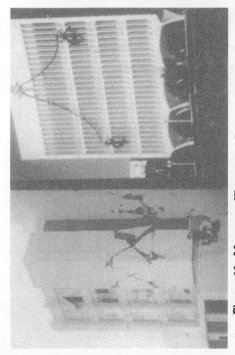

Figure 11. Managua, Nicaragua. 5-story Insurance Building and 5-story Enaluf Building.

Figure 14. Mexico City. Collapsed 14-story frame building.

Figure 16. Spitak, Armenia, USSR. Undamaged 5-story Large Panel Structure in the totally devastated city.

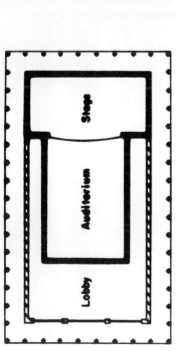

Figure 13. National Theater - Plan.

Figure 15. Leninakan, Armenia, USSR. Surviving 10-story Large Panel Structures amidst collapsed neighbors.

STRESS-STRAIN RELATIONSHIPS
FOR REINFORCED CONCRETE MEMBRANE ELEMENTS

THOMAS T. C. HSU, ABDELDJELIL BELARBI, AND XIAO-BO PANG
Professor/Ph.D. Candidate/ Ph.D. Candidate
Dept. of Civil and Environmental Engineering, University of Houston,
Houston, Texas, 77204, U. S. A.

ABSTRACT

Based on the tests of 34 full-size reinforced concrete panels subjected to membrane stresses, analytical expressions are derived for the stress-strain relationships of concrete in compression and tension, as well as steel reinforcement. Softened coefficients for concrete compression are found for both stress and strain at peak point of the curve. The stress-strain curve of concrete in tension is applicable to the whole range of pre-cracking, post-cracking and post-yielding. The stress-strain curve of mild steel bars stiffened by concrete does not exhibit a yield plateau and is valid up to strain hardening.

INTRODUCTION

Wall-type and shell-type reinforced concrete structures have received considerable attention in recent years. An element isolated from such a structure is subjected to membrane stresses. Since the understanding of the behavior of a reinforced concrete element is the key to the analysis of the whole structure, a softened truss model theory has been developed for the nonlinear analysis of such membrane elements[1]. An accurate prediction by the softened truss model depends strongly on the material laws of the concrete and steel in the elements. Using the universal panel tester constructed at the University of Houston, extensive tests of reinforced concrete panels 55 in. (1.4 m) square and 7 in. (17.8 cm) thick have been carried out to determine the membrane stress-strain relationships. Based on these tests, new material laws are reported in this paper.

BASIC EQUATIONS FOR MEMBRANE ELEMENTS

A reinforced concrete membrane element is subjected to in-plane normal stresses and shear stresses. The directions of the longitudinal and transverse steel bars are designated as the ℓ- and t-axes, respectively, constituting the ℓ-t coordinate system. Accordingly, the normal stresses are σ_ℓ and σ_t, and the shear stresses are $\tau_{\ell t}$, Figure 1 (a). The principal stresses and strains due to the applied stresses σ_ℓ, σ_t and $\tau_{\ell t}$ are oriented in the 2-1 axes, Fig. 1 (b). The angle between the direction of the principal compressive stress (2-axis) and the longitudinal steel (ℓ-axis) is denoted as α_2.

After the development of diagonal cracks in the 2-direction, Fig. 1 (a), a truss action is formed consisting of the diagonal concrete struts in compression and the steel bars in tension as shown in Figure 1 (c). The direction of the diagonal concrete struts are assumed to coincide with the direction of the principal compressive stress and strain in the concrete, which is inclined at an angle α to the longitudinal steel bars. This direction of the diagonal concrete struts is designated as the d-direction. Taking the direction perpendicular to the d-axis as the r-axis, we have a d-r coordinate system as shown in Figure 1 (d). The angle α deviates from the fixed angle α_2, and will be called the rotating angle. The principal stresses in the concrete are σ_d and σ_r.

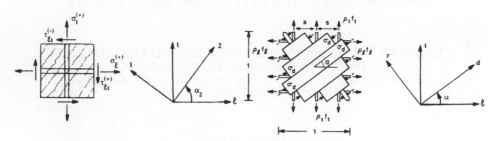

(a) Stresses in ℓ-t coordinate system (b) Principal 2-1 coordinate system (c) Truss model (d) Principal d-r coordinate system

Figure 1. Definitions of stresses and coordinate systems

Equilibrium Equations

From the three equilibrium conditions of the truss model, it can be shown[2] that the stresses in the concrete satisfy Mohr's stress circle. Assuming that the steel bars can resist only axial stresses, then the superposition of concrete stresses and steel stresses results in:

$$\sigma_\ell = \sigma_d \cos^2\alpha + \sigma_r \sin^2\alpha + \rho_\ell f_\ell \qquad \boxed{1}$$

$$\sigma_t = \sigma_d \sin^2\alpha + \sigma_r \cos^2\alpha + \rho_t f_t \qquad \boxed{2}$$

$$\tau_{\ell t} = (-\sigma_d + \sigma_r) \sin\alpha \cos\alpha \qquad \boxed{3}$$

where

$\sigma_\ell, \sigma_t, \tau_{\ell t}$	= normal and shear stresses on the element in the ℓ-t coordinate, respectively,
σ_d, σ_r	= principal stresses in the d and r directions, respectively,
α	= angle of inclination of the d-axis with respect to ℓ-axis,
ρ_ℓ, ρ_t	= reinforcement ratios in the ℓ and t directions, respectively,
f_ℓ, f_t	= steel stresses in the ℓ and t directions, respectively.

Compatibility Equations

From the compatibility condition of the truss model, it can also be shown[2] that the average strains (or smeared strains) satisfy Mohr's strain circle, giving:

$$\varepsilon_\ell = \varepsilon_d \cos^2\alpha + \varepsilon_r \sin^2\alpha \qquad \boxed{4}$$

$$\varepsilon_t = \varepsilon_d \sin^2\alpha + \varepsilon_r \cos^2\alpha \qquad \boxed{5}$$

$$\gamma_{\ell t} = 2(-\varepsilon_d + \varepsilon_r) \sin\alpha \cos\alpha \qquad \boxed{6}$$

where

$\varepsilon_\ell, \varepsilon_t, \gamma_{\ell t}$ = average normal and shear strains in the ℓ-t coordinate, respectively,

$\varepsilon_d, \varepsilon_r$ = average principal strains in the d and r directions, respectively .

The solution of the six equations above, (1) to (6), requires four stress-strain relationships for materials: two for concrete in the d and r directions and two for steel in the ℓ and t directions. These relationships are now introduced.

Material Laws for Concrete

The two stress-strain relationships for concrete can be expressed by four equations: three for compression and one for tension. The three equations for concrete in compression along the d direction are:

$$\sigma_d = f_1 (\varepsilon_r, \varepsilon_d, \zeta_{\sigma o}, \zeta_{\varepsilon o}) \qquad \boxed{7}$$

where $\zeta_{\sigma o}$ and $\zeta_{\varepsilon o}$ are the softened coefficients for stress and strain at peak point, respectively, expressed as:

$$\zeta_{\sigma o} = f_2 (\varepsilon_r, \varepsilon_d) \qquad \boxed{8}$$

$$\zeta_{\varepsilon o} = f_3 (\varepsilon_r, \varepsilon_d) \qquad \boxed{9}$$

f_1, f_2, and f_3 are functions to be determined from tests for concrete compression. The equation for concrete tension in the r direction is

$$\sigma_r = f_4 (\varepsilon_r, \varepsilon_d) \qquad \boxed{10}$$

where f_4 is a function to be determined from tests for concrete tension.

Material Laws for Steel

The two equations for reinforcing steel are:

$$f_\ell = f_5(\varepsilon_\ell) \qquad \boxed{11}$$

$$f_t = f_6(\varepsilon_t) \qquad \boxed{12}$$

where f_5 and f_6 are functions to be determined from tests for mild steel.

Solution Method

The twelve governing equations for a membrane element, Eqs. $\boxed{1}$ to $\boxed{12}$, contain fifteen unknown variables. These unknown variables include seven stresses (σ_ℓ, σ_t, $\tau_{\ell t}$, σ_d, σ_r, f_ℓ, f_t) and five strains (ε_ℓ, ε_t, $\gamma_{\ell t}$, ε_d, ε_r), as well as the the angle α and the two material coefficients $\zeta_{\sigma o}$ and $\zeta_{\varepsilon o}$. If three unknown variables are given (usually the three applied stresses, σ_ℓ, σ_t and $\tau_{\ell t}$), then the remaining twelve unknown variables can be solved by the twelve equations. An efficient algorithm to solve this set of twelve equations has been developed[3].

STRESS-STRAIN CURVES OF CONCRETE AND STEEL

Compressive Stress-Strain Curve of Concrete

Two softening effects were observed in the compression stress-strain relationships of concrete struts. The peak stress σ_p was found to be less than the standard cylinder strength of f_c', and the strain at peak stress, ε_p, may be less than the corresponding strain $\varepsilon_o = 0.002$. Define the softened coefficients $\zeta_{\sigma o}$ and $\zeta_{\varepsilon o}$, for stress and strain at peak point, respectively, as follows:

$$\sigma_p = \zeta_{\sigma o} f_c' \tag{13}$$

$$\varepsilon_p = \zeta_{\varepsilon o} \varepsilon_o \tag{14}$$

Assuming a parabolic curve both before and after the peak point, Eq. $\boxed{7}$ is expressed as follows:

$$\varepsilon_d / \zeta_{\varepsilon o} \varepsilon_o \leq 1 \qquad \sigma_d = \zeta_{\sigma o} f_c' \left[2 \left(\frac{\varepsilon_d}{\zeta_{\varepsilon o} \varepsilon_o} \right) - \left(\frac{\varepsilon_d}{\zeta_{\varepsilon o} \varepsilon_o} \right)^2 \right] \qquad \boxed{7a}$$

$$\varepsilon_d / \zeta_{\varepsilon o} \varepsilon_o > 1 \qquad \sigma_d = \zeta_{\sigma o} f_c' \left[1 - \left(\frac{\varepsilon_d / \zeta_{\varepsilon o} \varepsilon_o - 1}{2 / \zeta_{\varepsilon o} - 1} \right)^2 \right] \qquad \boxed{7b}$$

Eqs. $\boxed{7a}$ and $\boxed{7b}$ are plotted in Figure 2.

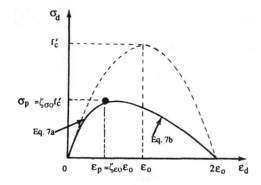

Figure 2. Softened stress-strain curve of concrete

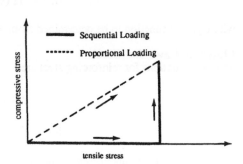

Figure 3. Two types of load paths used in tests

The softening effects are apparently caused by the diagonal cracking in the r-direction. Consequently, the softened coefficients $\zeta_{\sigma o}$ and $\zeta_{\varepsilon o}$ must be related to a parameter that measures the severity of cracking. Since the most important parameter to measure the severity of cracking is the tensile strain in the r-direction, ε_r, the peak-softening coefficient for stress is related to ε_r as follows:

$$\zeta_{\sigma o} = \frac{0.9}{\sqrt{1 + k_\sigma \varepsilon_r}} \qquad \boxed{8a}$$

and the peak softening coefficient for strain is

$$\zeta_{\varepsilon O} = \frac{1}{\sqrt{1 + k_\varepsilon \varepsilon_r}} \qquad \boxed{9a}$$

The constant 0.9 in the numerator of Eq. $\boxed{8a}$ takes into account the size effect, the shape effect and the loading-rate effect between testing of standard cylinders and testing of panels. The constants $k_{\sigma O}$ and $k_{\varepsilon O}$ in both Eqs. $\boxed{8a}$ and $\boxed{9a}$ depend on three factors: (1) the α_2 angle between the longitudinal bars and the applied principal compressive stress, (2) the α angle between the longitudinal bars and the principal compressive stress in concrete, and (3) the load path of the tensile and the compressive applied principal stresses.

Two α_2 angles were used in the tests, namely 90° and 45°. For each α_2 angle, two load paths were used, as shown in Fig. 3, namely, proportional loading (PL), and sequential loading with the tensile stresses applied first (SL). In one series of panels the ratio of the longitudinal steel to transverse steel is varied such that α_2-α varies from 0 to 19°. The coefficients k_σ and k_ε for the five cases are given in the following table:

	k_σ	k_ε
$\alpha_2 = \alpha = 90°$ and SL	250	0
$\alpha_2 = \alpha = 90°$ and PL	400	500
$\alpha_2 = \alpha = 45°$ and SL	400	160
$\alpha_2 = \alpha = 45°$ and PL	400	160
$\alpha_2 = 45°$, $\alpha < \alpha_2$ and PL	$400(1 + \frac{\alpha_2 - \alpha}{12})$	$160(1 + \frac{\alpha_2 - \alpha}{18})$

The stress-softened coefficients from Eq. $\boxed{8a}$ are compared to the tests in Figure 4, and the strain-softened coefficients from Eq. $\boxed{9a}$ in Figure 5. The formula proposed by Vecchio and Collins in 1986[4] are also shown in the two figures.

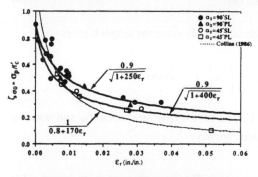

Figure 4. Stress-softened coefficient of concrete

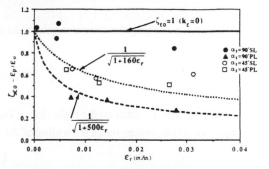

Figure 5. Strain-softened coefficient of concrete

Tensile Stress-Strain Curve of Concrete

A bar surrounded by concrete and subjected to a tensile force P is shown in Figure 6. At the two cracks indicated, the steel stress will be designated σ_{sO}. Between these two cracks, however, the steel stress $\sigma_s(r)$ at any section r will be less than σ_{sO}, and the difference will be carried by the concrete in tension. From the longitudinal equilibrium of forces at any section r

48

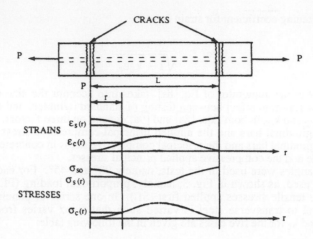

Figure 6. Stress and strain distributions along concrete-stiffened bars

between the two cracks

$$\sigma_{so} = \sigma_s(r) + \frac{1}{\rho} \sigma_c(r) \tag{15}$$

where ρ is the percentage of steel $= A_s/A_c$.

To relate the concrete stress σ_r and strain ε_r in the r-direction, we note that ε_r is an *average* strain measured along a length that crossed several cracks. Hence

$$\varepsilon_r = \frac{1}{L} \int_0^L \varepsilon_s(r) \, dr \tag{16}$$

Similarly, averaging the stresses on the right side of Eq. (15) along the length L gives

$$\sigma_{so} = \frac{1}{L} \int_0^L \sigma_s(r) \, dr + \frac{1}{\rho} \left(\frac{1}{L} \int_0^L \sigma_c(r) \, dr \right) \tag{17}$$

Denoting the first term on the right side of Eq. (17) as the *average* steel stress σ_s, and noticing $\varepsilon_s(r) = \sigma_s(r)/E_s$ before the first yielding of the steel, we have

$$\sigma_s = \frac{1}{L} \int_0^L \sigma_s(r) \, dr = E_s \varepsilon_r \tag{18}$$

If the parenthesis in the second term on the right side of Eq. (17) is defined as the average concrete stress in tension, σ_r, then Eq. (17) becomes

$$\sigma_{so} = \sigma_s + \frac{1}{\rho} \sigma_r \tag{19}$$

Substituting $\sigma_s = E_s\varepsilon_r$ (before yield of steel) into Eq. (15), we derived the relationship between the average concrete stress σ_r and the average strain ε_r:

$$\sigma_r = \rho\,(\sigma_{so} - E_s\varepsilon_r) = \rho\left(\frac{P}{A_s} - E_s\varepsilon_r\right) = \frac{P}{A_c} - \rho E_s\varepsilon_r \tag{20}$$

For an average strain ε_r and its corresponding load P measured in a panel test, an average concrete stress σ_r was calculated by Eq. (20). Plotting σ_r vs. ε_r consecutively for all load stages, we arrived at an experimental stress-strain curve for concrete stress in the r-direction. It was found that the best mathematical form to fit the experimental stress-strain curve is expressed by two curves as shown in Figure 7 (a) and (b):

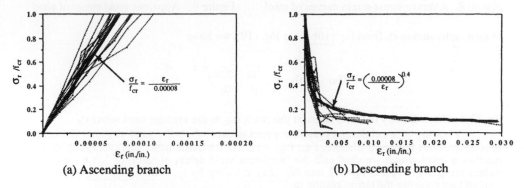

(a) Ascending branch   (b) Descending branch

Figure 7. Average concrete stress in the cracking direction

Ascending branch $(\varepsilon_r \leq \varepsilon_{cr})$ $$\sigma_r = E_c\,\varepsilon_r \qquad \boxed{10a}$$

Descending branch $(\varepsilon_r > \varepsilon_{cr})$ $$\sigma_r = f_{cr}\left(\frac{\varepsilon_{cr}}{\varepsilon_r}\right)^c \qquad \boxed{10b}$$

Eq. $\boxed{10b}$ was first proposed by Tamai et al.[5] at the University of Tokyo based on tension members. This equation was also found to be applicable to membrane elements as shown in Figure 7 (b), and the best experimental values for E_c, f_{cr}, ε_{cr} and c are:

E_c = 47,000 $\sqrt{f_c^r}$, where f_c^r and $\sqrt{f_c^r}$ are in psi

f_{cr} = tensile cracking strength of concrete, taken as 3.75 $\sqrt{f_c^r}$

ε_{cr} = strain at cracking of concrete, taken as 0.00008 in./in.

c = a constant, equal to 0.4.

Average Stress-Strain Curve of Steel

The stress-strain curve of mild steel is usually assumed to be elastic-perfectly plastic, based on tests of bare bars. However, when the mild steel bars are stiffened by concrete as in a membrane element, the post-cracking *average* stress-strain relationship is quite different, Figure 8. First, the yield stress is lowered; and second, the yield plateau is replaced by a sloped post-yield stress-strain curve.

Apparent Yield Stress of Steel, f_y^*: Inserting $\varepsilon_r = \sigma_s/E_s$ and c = 0.4 into Eq. $\boxed{10b}$ and,

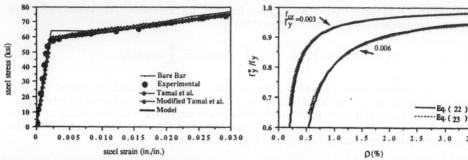

Figure 8. Average stress-strain curves of steel   Figure 9. Apparent yield stress of steel

in turn, substituting σ_r from Eq. $\boxed{10b}$ into Eq. (19), we have

$$\sigma_{so} = \sigma_s + \frac{f_{cr}}{\rho}\left(\frac{E_s\varepsilon_{cr}}{\sigma_s}\right)^{0.4} \tag{21}$$

Eq. (21) relates directly the steel stress at the crack σ_{so} to the average steel stress σ_s.

Yielding of a reinforced concrete panel occurs when the steel stress at the cracked section reaches the yield plateau, i.e., $\sigma_{so} = f_y$. At the same time, the average steel stress reaches a level which we shall call the "apparent yield stress of steel, f_y^*", i.e., $\sigma_s = f_y^*$. Substituting $\sigma_{so} = f_y$ and $\sigma_s = f_y^*$ into Eq. (21), dividing by f_y, noticing $f_{cr} = E_c\varepsilon_{cr}$ and $E_s/E_c = n$, and rearranging the terms, results in:

$$\left(\frac{f_y^*}{f_y}\right)^{0.4} - \left(\frac{f_y^*}{f_y}\right)^{1.4} = \frac{n^{0.4}}{\rho}\left(\frac{f_{cr}}{f_y}\right)^{1.4} \tag{22}$$

The apparent yield stress of steel f_y^* can be solved by Eq. (22) using trial and error method. It can be seen that the yield-stress ratio, f_y^*/f_y, is primarily a function of two variables. The primary variable is the steel percentage ρ, and the secondary variable is the cracking stress ratio, f_{cr}/f_y. The yield stress ratio f_y^*/f_y is plotted against the steel percentage ρ using the cracking stress ratio f_{cr}/f_y as a parameter in Figure 9. It can be seen that the family of curves can be closely approximated by a simple formula:

$$\frac{f_y^*}{f_y} = 1 - \frac{4.6}{\rho}\left(\frac{f_{cr}}{f_y}\right)^{1.5} \tag{23}$$

Eq. (23) is valid when ρ is greater than 0.5%, and f_{cr}/f_y less than 0.006. It has been derived for the case when the cracks are perpendicular to the steel bars, i. e. $\alpha_2 = 90°$.

When the steel bars cross the cracks at an angle of 45°, i.e. $\alpha_2 = 45°$, the yield stress ratio f_y^*/f_y is less than that given by Eq. (23) because of the kinking of steel bars at the cracks. For this case f_y^*/f_y can be expressed by:

$$\frac{f_y^*}{f_y} = 0.9 - \frac{4.6}{\rho}\left(\frac{f_{cr}}{f_y}\right)^{1.5} \tag{24}$$

The validity of Eqs. (23) and (24) is shown in Fig. 10.

Post-Yield Stress-Strain Curve of Steel: The averaging of the steel strains and the corresponding steel stresses along the length L between two cracks becomes mathematically

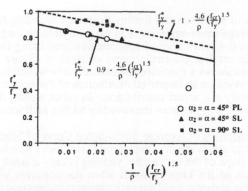

Fig. 10. Yield stress ratio f_y^*/f_y as a function of $(4.6/\rho)(f_{cr}/f_y)^{1.5}$

more complex. The integration process involved in the averaging requires numerical integration. In order to simplify this averaging process, two assumptions are made by Tamai et al.[5]:

(a) The stress distribution in the steel between two adjacent cracks is assumed to follow a full cosine curve.

(b) The average stress-strain relationship of concrete in tension, Eq. $\boxed{10b}$, is valid both before and after yielding. That is to say, Eq. $\boxed{10b}$, which had been calibrated to fit the test results before yielding, remains valid after yielding.

From the first assumption we can write

$$\sigma_s(r) = \sigma_s + (\sigma_{so} - \sigma_s)\cos\frac{2\pi r}{L} \tag{25}$$

Substituting $(\sigma_{so} - \sigma_s) = \sigma_r/\rho$ from Eq. (19) into Eq. (25) gives

$$\sigma_s(r) = \sigma_s + \frac{\sigma_r}{\rho}\cos\frac{2\pi r}{L} \tag{26}$$

Now, using the second assumption and substituting σ_r from Eq. $\boxed{10b}$ into Eq.(26) gives

$$\sigma_s(r) = \sigma_s + \frac{1}{\rho}f_{cr}\left(\frac{\varepsilon_{cr}}{\varepsilon_r}\right)^{0.4}\cos\frac{2\pi r}{L} \tag{27}$$

With these two assumptions, the averaging process is summarized as follows:

(1) Select a value of σ_s

(2) Assume an average strain ε_r

(3) Calculate the distribution of steel stress, $\sigma_s(r)$, from Eq. (27)

(4) Determine the corresponding distribution of steel strain, $\varepsilon_s(r)$, according to the stress-strain curve of bare bars.

(5) Calculate the average steel strain ε_r by numerical integration using Eq. (16)

(6) If ε_r calculated from Eq. (16) is not the same as that assumed, repeat steps (2) to (5) until the calculated ε_r is sufficiently close to the assumed value. The calculated ε_r and the selected value σ_s provide one point on the post-yield stress-strain curve.

(7) By selecting a series of σ_s values and finding their corresponding ε_r values from steps (2) to (6), the whole average stress-strain curve in the post-yielding range can be plotted.

A theoretical stress-strain curve of concrete obtained using the above procedures is compared to an experimental stress-strain curve of steel in Figure 8. The agreement is acceptable. Figure 8 also shows a modified theoretical curve which is even closer to the test points. This curve was based on an improved distribution of the assumed steel stresses, which approximates more closely the actual distribution. In other words, the first assumption of cosine distribution of steel stress has been improved by adding additional terms.

Mathematical Modelling of Average Stress-Strain Curve of Steel: As shown in Figure 8, the shape of the average stress-strain curve of mild steel looks like two straight lines joined by a curved knee. The slope of the curve after yielding is only a small fraction of that before yielding. The curvature of the knee is small when the apparent yield stress is low, but increases rapidly when the apparent yield strength ratio approaches unity.

The shape of a curve described above can best be represented by the Richard-Abbott equation:

$$f_s = \frac{(E_s - E_p)\varepsilon_s}{\left[1 + \left\{\frac{(E_s - E_p)\varepsilon_s}{f_o}\right\}^m\right]^{\frac{1}{m}}} + E_p\varepsilon_s \qquad \boxed{11/12}$$

where

f_s = stress in mild steel. f_s becomes f_ℓ or f_t when applied to the longitudinal and transverse steel, respectively

ε_s = strain in mild steel. ε_s becomes ε_ℓ or ε_t, when applied to the longitudinal and transverse steel, respectively

E_s = elastic modulus, or the slope of the first straight portion, taken as 28,000 ksi

E_p = modulus in plastic region, i.e., the slope of the second straight portion, taken as $(f_{0.05} - f_y)/(0.05 - \varepsilon_h)$

f_o = reference stress or vertical intercept of the second straight portion

m = curved knee-shape parameter

$f_{0.05}$ = stress of bare bar at a strain of 0.05 in the strain hardening region

ε_h = strain of bare bar at the beginning of the strain hardening region.

Eq. $\boxed{11/12}$ is plotted in Figure 11. Notice that there are two terms in the denominator of the first expression on the right side of the equation. If the second of the two terms is taken as zero, then Eq. $\boxed{11/12}$ becomes

$$f_s = E_s\varepsilon_s \qquad (28)$$

Eq. (28) is the first straight line through the origin with a slope of E_s.

If the first term in the denominator is taken as zero, then Eq. $\boxed{11/12}$ becomes

$$f_s = f_o + E_p\varepsilon_s \qquad (29)$$

Eq. (29) is the second straight line with a vertical intercept of f_o and a slope of E_p. The reference stress f_o are expressed as follows:

$$\frac{f_o}{f_y} = 1.0 - \frac{\varepsilon_h E_p}{f_y} \qquad \text{for } \alpha_2 = 90° \qquad (30)$$

$$\frac{f_o}{f_y} = 0.95(1.0 - \frac{\varepsilon_h E_p}{f_y})(1 - \frac{1}{4000\rho}) \qquad \text{for } \alpha_2 = 45° \qquad (31)$$

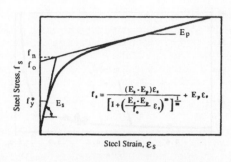

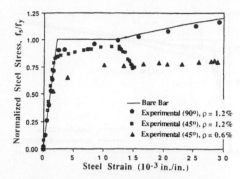

Figure 11. Graphical expression of equation for average stress-strain curve of steel

Figure 12. Reference stresses for 90° and 45° reinforcement

The validity of Eqs. (30) and (31) is shown in Fig. 12.

The two straight lines, Eqs. (28) and (29), serve as the asymptotic limits of the Richard-Abbott curve. The parameter m controls the curvature of the knee. The intersection of the two straight lines defines a knee point with a stress of f_n:

$$f_n = f_0 \frac{E_s}{E_s - E_p} \tag{32}$$

This knee point represents the limit of the knee when m approaches infinity.

The knee parameter m is observed to be a function of the apparent yield stress f_y^*, which, in turn, is a function of three variables, ρ, f_{cr} and f_y according to Eq. (23). A parametric study has been carried out to explore the effect of each of the three variables on the post-yield stress-strain curves of steel according to the 7-step iteration procedure described previously. A typical example is given in Fig. 13 which compares two such curves to the modeled curves characterized by the parameter m. In this way, m is related to the apparent yield stress f_y^* by a simple formula:

$$m = \frac{0.5}{1 - \frac{f_y^*}{f_n}} \tag{33}$$

Eq. (33) is plotted in Figure 14 and compared to the test results. Eq. (33) is valid when f_y^* is less than f_n. When f_y^* is equal to f_n, m becomes infinity, and the stress-strain curve becomes two straight lines.

CONCLUSIONS

(1) Softened truss model theory was developed for the nonlinear analysis of concrete membrane elements. The theory involves three equilibrium equations, three compatibility equations and six equations for material laws of concrete and steel. Based on the test results of 34 full-size reinforced concrete panels, stress-strain relationships have been determined for concrete in compression, concrete in tension, and steel reinforcement stiffened by concrete.

(2) The stress-strain relationship of concrete in compression can be expressed by Eqs. $\boxed{7a}$ and $\boxed{7b}$. The peak-softening coefficients are primarily a function of the tensile strain, ε_r. They also depend on the steel orientation, the load path and the angle between the applied principal stress and the concrete principal stress. Formulas for softening coefficients, Eqs. $\boxed{8a}$ and

54

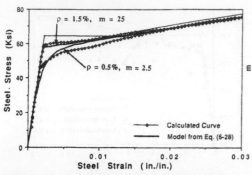

Figure 13. Modelling of Post-Yield Stress-Stress Curve of Steel

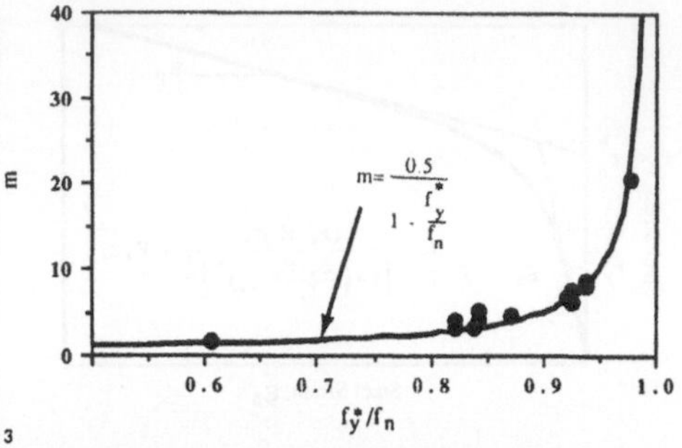

Figure 14. Parameter m as a function of Apparent Yield Stress f_y^*

9a , are developed for the cases of 90° vs 45° steel orientations, proportional vs sequential loadings and for the case of varying the angle between the two principal stresses.

(3) The concrete carries tensile stresses even after yielding of steel. The stress-strain curve for concrete in tension is expressed by Eqs. 10a and 10b .

(4) The apparent yield point of the average stress-strain curve of the steel bars is lower than that of the bare bar and can be determined accurately from Eqs. (23) and (24). Formula for calculating the average tensile stress-strain curve of steel in reinforced concrete membrane elements is proposed in Eq. 11/12 . The reference stress f_0 in the equation is given by Eqs. (30) and (31) and the parameter m is expressed by Eq. (33).

ACKNOWLEDGMENT

This research is sponsored by the National Science Foundation through Grant No. MSM-8815524.

REFERENCES

(1) Hsu, T. T. C., "Softened Truss Model Theory for Shear and Torsion," ACI Structural Journal, Vol. 85, No. 6, Nov. - Dec., 1988, pp. 624 - 635.

(2) Hsu, T. T. C., Torsion of Reinforced Concrete, Van Nostrand Reinhold, Inc., 1984, 544 pp.

(3) Hsu, T. T. C., "Nonlinear Analysis of Concrete Membrane Elements" Research Report UHCEE 90-7, Dept. of Civil and Environmental Engineering, University of Houston, Houston, Texas. (To be published in ACI Structural Journal.)

(4) Vecchio, F., and Collins, M. P., "The Modified Compression-Field Theory for Reinforced Concrete Elements Subjected to Shear," ACI Structural Journal, Proc. Vol. 83, No. 2, March-April 1986, pp. 219-231.

(5) Tamai, S., Shima, H., Izumo, J., and Okamura, H., "Average Stress-Strain Relationship in Post Yield Range of Steel Bar in Concrete," Concrete Library of JSCE, No. 11, June 1988, pp. 117-129. (Translation from Proceedings of JSCE, No. 378/V-6, Feb. 1987.)

HIGHER-STRENGTH CONCRETE PRESTRESSED DEEP BEAMS

JULIO A. RAMIREZ
Associate Professor of Civil Engineering
Purdue University
West Lafayette, IN 47907

ABSTRACT

This paper presents an evaluation of the ultimate behavior of prestressed higher-strength concrete deep beams. The experimental results of three prestressed concrete deep beams tested to failure under bending and shear are discussed. The beams are full-scale AASHTO Type I-bridge girders with concrete compressive strengths near 9000 psi (60 MPa).

INTRODUCTION

When the span-to-depth ratio of simply supported beams is less than 5, and they are loaded on one face and directly supported on the opposite face, it is customary to define these beams as deep. The usual assumption made for beams of linear distribution of strains over the depth of the section is not adequate for deep beams. The capacity of these members either in flexure or shear depends heavily on the detailing of loading and support.

To date, several studies have been conducted on the behavior and design of simply supported and continuous reinforced concrete deep beams with concrete strengths up to 6000 psi. However, prestressed higher-strength concrete deep beams have not been the subject of much dedicated effort. This paper reports the results of three tests of simply supported prestressed deep beams with concrete strengths up to 9000 psi (60 MPa) subjected to point loading.

SIGNIFICANCE OF RESEARCH

Due to the highly empirical nature of current design procedures for deep beams in the ACI Building Code, caution should be exercised in extrapolating these procedures to higher-strength concretes [1]. Higher-strength concretes have some characteristics and engineering properties that are different from those of lower-strength concretes. Internal changes resulting from short-term and sustained loads and environmental factors are known to be different. Because of the numerous advantages and continuously increasing use of higher-strength concretes, basic information on the performance of higher-strength concrete members is needed. This paper makes an immediate contribution in this regard.

EXPERIMENTAL PROGRAM

The test specimens consisted of full scale pretensioned AASHTO Type I and II beams with span-to-depth ratios between 4.08 and 4.39. The beams were fabricated at a local precast plant and, upon transfer of prestress force, were transported to the Structural Engineering Laboratory at Purdue University for testing [2]. The nominal dimensions of the beams are shown in Figures 1-3. Detailed specimen information is given in Tables 1 and 2. The mild reinforcement in the beams consisted of Grade 40 and Grade 60 bars. Stirrups and flange reinforcement consisted of Grade 40, #3 and #4 bars from separate heats. The stress-strain behavior in tension of the prestressing strand is given in Figure 4. The strand was instrumented with strain gages attached to a single wire and a 24-inch extensometer to measure axial strains. The area of the strand was determined by measuring the diameter of each wire and calculating the total area. The concrete material properties were monitored from transfer until the time of testing. Standard 6 x 12 inch test cylinders were used to measure compressive strength, split cylinder strength, and the modulus of elasticity. The modulus of rupture was determined with 6 x 6 x 18 inch flexure beams. Three flexure beams were tested at the time of transfer of prestress force and at test date.

The loading scheme used to test the beams is shown in Figure 5. A 600,000 lb. Baldwin test machine was used in all the tests. The load was applied incrementally. At each load increment, cracks were outlined and the corresponding shear force was marked on the surface of the beam. The test data was recorded with an automated data acquisition system as well as manual readings. Several types of transducers were used to record data from each test. Electrical resistance strain gages were attached to the prestressing strand and the mild reinforcement. Before placing the strain gages on the strand, it was tensioned to 5000 pounds. After the strand was instrumented, it was stressed to 33818 pounds. External gages were placed on the concrete surface to measure strains in the web and the compression flange. A combination of LVDT's, potentiometers, and mechanical dial gages were used to measure beam deflections. Mechanical dial gages were also used to detect possible strand slip at the ends of the beam during each test. Electrical strain gages were placed on the stirrup legs. The location of the gage on the stirrup leg was chosen to measure the stirrup strain following the expected direction of a potential failure shear crack.

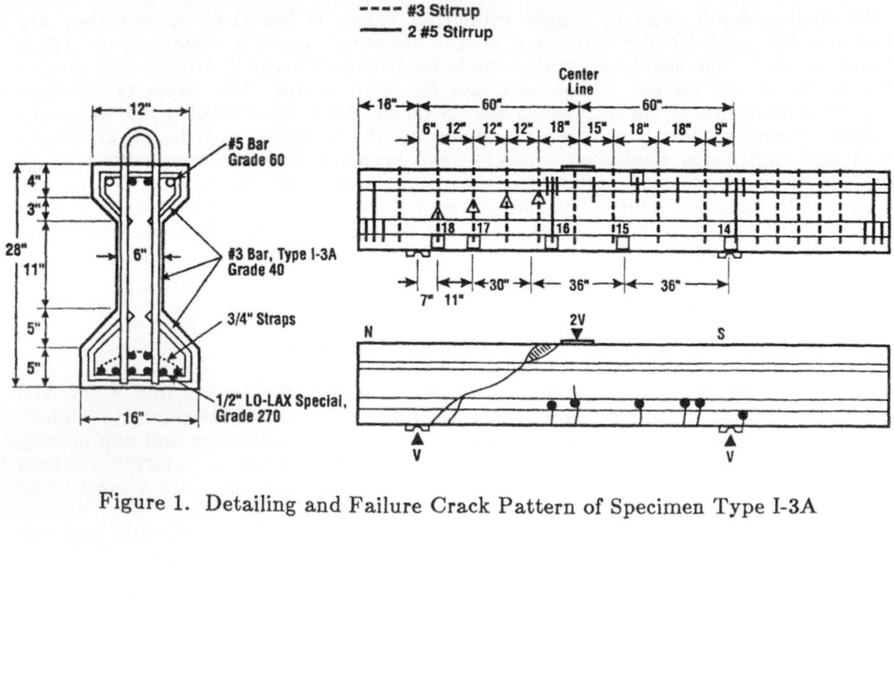

Figure 1. Detailing and Failure Crack Pattern of Specimen Type I-3A

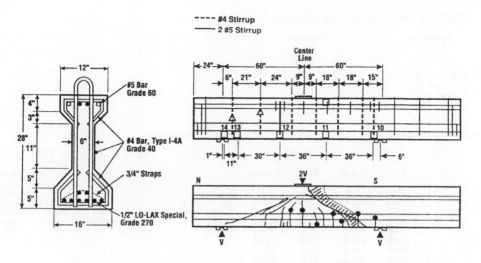

Figure 2. Detailing and Failure Crack Pattern of Specimen Type I-4A

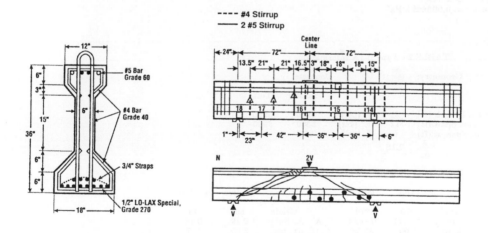

Figure 3. Detailing and Failure Crack Pattern of Specimen Type II-1A

TABLE 1 - Types I-3A and I-4A Information

Geometry: Concrete:

			Transfer	Test
Beam Length (ft)	17	f_c' (psi)	5840	8810
Test Span (ft)	10	E_c (ksi)	5620	5730
Shear Span, a (ft)	5	f_r (psi)	920	-
$\frac{a}{d_p}$	2.35			

Prestressing Strand: Mild Reinforcement:

	TYPE I-3A		TYPE I-4A				TYPE I - 3A		TYPE I-4A	
	Top	Bottom	Top	Bottom			#5 Bar	#3 Bar	#5 Bar	#4 Bar
Grade	270	270	270	270		Grade	60	40	60	60
A_{ps} (in^2)	0.1633	0.1633	0.1633	0.1633		A_s, A_v (in^2)	0.31	0.104	0.31	0.19
d_p,d_p (in)	2.00	25.5	2.00	26.00		d', rf$_y$ (in, psi)	2.00	80	2.00	165
E_{ps} (ksi)	27920	27920	27920	27920		E_s (ksi)	29020	29150	29020	29500
f_{pu} (ksi)	282.0	282.0	282.0	282.0		f_y (ksi)	64	46	64	52
f_{si} (ksi)	207.6	193.3	207.6	193.3						
f_{se} (ksi)	200.9	184.6	199.9	187.8						
P_{e1},P_{e2} (kips)	65.6	241.2	65.3	245.4						

SI Equivalents

1 in   = 25.4 mm
1 in^2 = 645.2 mm^2
l lb   = 4.448N
1 psi   = 0.006895 MPa

TABLE 2 - Type II-1A Information

Geometry: Concrete:

			Transfer	Test
Beam Length (ft)	21	f_c' (psi)	5980	8950
Test Span (ft)	12	E_c (ksi)	5580	5900
Shear Span, a (ft)	6	f_r (psi)	920	-
$\frac{a}{d_p}$	2.16			

Prestressing Strand: Mild Reinforcement:

	Top	Bottom		#5 Bar	#4 Bar
Grade	270	270	Grade	60	40
A_{ps} (in^2)	0.1633	0.1633	A_s', A_v (in^2)	0.29	0.19
d_p,d_p (in)	2.00	33.33	d', rf$_y$ (in, psi)	2.00	137
E_{ps} (ksi)	27920	27920	E_s (ksi)	28570	29500
f_{pu} (ksi)	282.0	282.0	f_y (ksi)	62	52
f_{si} (ksi)	207.6	193.5			
f_{se} (ksi)	200.9	183.5			
P_{e1},P_{e2} (kips)	35.6	245.4			

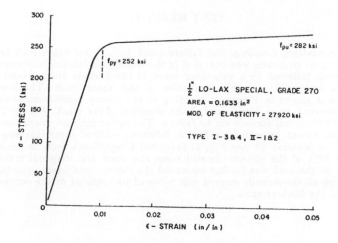

Figure 4. Stress-Strain Curve for Prestressing Strand

Plot labels:
- $f_{pu} = 282$ ksi
- $f_{py} = 252$ ksi
- $\frac{1}{2}$" LO-LAX SPECIAL, GRADE 270
- AREA = 0.1633 in²
- MOD. OF ELASTICITY = 27920 ksi
- TYPE I – 3 & 4, II – 1 & 2
- σ – STRESS (ksi)
- ε – STRAIN (in/in)

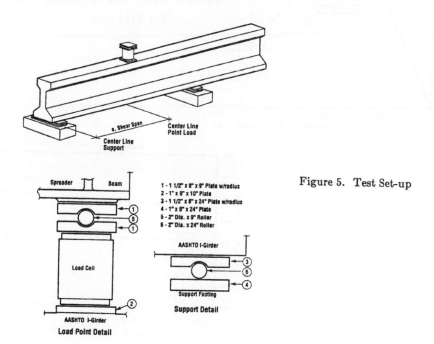

Figure 5. Test Set-up

a. Shear Span
Center Line Point Load
Center Line Support

Load Point Detail
Spreader Beam
Load Cell
AASHTO I-Girder

1 - 1 1/2" x 8" x 9" Plate w/radius
2 - 1" x 9" x 10" Plate
3 - 1 1/2" x 8" x 24" Plate w/radius
4 - 1" x 8" x 24" Plate
5 - 2" Dia. x 9" Roller
6 - 2" Dia. x 24" Roller

Support Detail
AASHTO I-Girder
Support Footing

TEST RESULTS

Beam I-3A

The instrumentation, detailing, and failure crack pattern for beam I-3A are shown in Figure 1. An 18 inch overhang was provided in the shear span labeled N. Flexural cracking was first observed, followed by a web-shear crack in the N shear span. About half of the strands showed some slip upon the formation of this shear crack. The load-centerline deflection graph is shown in Figure 6. Yielding of the stirrup reinforcement on the shear span N was observed upon formation of the diagonal shear crack. Figure 7 shows the strains in the longitudinal strands up to failure. The major inclined crack developed at a shear of 98 kips. As can be seen from Fig. 7, following inclined shear cracking the instrumented strands at location 18 (see Fig. 1) indicated a significant drop in the strain level. Approximately 50% of the strands showed some slip upon first diagonal cracking on the N-shear span. As the load was further increased the strains once again increased and at a shear of 113 kips all the strands showed slip followed by crushing of the concrete near the loading plate on the N-shear span.

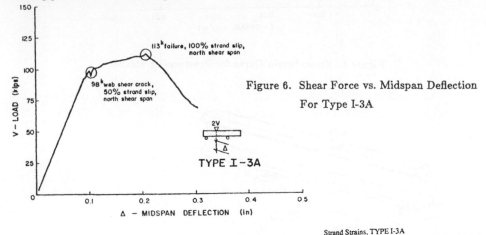

Figure 6. Shear Force vs. Midspan Deflection

For Type I-3A

Figure 7. Strains in Longitudinal Strands

for Type I-3A

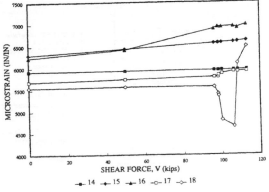

Beam I-4A

The detailing, instrumentation and failure crack pattern of Beam I-4A are shown in Figure 2. A 24 inch overhang was provided beyond the support on the N-shear span and more than 2 ft. on the S-shear span. Major inclined cracks developed in both shear spans. The first diagonal crack opened in the S-shear span at a shear of 118 kips followed by a diagonal crack in the N shear span at a shear of 120 kips. In this test the longitudinal strands showed no signs of slip up to failure. Ultimate failure was started by initial spalling of the concrete under the edge of the loaded plate on the S-shear span followed by crushing of the compression strut on the same shear span as shown by the shaded region in Figure 2. The load-mispan deflection graph for this test is shown in Figure 8. Yielding of the stirrup reinforcement was observed upon the formation of the inclined shear crack. After inclined cracking the beam started to show tied-arch behavior. This is shown by the strand strain diagrams in Figure 9 at locations 10 and 14 near the supports (see Fig. 2).

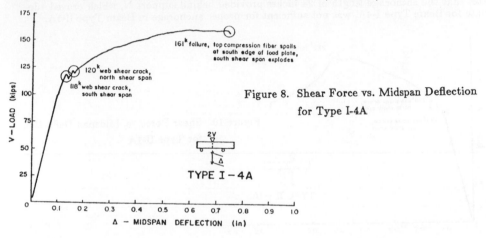

Figure 8. Shear Force vs. Midspan Deflection for Type I-4A

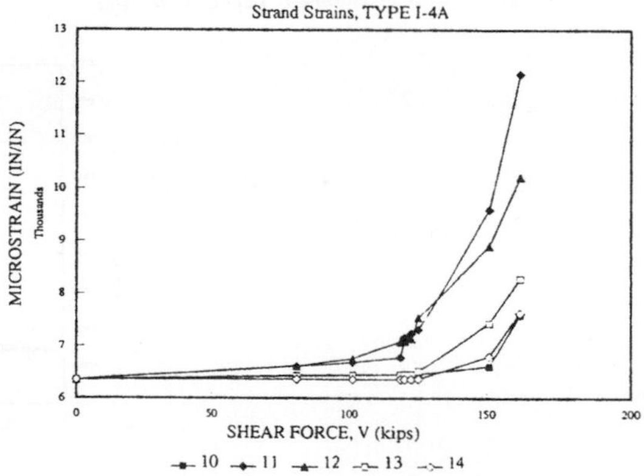

Figure 9. Strains in Longitudinal strands for Type I-4A.

Beam II-1A

Figure 3 shows the detailing, instrumentation and crack pattern at failure for Beam II-1A. A 24-inch overhang was provided on the shear span N. Major inclined shear cracks developed, first in the S-shear span at a shear of 150 kips. This was followed by a second web-shear crack on the N-shear span at a shear of 158 kips. The strands showed initially no signs of slip. Ultimate failure followed initial spalling of the concrete near the edge of the bearing plate at a shear of 222 kips. This was followed by crushing of the concrete on the top flange on the N-shear span as shown by the shaded region in Figure 3. The load-midspan deflection plot is shown in Figure 10. Yielding of the stirrups in the shear span immediately followed first web-shear cracking. After inclined cracking the beam showed tied-arch behavior as indicated by the strain gage readings on the longitudinal steel. The longitudinal steel strains, shown in Figure 11 for location 18 (see Fig. 3) at the base of the failure shear crack near the support, decreased as the load approached failure. This indicates that the anchorage length of 24 inches provided behind support N, which proved adequate for Beam Type I-4A, was not sufficient for proper anchorage in Beam Type II-1A.

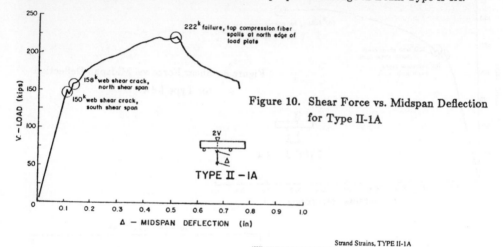

Figure 10. Shear Force vs. Midspan Deflection
for Type II-1A

Figure 11. Strains in Longitudinal Strands
for Type II-1A.

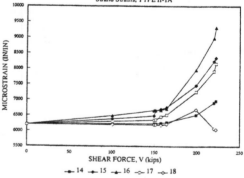

COMPARISON OF TESTS WITH ACI BUILDING CODE

Section 11.8 of the ACI Building Code [3] defines a beam with l_n/d less than 5 as a deep beam for purposes of design for shear. All the test specimens fitted this definition. However, the provisions in Sec. 11.8 for simply supported deep flexural members are intended to be used in reinforced concrete members [4,5].

As an alternative, the nominal shear capacities computed using the ACI Code Secs. 11.4 and 11.5 for shear design of prestressed members are given in Table 3. As can be seen from the ratio of test to code value in column (9), there is a clear conservatism in the use of these provisions in prestressed deep beams even in the case of specimen Type I-3A. The predicted maximum shears based on the flexural strength of the cross section are given in column (8). As can be seen from the values in column (10), the improper strand anchorage in specimen TYPE I-3A resulted in an unconservative estimate of the ultimate capacity. In the other two specimens the actual shear capacity was greater than the predicted maximum shears based on flexural strength. The incorrect predictions by the ACI Code equations result from the fact that they are based on the wrong behavioral model for deep beams. As an alternative, the strut-tie approach [6] offers a much improved behavioral model for the design of both reinforced and prestressed concrete deep beams.

TABLE 3. Shears at Failure

(1)	(2)	(3)	(4)	(5)	(6)	(7)	(8)	(9)	(10)
Beam	V_{cr} (kips)	$V_{Failure}$ (kips)	V_{ci} (kips)	V_{cw} (kips)	V_s (kips)	V_{nACI} (kips)	V_{FLEX} (kips)	(3)/(7)	(3)/(8)
Type I-3A	98	113	85	101.3	12.2	97.2	134.6	1.16	0.84
Type I-4A	118	161	86.2	101.9	25.2	111.4	134.6	1.45	1.20
Type II-1A	150	222	129.5	135.3	27.4	156.9	186.9	1.41	1.19

V_{FLEX}: Predicted maximum shear force based on flexural strength based on plane sections and rectangular stress block assumptions.

SUMMARY

The main type of behavior observed for these beams with low amount of stirrups approached tied-arch action at failure. Failures in the case of specimens with sufficient anchorage of the longitudinal tension chord were due to crushing of the concrete in the flexural compression zone. The use of higher-strength concrete will tend to strengthen this critical region as well as the web section, thus leading to higher failure loads. This, however, increases the importance of adequate detailing of the reinforcement. In particular the longitudinal tension chord as the tied-arch mechanism becomes the principal load carrying system at failure in deep beams. Also of considerable importance is the adequate detailing of the bearing plate under point loads and at supports as smaller plate sizes would lead to larger stresses for a given load level.

ACKNOWLEDGMENTS

The research was made possible through the support of the Indiana Department of Transportation and the Federal Highway Administration under the research project, Indiana HPR-2376-(024). Any opinions, findings, and conclusions expressed in this paper are those of the authors and do not necessarily reflect the views of the sponsor.

REFERENCES

1. ACI Committee 363, "State of the Art Report on High-Strength Concrete," (ACI 363R-84), ACI Journal, Proceedings, Vol. 81, No. 4, July-August, 1984, pp. 364-411.

2. Kaufman, M.K., "The Ultimate Strength and Behavior of High Strength Concrete Prestressed I-Beams, Ph.D. Thesis, Purdue University, School of Civil Engineering, December 1989, 230 pp.

3. ACI Committee 318, Building Code Requirements for Reinforced Concrete (ACI 318-89) and Commentary - ACI 318R-89, American Concrete Institute, Detroit, Michigan, November 1989, 353 pp.

4. Rogowsky D.M., and McGregor, J.G., "The Design of Reinforced Concrete Deep Beams," Concrete International: Design & Construction, V. 8, No. 8, Aug. 1986, pp. 49-58.

5. Rogowsky, D.M., and McGregor, J.G., Ong, S.Y., "Tests of Reinforced Concrete Deep Beams," Journal of American Concrete Institute, Proceedings, V. 83, No. 4, July-August 1986, pp. 614-623.

6. Schlaich, J., Schafer, I., and Jennewein, M., "Towards a Consistent Design of Structural Concrete," Journal of the Prestressed Concrete Institute, Vol. 32, No. 3, May-June 1987, pp. 74-150.

ASEISMIC BEHAVIOR OF RC AND SRC SHORT COLUMNS

Zhan-Qin Lu*, Jia-Kui Chen**

ABSTRACT

Experiments on 94 reinforced concrete (RC) short columns, 32 steel fiber reinforced concrete (SFRC) short columns and 53 steel reinforced concrete (SRC) short columns under cyclic loading have been conducted during the past years at the Southwest Jiaotong University in China. The aseismic behavior of short columns has been the object of the experimental research. In this paper, the study on the deformation capability of reinforced concrete short columns is presented. Based on the experimental results, the relationships between the deformation capability of reinforced concrete short columns and the various parameters, such as shear span ratio, axial load level ratio, web reinforcement ratio and longitudinal reinforcement ratio, are clarified, and design suggestions are also presented.

INTRODUCTION

The current seismic design codes emphasize the avoidance of using RC short columns with shear span ratio $\lambda \leq 2$ because of its poor aseismic performance. However, existance of short columns is inevitable, such as the columns of ground floor in multistorey buildings. Hence it is essential to study and understand the response of short columns subjected to reversed cyclic horizontal loading in the inelastic range.

According to the provisions of the Chinese Building Seismic Design code GBJ11-89[1], the computed interstorey elastoplastic displacement, Δu_p, responding to a maximum credible earthquake, shall be controlled by

$$\Delta u_p \leq [\theta_p] \, H \tag{1}$$

where H = critical storey height

[θ_p] = maximum allowable computed interstorey displacement angle, which shall be limited to $\frac{1}{50} \sim \frac{1}{70}$ for frame structures.

Therefore, it is required that the short columns must possess sufficient deformation capability, with $R_u \geq [\theta_p]$. The ultimate value of deformation angle R_u is considered as an index representing the deformation capability of reinforced concrete columns. The value means the useful ultimate value of deformation angle and is defined as the deformation angle when the first load of cyclic loading at a given deflection amplitude deteriorated less than 85% of maximum load or when the fifth load deteriorated less than 50% of maximum load.

In this paper, the ultimate value of deformation angle, together with the effects of main

*   Professor, Southwest Jiaotong University, China.
**   Associate Professor, Southwest Jiaotong University, China.

parameters, is statistically examined by the test results.

OUTLINE OF THE TESTS

The dimensions and parameters of specimens are shown in Table 1. Table 2 gives parts of the specimen details and test results. The specimens were subjected to constant axial compressive load and reversed cyclic horizontal load. Tests were carried out with a special loading system. The loading arrangement was shown schematically in Fig.1. The horizontal loading history was deformation-controlled with incremental displacement angle amplitudes of 0.5% or 1.0%.

TABLE 1
Summary of specimens

SPECIMEN	b×h	λ	n	ρ_{sv} (%)	ρ (%)	No. of tests
RC	20×20cm	1.5~2	0.2~1.06	0~1.03	0.78~3.14	94
SFRC	20×20cm	1~2	0.1~0.8	0.4~1.0	2.0~4.1	32
SRC	20×20cm	1~2	0.17~1.07	0~1.0	1.13~3.14	53

Note: b, h — width and depth of section, respectively $\lambda = \dfrac{H}{2h}$ — shear span ratio
      H — height of column
      N — axial compressive load $n = \dfrac{N}{f_c bh}$ — axial load level ratio
      f_c — compressive strength of concrete
    A_{sv} — area of web reinforcement spaced at $\rho_{sv} = \dfrac{A_{sv}}{bs}$ — web reinforcement ratio
      a distance s along column
    A_s — total area of lognitudinal bars $\rho = \dfrac{A_s}{bh}$ — longitudinal reinforcement ratio

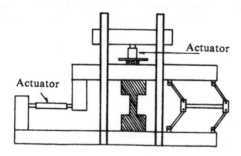

Fig. 1 Loading arrangement

MODES OF FAILURE

Test results indicate that the failure modes of RC short columns can be classified as follows:
1. Shear Tension Failure.
    In the case of specimens with lower axial load and insufficient hoop, diagonal cracks occur

TABLE 2
Specimen Details and Test Results

NO.	Specimens	λ	n	ρ_{sv} (%)	ρ (%)	R_u (%)	Modes of failure
1	SC–1D1	2.0	0.2	0	1.54	1.0	ST · F
2	SC–1D2	2.0	0.4	0	1.54	1.45	ST · F
3	SC–1D3	2.0	0.55	0	1.54	1.0	ST · F
4	SC–1D4	2.0	0.65	0	1.54	1.0	ST · F
5	SC–1D5	2.0	0.75	0	1.54	1.0	F
6	SC–1D6	2.0	0.85	0	1.54	0.50	F
7	SC–1D7	2.0	1.0	0	1.54	0.5	F
8	SC–2D1	2.0	0.2	0.2	1.54	2.5	B · F
9	SC–2D2	2.0	0.4	0.2	1.54	2	B · F
10	SC–2D3	2.0	0.55	0.2	1.54	1.25	F
11	SC–2D4	2.0	0.65	0.2	1.54	1.0	F
12	SC–2D5	2.0	0.75	0.2	1.54	0.5	F
13	SC–2D6	2.0	0.85	0.2	1.54	0.5	F
14	SC–2D7	2.0	1.0	0.2	1.54	0.5	F
15	SC–3D1	2.0	0.2	0.565	1.54	4.5	SF
16	SC–3D2	2.0	0.4	0.565	1.54	3.0	B · F
17	SC–3D3	2.0	0.55	0.565	1.54	2.5	F
18	SC–3D4	2.0	0.65	0.565	1.54	1.5	F
19	SC–3D5	2.0	0.75	0.565	1.54	1.0	F
20	SC–3D6	2.0	0.85	0.565	1.54	0.5	F
21	SC–3D7	2.0	1.0	0.565	1.54	0.5	F
22	SC–4D1	2.0	0.2	1.0	1.54	5.5	F
23	SC–4D2	2.0	0.4	1.0	1.54	3.5	F
24	SC–4D3	2.0	0.55	1.0	1.54	3	F
25	SC–4D4	2.0	0.65	1.0	1.54	3	F
26	SC–4D5	2.0	0.75	1.0	1.54	2.5	F
27	SC–4D6	2.0	0.85	1.0	1.54	1.5	F
28	SC112	2.0	0.40	0.707	2.01	2.3	B
29	SC113	2.0	0.50	0.704	2.01	1.8	HCS
30	SC114	2.0	0.65	0.711	2.01	1.25	HCS
31	SC211	2.0	0.20	0.707	2.54	3.2	B
32	SC212	2.0	0.40	0.765	2.54	2.1	B
33	SC213	2.0	0.50	0.707	2.54	1.93	B
34	SC214	2.0	0.65	0.700	2.54	1.5	HCS
35	SC311	2.0	0.20	0.711	3.14	2.9	B
36	SC312	2.0	0.40	0.704	3.14	2.5	B
37	SC313	2.0	0.50	0.711	3.14	2	B
38	SC314	2.0	0.65	0.707	3.14	1.5	HCS
39	SC321	2.0	0.20	0.985	3.14	3.2	B
40	SC322	2.0	0.40	0.985	3.14	2.7	B
41	SC323	2.0	0.50	0.995	3.14	2.3	B
42	SC324	2.0	0.65	0.985	3.14	1.5	HCS
43	SC331	2.0	0.20	0.401	3.14	2.3	B
44	SC332	2.0	0.40	0.397	3.14	1.7	B

TABLE 2
Continued

NO.	Specimens	λ	n	ρ_{sv} (%)	ρ (%)	R_u (%)	Modes of failure
45	SC333	2.0	0.50	0.403	3.14	1.9	B
46	SC312′	2.0	0.40	0.704	3.14	3.0	B
47	SC314′	2.0	0.65	0.700	3.14	3.2	F
48	SC312w	2.0	0.40	0.704	3.14	2.5	B
49	SC314w	2.0	0.65	0.700	3.14	1.5	HCS
50	SC312w′	2.0	0.40	0.700	3.14	5	F
51	SC322w	2.0	0.41	0.995	3.14	3.3	B
52	SC332w	2.0	0.40	0.395	3.14	2.2	B
53	DZ11−1	1.5	0.53	0.220	2.308	1.0	ST
54	DZ11−2	1.5	0.32	0.220	2.31	1.6	B
55	DZ11−3	1.5	0.11	0.220	2.31	2.1	B
56	DZ11−4	1.5	0.74	0.220	2.31	1.1	HCS
57	DZ12−1	1.5	0.71	0.378	2.31	1.2	HCS
58	DZ12−2	1.5	0.10	0.378	2.31	2.7	B
59	DZ12−3	1.5	0.30	0.378	2.31	1.9	B
60	DZ12−4	1.5	0.50	0.371	2.31	1.3	SC
61	DZ13−1	1.5	0.32	0.647	2.31	1.9	B
62	DZ13−2	1.5	0.74	0.644	2.31	1.3	HCS
63	DZ21−1	1.5	0.87	0.215	3.02	1.4	HCS
64	DZ21−2	1.5	0.11	0.214	3.02	2.6	B
65	DZ21−3	1.5	0.31	0.215	3.02	1.5	ST
66	DZ21−4	1.5	0.52	0.215	3.02	1.2	ST
67	DZ21−5	1.5	0.74	0.215	3.02	1.2	HCS
68	DZ21−6	1.5	1.06	0.215	3.02	0.7	F
69	DZ22−1	1.5	0.10	0.371	3.02	2.8	B
70	DZ22−2	1.5	0.62	0.371	3.02	1.3	ST
71	DZ22−3	1.5	0.74	0.371	3.02	1.1	HCS
72	DZ22−4	1.5	0.53	0.371	3.02	1.3	ST
73	DZ22−5	1.5	1.06	0.370	3.02	0.6	F
74	DZ22−6	1.5	0.44	0.370	3.02	1.7	B
75	DZ23−1	1.5	0.31	0.600	3.02	2.2	B
76	DZ23−2[#]	1.5	0.31	0.600	3.02	2.8	SC
77	DZ23−3	1.5	0.52	0.603	3.02	1.8	SC
78	DZ23−4	1.5	0.73	0.594	3.02	1.0	HCS
79	DZ24−1	1.5	0.50	1.032	3.02	1.9	B
80	DZ24−2	1.5	0.70	1.032	3.02	1.6	HCS
81	DZ25−1	1.5	0.5	0.8	2.3	2.0	SC
82	DZ25−2	1.5	0.6	0.8	2.3	1.62	SC
83	DZ25−3	1.5	0.7	0.8	2.3	1.21	HCS
84	DZ25−4	1.5	0.8	0.8	2.3	0.81	HCS
85	DZ31−1	1.5	0.3	0.20	1.54	2.00	B
86	DZ31−2	1.5	0.5	0.20	1.54	1.13	ST
87	DZ31−3	1.5	0.7	0.20	1.54	0.68	ST
88	DZ32−1	1.5	0.3	0.60	1.54	2.50	B

TABLE 2

Continued

NO.	Specimens	λ	n	ρ_{sv} (%)	ρ (%)	R_u (%)	Modes of failure
89	DZ32−2	1.5	0.5	0.60	1.54	1.80	SC
90	DZ32−3	1.5	07	0.60	1.54	1.07	HCS
91	DZ41−1	1.5	0.3	0.20	0.78	1.60	SF
92	DZ41−2	1.5	0.5	0.20	0.78	1.13	ST
93	DZ42−1	1.5	0.3	0.60	0.78	2.50	B · F
94	DZ42−2	1.5	0.5	0.60	0.78	1.80	SC
95	S1−25	1.0	0.25	0.3	2.40	5.7	S
96	SS1−25[#]	1.0	0.25	0.3	2.40	15.0	S
97	S1−25−1[*]	1.0	0.25	0.3	2.40	7.0	S
98	S2−40	1.0	0.4	0.3	2.40	3.0	S
99	SS2−40[#]	1.0	0.4	0.3	2.40	6.0	S
100	S2−55	1.0	0.55	0.3	2.40	3.0	S
101	SS2−55[#]	1.0	0.55	0.3	2.40	6.0	S
102	S3−65	1.0	0.65	0.3	2.40	3.9	S
103	S3−75	1.0	0.75	0.3	2.40	3.0	S
104	S33−75	1.0	0.75	0.6	2.40	4.7	S
105	S33−90	1.0	0.9	0.6	2.40	4.0	S
106	S44−40−1[*]	1.0	0.4	0.6	2.40	6.0	S
107	S40−65	1.0	0.65	0.0	2.40	2.4	S
108	S40−40−1[*]	1.0	0.4	0.0	2.40	3.0	S

Note: 1. The specimens form No.1 to No.94 are RC short columns. Others are SRC short columns.

2. B — Shear bond failure prior to flexural yielding

B · F — Shear bond failure after flexural yielding

F — Flexure failure

HCS — High compression shear failure

SC — Shear compression failure

ST — Shear tension failure

ST · F — Shear tension failure after flexural yielding

S — Shear failure of SRC short columns

SF — Shear flexure failure

suddenly along the diagonal of the specimen. Because of lower web reinforcement content, the hoop steel stress increases drastically and the hoop steel yields or even fractures. In this failure type, R_u is less than 1/70.

2. Shear Compression Failure.

In the case of specimens with moderate n and larger ρ_{sv}, shear compression failure will take place due to the combined stress in the compressed region of the concrete reaching its maximum capacity. The deformation capability of columns in shear compression failure is superior to that of columns in shear tension failure. For $n \leq 0.5$, the value of R_u can reach 1/70, even can reach 1/50. The typical crack pattern and hysteresis loops of this kind of failure are shown in Fig. 2.

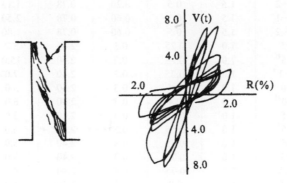

Fig. 2 Shear compression failure

3. High Compression Shear Failure.

This kind of failure occures mostly in the specimens with higher n (n> 0.6~0.7). The failure initiates with excessive longitudinal cracks, then the cover of concrete spalls off and longitudinal bars buckle, and finally the specimen fails by slipping along a major diagonal crack in its mid-height region. Because of failing in brittle fashion, the value of R_u is low, which does not satisfy the design requirement. However, if the $\rho_{sv} > 1\%$ and composite spiral steel is used, the value of R_u can reach 1/70 or 1/50 when n=0.7.

4. Shear Bond Failure.

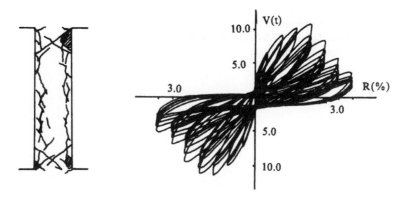

Fig. 3 Shear bond failure

In the condition of $\lambda=1.5\sim2.0$, large value of ρ, large diameter of bars and low strength of concrete, the shear bond failure will likely take place. In this type of failure, diagonal cracks appear first at the ends of columns, then bond cracks initiate at a distance h from the end of the column. At the later stages of loading, bond cracks extend through the entire height of the column. Shear bond failure may occur prior to the tension yielding of longitudinal bars or after its tension yielding. The former case should be eliminated. In the latter case, the value of Ru can reach $\frac{1}{70}$, sometimes even reach $\frac{1}{50}$. The typical crack pattern and the hysteresis loops of shear bond failure are shown in Fig. 3.
5. Shear Flexure Failure.

This type of failure may occur when ρ and n are low. The failure is initiated by yielding of the longitudinal bars in tension. With further increase in deformation, the crushing commences in the compressed region of the concrete and the specimen reaches its failure. The specimens can withstand large deformation. The typical crack pattern and hysteresis loops of the failure are shown in Fig. 4.

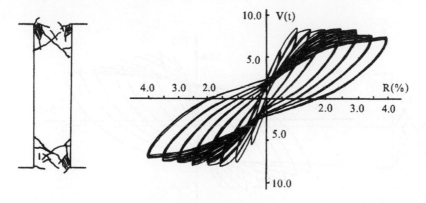

Fig. 4 Shear flexure failure

The modes of failure of SFRC specimens are similar to those of RC short columns. However, because the tensile strength of SFRC is higher than that of RC, the possibility of occuring of bond failure is reduced. For example, to RC short columns, the bond failure occurs when ρ is nearly 2%, but to SFRC short column, it occurs when ρ is nearly 3%. The test results indicate that the deformation capability of SFRC short column is improved to some extent relative to RC short columns, as shown in Fig. 5.

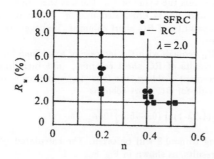

Fig. 5 Comparison of deformation capability of SFRC and RC columns

Like RC short columns, all the tested SRC short columns also collapse in shear behavior, but their deformation capability is fairly good comparing with that of RC short columns. The failure types of SRC short columns are classified as follows.

When $\lambda=1.5$ or 2.0 , the failure type is in shear bond behavior. At beginning of cracking, a series of short and steep diagonal cracks appear along the flange of encased steel on the face of the specimens. As the applied load increases, these short cracks develop into two critical longitudinal cracks, causing the spalling of the cover of concrete. Finally, with the disintegration of the core concrete and buckling of the encased steel, the failure is reached.

When $\lambda=1.0$, the failure type is in shear behavior. In the initial stage of loading, a series of short diagonal cracks appear along the diagonal direction of specimens. With increased loading, more cracks form and the crack width enlarges, the web of the encased I steel yields. After peak loading, the cover of concrete spalls off and the spalling extends slowly towards the core concrete, finally resulting in the collapse of specimen.

The typical crack pattern and hysteresis loops of SRC short columns are shown in Fig. 6.

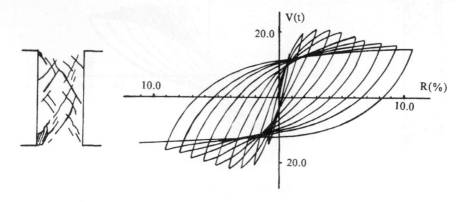

Fig. 6 Typical crack pattern and hysteresis loops of SRC short columns

DEFORMATION CAPABILITY ANALYSIS OF SHORT COLUMNS

The deformation capability of RC short columns is affected by many parameters. In this paper, only four primary parameters, axial load level ratio n, shear span ratio λ, longitudinal reinforcement ratio ρ and web reinforcement ratio ρ_{sv} are considered.

The test results indicate that the less the value of λ is, the less the value of R_u is.

The relationship between ρ_{sv} and R_u is illustrated in Fig. 7. An increase in ρ_{sv} causes an improvement on the deformation capability of RC short columns. However, the improvement will be little after ρ_{sv} reaches a certain limit.

As Fig. 8 shows, the value of R_u reduces with the increase of axial load level ratio n.

Based on the test statistical results of RC short columns, not including those of shear bond failure prior to longitudinal reinforcement yielding, the following equation is proposed for predicting the ultimate value of deformation angle for RC short columns

$$R_u = [\lambda-0.7-(1.2\lambda-1)n] \, (0.6+5.0\sqrt{100\rho_{sv}})\% \qquad (1)$$

If n>0.8, the R_u corresponding to n=0.8 will be taken. The calculated results according to Eq. (1) agree well with the experiment results, as shown in Fig. 8.

In the range of λ from 1.5 to 2.0, when value of ρ increases to a certain extent, shear bond failure prior to longitudinal reinforcement yielding may occurs. In order to prevent this type of

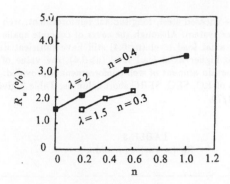

Fig. 7 Variation of R_u with ρ_{sv} for RC columns

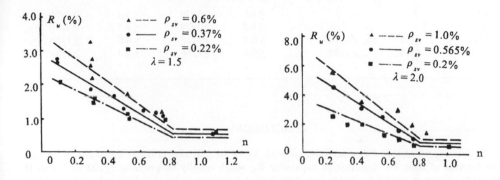

Fig. 8 Comparison of Eq.(1) and test results

failure, according to Fig. 9, the amount of longitudinal reinforcement ρ should be limited to

$$\rho \frac{f_y}{f_c} \leq 0.2\lambda \tag{2}$$

where f_y is the tension strength of longitudinal bar.

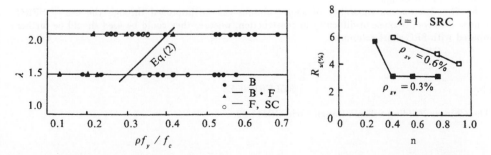

Fig. 9 Distribution of types of failure
for RC short columns

Fig. 10 R_u vs. n for SRC
short columns

For SRC short columns, the encased steel, longitudinal reinforcement, web reinforcement as well as core concrete form a resistance system. Although the cover of concrete spalls after peak load, SRC short columns subjected to low axial load level (n<0.4) still have sufficient deformation capability. Even for specimens subjected to higher axial load level (n>0.4), the value of R_u can be far more greater than 1/50, so long as a certain amount of web reinforcement is provided, as shown in Fig. 10.

According to the test results of RC, SRC, SFRC short columns, Table 3 gives the values of λ and n that will satisfy $R_u \geq 1/50$ or 1/70.

TABLE 3

R_u for various λ and n

Short Column	$\lambda = 1.0$		$\lambda = 1.5$		$\lambda = 2.0$	
	n	R_u	n	R_u	n	R_u
RC			≤ 0.5	$\geq 1/70$	≤ 0.4	$\geq 1/50$
RC			≤ 0.2	$\geq 1/50$	≤ 0.6	$\geq 1/70$
SFRC	≤ 0.2	$\geq 1/70$	≤ 0.4	$\geq 1/50$	≤ 0.6	$\geq 1/50$
SRC	≤ 0.9	$\geq 1/50$	≤ 0.9	$\geq 1/50$	≤ 0.9	$\geq 1/50$

CONCLUSIONS

1. It is suggested that RC short columns with $\lambda \leq 1.0$ be not used in frame structures in seismic zones because of their low deformation capability R_u which usually ranges from 1/100 to 1/300. If such short column is needed by design considerations, it is recommended that SRC columns should be used.

2. The minimum value of ρ_{sv} not less than 0.5% is suggested to prevent a sudden brittle shear tension failure.

3. In order to avoid high compression shear failure, it is recommended that the maximum n shall not exceed 0.6 to assure neccessary ductility.

4. In order to prevent shear bond failure prior to longitudinal reinforcement yielding, the amount of longitudinal reinforcement ρ should be limited by Eq. (2). For RC short columns satisfying the requirement of Eq. (2), the failure is usually shear flexure failure, shear compression failure, or shear bond failure after longitudinal reinforcement yielding. They possess certain ductility and can be used in seismic building design.

5. The deformation capability of short columns can be improved through the use of SFRC columns. However, because of difficulty in construction, whether they could be used should be further compared with SRC short columns.

REFERENCES

1. The Chinese Building Seismic Design Code GBJ11-89.

BEHAVIOR OF FIBER REINFORCED CONCRETE UNDER SHEAR

P. BALAGURU
Professor of Civil Engineering
Rutgers, The State Univ. of N.J.
Piscataway, New Jersey

SURENDRA P. SHAH, DIRECTOR
NSF Center for Advanced
Cement Based Materials
Northwestern University
Evanston, Illinois

ABSTRACT

A number of researchers have investigated the use of steel fibers to improve shear behavior of beams. These studies deal with: regular beams with length/depth of about 5, deep beams with and without openings in which shear is a controlling factor, beam-column connections in which the beams failed in shear, beam-column connections in which columns were subjected to shear loads (simulating earthquake movement), corbels, direct shear test using double L specimens, and behavior under biaxial loading condition. Responses under both monotonic and cyclic loads were investigated. Fiber reinforced concrete beams have also been tested under torsion. A review of the results available in the literature is presented in this paper. Emphasis is placed on the fiber contribution to ductility of structural components subjected to cyclic loading.

INTRODUCTION

Fiber reinforced concrete is defined as portland cement concrete reinforced with short randomly oriented fibers. The fibers could be steel, polymeric such as nylon or polypropylene, carbon, glass, or naturally occurring fibers such as wood or sisal fibers. In most, if not all, structural applications only steel fibers have been evaluated. Hence, only steel fiber reinforced concrete is discussed in this paper.

Fiber reinforced concrete has excellent potential for use in structural members subjected to shear. The random distribution of fibers provides a close spacing that cannot be duplicated with bar reinforcement. The fibers also intercept and bridge cracks in all the directions. This process not only increases shear capacity but also provides substantial post peak resistance and ductility. Laboratory tests have established that both the strength and ductility can be enhanced by using fibers. The test variables investigated include: shear span to depth ratio (a/d ratio), fiber type, fiber volume fraction, and compressive strength of concrete. Shear span, 'a' is defined as the distance between the load and the support and 'd' is the depth of the beam measured from extreme compression fiber to the center of gravity of tension steel. Fibers have been used just by themselves and to augment stirrup reinforcement. When used in conjunction with stirrups, fibers make it possible to increase the spacing of stirrups thus reducing reinforcement congestion in high shear areas.

Fiber reinforced concrete has been evaluated for application in various types of structural elements. These include regular beams with a/d ratios around 5, deep beams, corbels and connections. In addition, FRC beams have also been subjected to torsion, and combined bending, shear and torsion. Columns subjected to lateral shear loads have also been investigated. The salient features of the test results are briefly discussed in this paper.

FRC BEAMS WITH AND WITHOUT SHEAR REINFORCEMENT

Tests conducted by Jindal show that shear strength of mortar beams could be doubled by using steel fibers [1]. He used a cement mortar with a water-cement ratio of 0.6 and mild steel or brass coated fibers with aspect (length/diameter) ratios ranging from 10 to 100. The fiber volume fraction was 1 percent. The 4x6 in. (100x150mm) beams had two - 0.5 in. (12mm) diameter steel bars for flexural reinforcement. They were tested over a simply supported span of 60in. (152cm). The loads were positioned to obtain shear span to depth ratios (a/d) varying from 2.0 to 4.8. The plain concrete (mortar) beams failed in shear for a/d ratios of 2.0, 2.5 and 3.6. The maximum shear stress was 433 psi(2.8 MPa). The mild steel fibers provided maximum increase of 44 percent and the brass coated fibers recorded 105 percent increase. The increase in aspect (length/diameter) ratio of fibers resulted in increase of shear strength up to a value of 75. Fibers with aspect ratio of 100 had lower strengths than fibers with aspect ratio 75. Higher aspect ratios provide higher bond strength, thus generating more resistance. When fibers are too long, they tend to ball and reduce compaction resulting in poor distribution. This could be the reason for obtaining lower strengths for fibers with aspect ratio of 100.

The primary variables for results reported by Batson et al are: fiber size, shape, and volume fraction and shear span to depth ratio (a/d) [2]. The beams had an overall dimension of 4x6x78in. (100x150x2000mm). Conventional flexural reinforcement were provided to induce shear failure. Plain matrix failed in shear for beams with a/d ratio of 4.8. When sufficient volume fraction of fibers were added, the failure mode changed from shear failure to moment failure. For each volume fraction of fibers, the shear span to depth (a/d) ratios were decreased so as to induce shear failure. The variation of shear strength with respect to shear span to depth ratio are shown in Fig. 1 [2]. A volume fraction of 1.76 percent provides considerable increase in shear strength. The results however show considerable scatter when a/d ratios are less than 3.0.

Bollana compared the performance of straight and hooked-end fibers [3]. Hooked-end fibers were found to provide much better contribution. A volume fraction of 0.75 and 1.00 percent hooked fibers recorded failure loads of 14.8 and 16.4 kips (65.6 and 72.7 kN) as compared to 14.0 and 15.0 kips (62.1 and 66.76 kN) recorded by beams with straight fibers. The shear capacity for beams with stirrups was 16.2 kips (72.1 kN) and the plain concrete had a capacity of 8.0 kips (35.6 kN).

Tests conducted using life size, 12x21.5x276 in. (300x540x7000mm), beams showed that steel fibers with hooked ends can be used as effective shear reinforcement [4]. The beams with 1.1 volume percent of hooked end steel fibers provided an increase of 67 percent shear capacity. The beams had no stirrup reinforcement.

The contribution of fibers were also evaluated for T beams [5]. The variables were fiber content, longitudinal (flexural) reinforcement ratio, and stirrup spacing. The shear span to depth (a/d) ratio was kept constant at 4.5. Ultimate shear strength was found to increase almost linearly for fiber volume fractions ranging from 0.8 to 1.2 percent. The addition of hooked end fibers were found to augment nominal amount of stirrups and enhance the shear strength contribution of dowel action.

Fibers were also found to be effective in I cross sections [6]. Fiber addition prevented shear failure for shear span to depth (a/d) ratios between 3.24 and 4.68.

Overall, it can be said that fibers are effective in providing improvement in shear capacity. Deformed fibers are more efficient than straight fibers. A minimum of 1 volume percent fibers can replace stirrups in some instances. However, the authors feel that the fibers should be used to augment nominal stirrup reinforcement reducing reinforcement congestion and at the same time providing improved ductility, rather than replacing stirrup reinforcement.

BEHAVIOR UNDER TORSION

Since torsion produces shear stresses, torsion behavior of FRC is briefly reviewed. Behavior of fiber reinforced concrete beams under pure torsion had been investigated by a few researchers [7-11]. The primary test variables were fiber type, volume fraction, aspect (length/diameter) ratio of fibers, and the amount of hoop reinforcement.

When fibers are added to plain concrete, both torsional strength and energy absorbing capacity were found to increase [11]. When straight steel fibers are used without incorporating longitudinal or hoop reinforcement the torsional strength increases from 10 to 26 percent for volume fractions ranging from 1 to 3 percent. The increase in strength with respect to aspect (length/diameter) ratio is not substantial. However, the area under the Torque – Twist curve increases substantially with increase in aspect ratio. Hence, the fiber reinforced composite can be assumed to be more ductile. The improvement in ductility with increase in fiber volume fraction is also more substantial than strength increase as shown in Fig. 2 [11]. Tests conducted using fibers in beams reinforced with longitudinal and hoop reinforcement indicate that both strength and ductility can be improved considerably using hooked end steel fibers [7]. The beams were 6x12in. (150x300mm) in cross-section and reinforced with four longitudinal bars. Some of the beams had hoop reinforcement spaced at 3.5 and 7.0in. (87 and 175mm). The improvement provided by fibers is much more substantial if hoop reinforcement is present, [7]. The fibers seem to provide synergistic effect in the presence of hoop reinforcement.

BEHAVIOR UNDER COMBINED BENDING, SHEAR AND TORSION

Very rarely, structural elements are subjected to pure bending, shear or

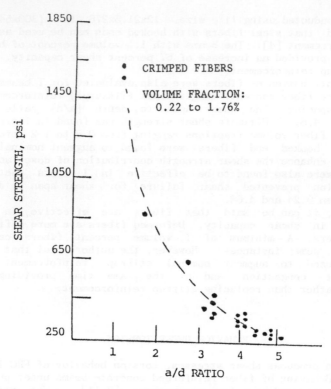

Figure 1. Shear Strength Vs a/d Ratio for Steel FRC [2]

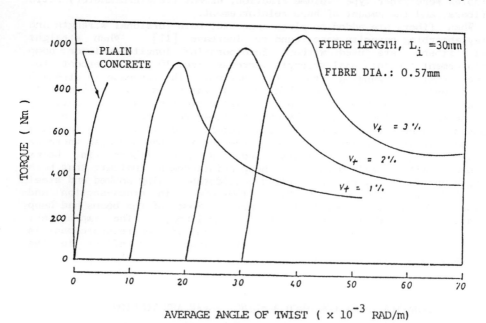

AVERAGE ANGLE OF TWIST (x 10^{-3} RAD/m)

Figure 2. Torque – Twist Curves for Steel FRC [11]

torsion. In most cases bending and shear force occur together. In some instances the element will also be subjected to torsion in addition to bending and shear. In the case of torsion, the members could be subjected to either equilibrium or compatibility torsion. When fibers are added to members subjected to compatibility torsion, they were found to reduce stiffness degradation. The cracks were more fine and more uniformly distributed, thus reducing sudden drops in stiffness at areas with wider cracks. The members could also sustain more rotation without losing their load carrying capacity [12].

The contribution of fibers to structural elements subjected to equilibrium torsion, moment and shear were investigated using both straight and hooked end steel fibers [13,14]. The performance of straight fibers was evaluated using 0.46mm diameter 36.8mm long (aspect ratio, ℓ/d = 80) at volume fractions of 0.5, 1.0, and 1.5 percent. Control specimens with no fibers were also tested. The rectangular beams with overall dimensions of 125x300x2500mm were reinforced with both longitudinal and hoop reinforcement. The testing was conducted under pure torsion, torsion and bending and torsion, bending, and shear.

The fiber addition provided an increase of 33 and 44 percent respectively for cracking and ultimate torque when only torsion was applied. Under combined bending and torsion the increases were 53 and 21 percent for cracking and ultimate torque. The increases were 82 and 65 percent when the beams were subjected to combined bending, shear and torsion. The higher increase under combined loading shows the effectiveness of randomly distributed fibers under complex loading patterns inducing tensile stresses in all three directions. Continuous reinforcement cannot be effectively placed to counteract the principle tensile stresses developed by the combined loading. The specimens with fibers were found to develop more uniformly distributed cracks with lower crack widths.

The addition of fibers also increases the rotational capacities of the elements. The fibers are more effective under combined loading than in pure torsion. The torsional capacity and maximum rotation increases substantially for beams with 1.5 volume percent fibers under torsion, bending and shear as compared to other two loading patterns involving just torsion or torsion and bending.

Fibers with hooked ends seem to increase the ductility much more than straight fibers. Tests conducted using 30mm long, 0.5mm diameter fibers with hooked ends were found to be very effective even at a volume fraction of 0.5 percent [13]. The rotation capacities of beams even with 0.5 percent fibers are about 5 times the rotation capacity of plain concrete. As in the case of straight fibers, hooked fibers also seem to provide best results under combined loads [14]. The improved ductility is important for members subjected to cyclic loading resulting from phenomenon such as earthquakes.

DEEP BEAMS AND CORBELS

Deep beams and corbels can be considered as special beams where shear force is much more predominant than bending moment. Fibers could be effectively used in these structural members, specially to improve the ductility.

The behavior of deep beams is very much influenced by the shear span to depth (a/d) ratio. Failure patterns change considerably around a/d ratios of 1.5. Fibers have been evaluated for a/d ratios ranging from 0.3

to 2.4 [15-17]. The overall dimensions of the beam cross section was 50x200mm. The fiber volume fraction varied from 1.2 to 1.8 percent. Fibers make a definite contribution to the strength development both before and after cracking [15]. The improvement in first crack load is higher than failure load for all the three shear span to depth (a/d) ratios. The magnitude of increase does not seem to vary with a/d ratio.

The addition of hooked end fibers was found to improve the performance both in strength and ductility for deep beams with and without opeing.

Corbels can be considered as special type of cantilever deep beams. These are often used to transfer heavy beam reactions to columns. They are typically reinforced with flexural reinforcement at the top and stirrups to resist shear force. Fibers have been tried as a replacement for the stirrup reinforcement. The behavior of corbels containing shear reinforcement in the form of stirrups was compared to the behavior of corbels containing only fibers [18, 19]. The fibers used were double-indented duoform steel fibers. The nominal size was 60mm by 0.65mm (2.4x0.0226in). The primary test variables were a/h ratio, shear reinforcement in the form of stirrups, and shear reinforcement in the form of fibers.

The test results showed that the failure is castastrophic when no shear reinforcement is provided. Addition of fibers was found to change the mode of failure and improve the ductility. Corbels with fibers exhibited good elasto-plastic behvior at fiber volume fractions greater than twice the main bar reinforcement. It was also found that the contribution of stirrup reinforcement can be duplicated by fibers.

BEAM-COLUMN CONNECTIONS

One of the most promising area of structural application for fiber reinforced concrete is structural connections. The fibers provide the necessary ductility to induce flexural failure rather than sudden shear failure. Investigations have been carried out using both straight and hooked end steel fibers [21-24]. All the typical connections in a framed structures namely: cross-type, T-type, and knee type have been studied. Since cyclic loading is of primary importance, all the investigations were carried out using low cycle high amplitude loading.

The contribution of fibers can be demonstrated using moment-rotation curves or energy absorption measured using hysteresis loops generated under cyclic loading. Typical load-deflection behavior under cyclic loading for plain and fibrous concretes are shown in Fig. 3 [23]. From this figure it can be seen that fiber reinforced specimens can sustain a higher percentage of initial load after the cyclic load applications. Fiber reinforced specimens also sustained higher deformations. The fibers used were 30mm long - 0.5mm diameter hooked end steel fibers. Fiber volume fraction was 1.5 percent.

The fiber reinforced specimens were also found to show less damage at junction, higher initial stiffness, less cracking and higher moment and shear capacities. All the investigators concluded that fibers make a definite contribution and design methods and specifications should be developed for using fibers in the actual field construction.

COLUMNS

Addition of fibers to columns that are subjected only to compressive loads provide improvement in ductility at failure [25]. The increase in compressive strength is not substantial. On the other hand if the columns are subjected to lateral shear force, the fibers provide a substantial contribution to ductility [25,26]

The effectiveness of the fibers in resisting later shear forces was demonstrated using a short column subjected to axial force and lateral shear [26]. The primary test variables were axial load, amount of the reinforcement and volume fraction of fibers. Typical results obtained for 0, 1 and 2 percent volume fraction of fibers are shown in Fig. 4 [26]. From this figure it can be seen that fiber reinforced specimens have a much higher load retaining capacity under cyclic (shear) loading. It was also observed that cracks were distributed for larger portion of column resulting in larger ductility. Improvement in ductility for fiber volume fraction from 0 to 1 percent was found to be much higher than the improvement provided by the increase from 1 to 2 percent.

PRESTRESSED CONCRETE BEAMS

Limited results available show that fibers can be used to improve the shear performance of prestressed concrete beams [27]. Tests conducted using T beam cross-section and high strength concrete show that first crack moment, and ductility can be increased by adding fibers. All the beams tested failed in shear mode. The 50mm long 0.5mm diameter hooked end steel fibers provided consistent improvement both in ductility and increase in first crack moment in addition to a slight increase in shear capacity.

SUMMARY

In summary, fiber reinforced concrete has excellent potential for use in structural components subjected to shear. The contribution of fibers can be effectively used to improve the ductility of beam column connections and deep beams. Fibers can also be used in conjunction with stirrup reinforcement to reduce reinforcement congestion.

ACKNOWLEDGEMENT

This summary paper is based on the work done while the first author was a visiting scholar at the NSF Center for Advanced Cement Based Materials, Northwestern University. The sponsorship provided by the Center is gratefully acknowledged. The first author also gratefully acknowledges the facilities provided by the Department of Civil Engineering of Northwestern University.

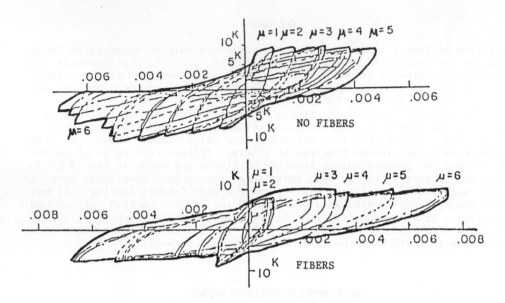

Figure 3. Beam Tip Load Vs Rotation of Beam at Column Face. [23]

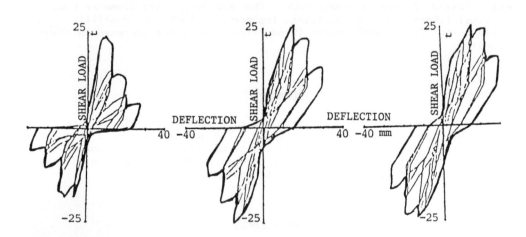

Figure 4. Shear Load - Deflection Curves of Columns. [26]

REFERENCES

1. Jindal, R.L., "Shear and Moment Capacities of Steel Fiber Reinforced Concrete Beams," ACI SP81, 1984, pp. 1-16.

2. Batson, G., Jenkins, E., and Spatney, R., "Steel Fibers as Shear Reinforcement in Beams," ACI Journal, Vol. 69, No. 10, October 1972, pp. 640-644.

3. Bollana, R.B., "Steel Fibers as Shear Reinforcement in Two Span Continuous Reinforced Concrete Beams," M.S. Thesis, Clarkson University, Potsdam, New York, U.S.A., May 1980.

4. Williamson, R.R., "Steel Fibers as Web Reinforcement in Reinforced Concrete," Proceedings, U.S. Army Science Conference, Vol. 3, 1978, West Point, New York.

5. Swamy, R.N., and Bahia, H.M., "The Effectiveness of Steel Fibers as Shear Reinforcement," Concrete International, March 1985, pp. 35-40.

6. Muhidin, N.A., and Regan, P.E., "Chopped Steel Fibers as Shear Reinforcement in Concrete Beams," Proceedings, Conference on Fiber Reinforced Materials: Design and Engineering Applications; Institute of Civil Engineers, London, 1977, pp. 149.

7. Craig, J.R., Dunya, W., Riaz, J., and Shirazi, H., "Torsional Behavior of Reinforced Fibrous Concrete Beams," ACI-SP81, 1984, pp. 17-49.

8. Narayanan, R., and Toorani-Goloosalar, Z., "Fiber Reinforced Concrete in Pure Torsion and in Combined Bending and Torsion," Proceedings, ICE, London, Part 2, Vol 67, December 1979, pp. 987-1001.

9. Mindess, S., "Torsional Tests of Steel Fiber Reinforced Concrete," The International Journal of Cement Composites, Vol. 2, No. 2, May 1980, pp. 85-89.

10. Narayanan, R., and Green, K.R.," Fiber Reinforced Concrete Beams in Pure Torsion," Proceedings, Institution of Civil Engineers, London, Part 2, Vol. 69, September 1980, pp. 1043-1044.

11. Mansur, M.A., and Paramasivam, P., "Steel Fiber Reinforced Concrete Beams in Pure Torsion," The International Journal of Cement Composites, Vol. 4, No. 1, February 1982, pp. 39-45.

12. Batson, G., "Use of Steel fibers for Shear Reinforcement and Ductility," Steel Fiber Concrete, Elsevier Applied Science Publishers, 1985, pp. 377-419.

13. Al-Ausi, M.A., Abdul-Wahab, H.M.S., and Khidair, R.M., "Effect of Fibers on the Strength of Reinforced Concrete Beams Under Combined Loading," Fiber Reinforced Cements and Concretes: Recent Developments, Elsevier Applied Science, 1989, pp. 664-675.

14. Kaushik, S.K., and Sasturkar, P.J., "Simply Supported Steel Fiber Reinforced Concrete Beams Under Combined Torsion, Bending and Shear,"

Ibid, pp. 687-698.

15. Roberts, T.M., and Ho, N.L., "Shear Failure of Deep Fiber Reinforced Concrete Beams," The International Journal of Cement Composites, Vol. 4, No. 3, August 1982, pp. 145-152.

16. Shanmugam, N.E., and Swaddiwudhipong, S., "The Ultimate Load Behavior of Fiber Reinforced Concrete Deep Beams," The Indian Concrete Journal, Vol. 58, No. 8, August 1984, pp. 207-211.

17. Shanmugam, N.E., and Swaddiwudhipong, S., "Behavior of Fiber Reinforced Concrete Deep Beams Containing Openings," Fiber Reinforced Cements and Concretes: Recent Developments, Elsevier Applied Science, 1989, pp. 479-488.

18. Fattuhi, N.I., and Hughes, B.P., "Reinforced Steel Fiber Concrete Corbels with Various Shear Span-to-Depth Ratios," ACI Structural Journal, Vol. 86, No. 6, Nov-Dec. 1989, pp. 590-596.

19. Fattuhi, N.I., and Hughes, B.P., "Ductility of Reinforced Concrete Corbels Containing Either Steel Fibers or Stirrups," ACI Materials Journal, Vol. 86, No. 6, Nov-Dec. 1989, pp. 644-651.

20. Hughes, B.P., and Fattuhi, N.I., "Reinforced Steel and Polypropylene Fiber Concrete Tests," The Structural Engineer, London, Vol. 67, No. 4, February 1989, pp. 68-72.

21. Henager, C.H., "Steel Fibrous, Ductile Concrete Joint for Seismic-Resistant Structures," ACI-SP53, 1974, pp. 371-386.

22. Jindal, R.L., and Hassan, K.A., "Behavior of Steel Fiber Reinforced Concrete Beam-Column Connections," ACI-SP81, 1984, pp. 107-123.

23. Craig, J.R., Mahadev, S., Patel, C.C., Viteri, M., and Kertesz, C., "Behavior of Joints Using Reinforced Fibrous Concrete," ACI-SP81, 1984, pp. 125-167.

24. Sood, V., and Gupta, S., "Behavior of Steel Fibrous Concrete Beam-Column Connections," ACI-SP105, 1987, pp. 437-474.

25. Craig, J.R., McConnell, J., Germann, H., Dib, N., and Kashani, F., "Behavior of Reinforced Fibrous Concrete Columns," ACI-SP81, 1984, pp. 69-105.

26. Yashiro, H., Tanaka, Y., Ro, Y., and Hirose, K., "Study of Shear Failure of Steel Fiber Reinforced Concrete Short Columns in Consideration of Arrangement of Ties," Fiber Reinforced Cements and Concretes: Recent Developments, Elsevier Applied Science, 1989, pp. 489-498.

27. Balaguru, P., and Ezeldin, A., "Behavior of Partially Prestressed Beams Made with High Strength Fiber Reinforced Concrete," ACI-SP105, 1987, pp. 419-434.

SHEAR WALL TESTS

JOHANNES MAIER*
Wettingen, Switzerland

ABSTRACT

This article presents the experiments on reinforced concrete shear walls carried out as part of a research project at the Swiss Federal Institute of Technology in Zürich.

The square test specimens were loaded with normal and shear force at the upper edge. The lower edge was firmly connected to the test floor. The following quantities were chosen as test parameters: Shape of the cross-section, opening in the compression zone of the wall, arrangement and amount of reinforcement, level of normal force loading and monotonous or cyclical increase of the shear force.

The article describes the geometry of the specimens and the loading and measuring system. Photographs show the crack patterns and diagrams present the load-displacement relations. Strain measurements are used to plot the strain field of the concrete surface.

INTRODUCTION

The walls of a multistorey building are frequently the most important part of the structural system for the lateral loads originating from wind or earthquake. Usually the walls also support a substantial part of the vertical dead and live loads.

Until quite recently structural walls were designed on the basis of a pure linear-elastic analysis. As long as the ratio between vertical and lateral load is high, the walls remain uncracked and the elastic analysis gives a good estimate of the stresses. But when cracking starts, the stresses will be redistributed. During the last two decades methods based on the theory of plasticity, which allows a stress redistribution, have become important for determining the ultimate strength of beams and slabs. The question, if this theory is also applicable to wall structures, was therefore close to hand. The research project *Reinforced Concrete Wall Structures*, carried out at the Swiss Federal Institute of Technology in Zürich, tried to find an answer. Besides the theoretical investigations [1], on which the article *Interactive Plastic Design of Walls* is based, the research project had an experimental part [2]: Tests on square wall elements were used to prove the sufficient amount of ductility required for the application of the theory of plasticity.

*Former research engineer at the Institute of Structural Engineering, ETH-Hönggerberg, CH-8093 Zürich, Switzerland.

TEST PARAMETERS

The test program consisted of experiments on ten square shear walls (specimen S1 to S10) loaded with normal and shear force at the upper edge. Whilst the normal force was constant throughout a test, the shear force was increased until failure occurred.

To facilitate the latter interpretation, only one parameter was changed from one test to the other. Starting from test specimen S1—which had a flanged cross-section, no openings, a medium amount of evenly spaced vertical and horizontal reinforcement and which was loaded with a low normal force and a monotonous increasing shear force—the following diagram can be developed to show the relations between the tests:

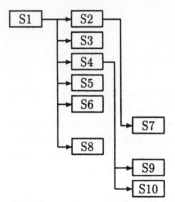

- high level of the normal force
- high amount of vertical reinforcement
- no flanges
- cyclical increase of the shear force
- low amount of horizontal reinforcement
- high level of the normal force and cyclical shear force
- opening in the compression zone
- no flanges and no horizontal reinforcement
- no flanges but additional vertical reinforcement

TEST SPECIMENS

The test specimens were composed of the actual wall and thick slabs at the upper and lower edge. While the upper slab was used for the application of the loads, the lower slab served as a rigid connection to the test floor. The wall had a length $l = 1.18$ m and a height $h = 1.20$ m, its thickness t was 0.10 m. Figure 1 shows the dimensions $l_{1,2}$ and $c_{1,2}$

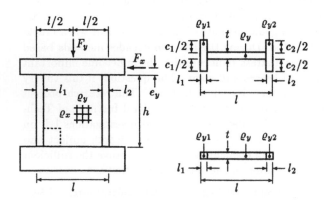

Test	Cross-section	$c_{1,2}$ [mm]	$l_{1,2}$ [mm]
S1	⊢⊣	300	100
S2	⊢⊣	300	100
S3	⊢⊣	300	100
S4	—	0	0
S5	⊢⊣	300	100
S6	⊢⊣	300	100
S7	⊢⊣	300	100
S8	⊢⊣	300	100
S9	—	0	0
S10	—	0	240

Figure 1: Geometry of the test specimens.

of the flanges and with dashed lines the position of the opening of test specimen S8. This opening had a length of 0.27 m and a height of 0.30 m. The heavy reinforced slabs had a length of 1.70 m and a width of 0.70 m. The thickness of the upper slab was 0.24 m and that of the lower amounted to 0.38 m. To avoid premature cracking at the joints between slabs and wall, the specimens were cast in one upright standing piece. The same formwork could be adjusted to all ten specimens.

The concrete mix had a maximum aggregate size of 16 mm, the cement content was 325 kg/m^2 and the water-cement ratio was about 0.5. Superplasticizer allowed a better workability. Testing started at the earliest 17 days after casting.

The reinforcement was made of hot rolled steel with a yield plateau until at least $15 \cdot 10^{-3}$ strain. The standard diameter d of the reinforcement was 8 mm and the distance s between the bars was usually 100 mm. The bars, arranged in vertical and horizontal layers on both sides of the wall, had a minimal concrete cover of 10 mm. Figure 2 shows the cross-sectional layout of the reinforcement for all specimens and a detail of the special reinforcement around the opening of specimen S8.

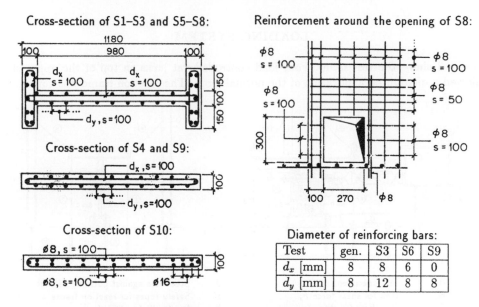

Figure 2: Reinforcement of the test specimens.

Both horizontal and vertical reinforcement had the form of closed stirrups. The vertical bars extended through both slabs, to allow a sufficient anchor length outside of the wall. With the exception of specimen S10, the horizontal bars of the inner part of the walls were always inside the vertical ones. In the flanges and at the edges, however, the horizontal bars enclosed the vertical steel to prevent premature buckling of the latter. Table 1 presents the resulting reinforcement ratios $\varrho_{x,y}$ together with the strength properties of steel and concrete (yield strength f_{sy} respectively strength of direct tension test specimens f_{ct} and compression strength of cylinders f_{cc}). The concrete modulus of elasticity was between 27.7 and 34.1 GPa. The measuring of the concrete material properties took place at the time of the respective wall test.

TABLE 1: Reinforcement ratios and strength properties of the test specimens.

Test	ϱ_x [10^{-3}]	f_{syx} [MPa]	ϱ_y [10^{-3}]	f_{syy} [MPa]	$\varrho_{y1,2}$ [10^{-3}]	$f_{syy1,2}$ [MPa]	f_{ct} [MPa]	f_{cc} [MPa]
S1	10.28	574	10.45	574	12.89	574	3.56	36.9
S2	10.28	574	10.45	574	12.89	574	—	34.6
S3	10.28	574	22.25	530	27.44	530	3.80	36.7
S4	10.28	574	10.45	574	0.00	—	3.44	32.9
S5	10.28	574	10.45	574	12.89	574	3.75	37.3
S6	5.68	537	10.21	575[1]	12.59	575[1]	4.17	35.6
S7	10.06	555	10.23	555	12.62	555	3.66	34.1
S8	10.06[2]	555	10.23	555	12.62	555	3.55	32.1
S9	0.00	—	9.88	560	0.00	—	3.19	29.2
S10	9.80	496	9.90	496	56.90[3]	518	3.16	31.0

LOADING SYSTEM

Figure 3 shows the test setup. Four load controlled flat jacks on top of the upper slab allowed a distributed application of the normal force. The reaction frame was connected

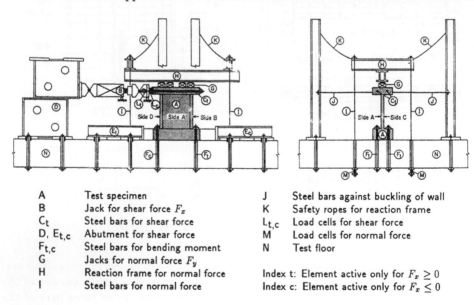

A	Test specimen	J	Steel bars against buckling of wall
B	Jack for shear force F_x	K	Safety ropes for reaction frame
C_t	Steel bars for shear force	$L_{t,c}$	Load cells for shear force
D, $E_{t,c}$	Abutment for shear force	M	Load cells for normal force
$F_{t,c}$	Steel bars for bending moment	N	Test floor
G	Jacks for normal force F_y		
H	Reaction frame for normal force	Index t: Element active only for $F_x \geq 0$	
I	Steel bars for normal force	Index c: Element active only for $F_x \leq 0$	

Figure 3: Test setup.

---

[1]The vertical reinforcement was made of cold worked steel bars with the proportionality limit $f_{s2.0} = 479$ MPa and the tension strength $f_{st} = 764$ MPa.

[2]Extra horizontal reinforcement was added in a 0.3 m high zone above the opening over the entire wall length: $\varrho_x = 16.77 \cdot 10^{-3}$.

[3]The left edge did not have any additional vertical reinforcement, i.e. $\varrho_{y1} = 0$.

by four thin high strength steel bars to the test floor and could move together with the upper slab in the longitudinal direction of the wall.

The shear force was produced by a single tension-compression jack, for which the servo-hydraulic loading system allowed a deformation control. Therefore the tests could be continued even after reaching the maximum shear force. Positive shear forces were applied by means of two high strength steel bars passing through the upper slab in plastic ducts to avoid any additional bending moment. The negative shear forces for the cyclic tests S5 and S7 were applied directly to the specimen on the opposite side of the slab. The vertical eccentricity e_y of the lateral load was 0.12 m in both cases.

Abutments at each end of the lower slab served as a support for the shear force. Prestressed vertical bars going through the lower slab transferred the in-plane bending moment from the base of the wall to the test floor.

MEASUREMENTS AND TEST PROCEDURE

In addition to the usual registration of the applied forces by load cells and of the global deformations with inductive displacement transducers, extensive measurements of local deformations on the concrete surface and on the reinforcing bars were carried out to record the deformational behaviour. Hand held extensometers with an internal inductive displacement transducer were used for the latter measurements. Aluminium targets glued to the concrete and steel helped to position the extensometers. Different base lengths of the extensometers allowed a continuous measuring net on the concrete surface.

The crack pattern was photographed and crack widths were noted. For the tests with a cyclical increase of the shear force two different colors were used to mark the cracks: Red color marked the cracks coming from a positive shear force and blue color those originating from a negative. The use of a blue and a red color filter enabled the obliteration of the crack marks with the opposite color when taking photographs on a black and white film.

Loading started with the application of the normal force. Afterwards the horizontal deflection of the wall was increased stepwise by the jack for the shear force. This deflection was kept constant during the measuring time. For a complete reading—more than 600 measurements—about one hour was needed. The loads were registered at the beginning and at the end of each measuring period, to show the reduction of the shear force due to creep. The tests S5 and S7 had ten full cycles on each deformation level. But the complete measuring program was only carried out for maximum and minimum deformation of the first and the last cycle.

RESULTS

Table 2 gives—together with the values of applied normal force F_y and resulting vertical deformation δ_y—the shear force F_x and the horizontal displacement δ_x for the cracking, maximum and final load stage. The diagrams of Figure 4 show the relationship between horizontal displacement and applied shear force.

The small difference between test S1 and S6 demonstrates the negligible influence of the amount of horizontal reinforcement on the maximum shear force. However, the

TABLE 2: Applied forces and resulting deformations.

Test	Normal Force		Cracking Load		Maximum Load		Final Load	
	F_y [kN]	δ_y [mm]	F_x [kN]	δ_x [mm]	F_x [kN]	δ_x [mm]	F_x [kN]	δ_x [mm]
S1	469	0.12	209	0.46	680	22.23	591	30.02
S2	1689	0.41	599	1.72	928	10.41	907	12.12
S3	460	0.07	164	0.48	977	14.72	976	15.92
S4	298	0.07	117	0.56	392	11.35	368	14.65
S5	452	0.07	252	0.98	701/−664	21.03/−18.63	505	21.35
S6	452	0.07	190	0.84	667	18.69	596	24.18
S7	1693	0.36	595	1.88	836/−874	7.92/ −7.18	335	9.98
S8	452	0.14	162	0.63	510	8.74	390	15.46
S9	296	0.10	108	0.71	342	8.99	292	12.22
S10	296	0.07	149	1.07	670	14.11	643	16.06

deformation capacity of S6 was reduced as a function of its lower horizontal reinforcement ratio. Even test S9 without any horizontal reinforcing bars reached a maximum shear force, which was only 13 % less than that of the related specimen S4. The cyclic test S5 with a low level of the normal force exhibited no influence of the load reversal when compared with the corresponding monotonous test S1; even the deformation capacity did not show any significant reduction. The cyclic loading of test S7 has to account for the 10 % decrease of the maximum positive shear force compared to specimen S2 loaded with the same, high normal force; for the extreme negative shear force a reduction of

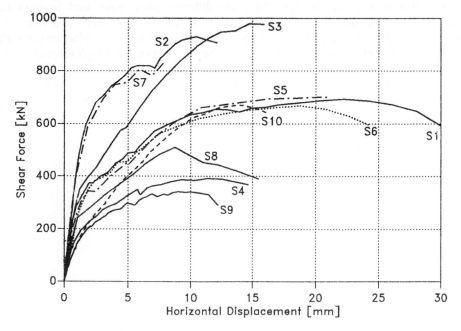

Figure 4: Measured relationship between horizontal displacement and applied shear force.

only 6 % was measured. The last test S10 demonstrated the possibility of achieving the same loadbearing capacity as in the basic test S1 even without flanges but with a heavy vertical reinforcement on the tension edge. Similar to test S3, this high reinforcement ratio reduced the maximum attainable horizontal deformation.

In general, failure occurred in the compression corner of the walls due to lateral splitting of the concrete and spalling of the cover. The specimens tested with a high normal force load exhibited a horizontal fracture zone over their entire length. Photographs 7 to 12 present the final crack pattern and the appearance of the fracture zone for test specimens S1, S2, S5 and S7 to S9.

The cracks on the tension edge were always initialized by one of the horizontal reinforcing bars and they did not, even for those specimens without flanges, deviate from the horizontal direction for the first 0.1 m. The high normal force of the tests S2 and S7 produced steeper cracks, and here, contrary to all other tests, the first cracks were not observed on the tension edge, but in the interior of the wall.

The results from tests S2 and S7 also show the importance of the level of the normal force loading on the crack width values. For these two tests, cracking did not start before 75 % of the maximum shear load and the cracks never opened more than 0.30 mm. Test S10 had just about the same maximum crack width. But in this case, as for all other tests with a low level of normal force, the first cracks were observed between 25 and 50 % of the maximum shear force. At half the maximum shear force, the crack widths were always below 0.12 mm. Until 75 % of the maximum this limit had only increased to 0.27 mm. Between 80 and 90 % of the maximum shear force, however, the cracks started to open up. For tests S1, S3, S4 and S5 extreme values over 1 mm were measured.

The local deformation measurements allowed plots of the displacement and strain field of the concrete surface. They give, together with the crack photographs, an idea of the deformational behaviour and they can be used for testing of new material models for finite element programs. Figures 5 and 6 show as an example the final strain field for the specimens S1 and S2.

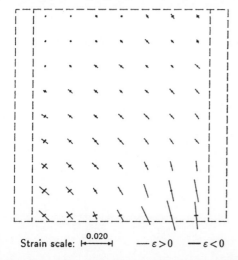

Strain scale: $\longmapsto$ 0.020 $\quad$ — $\varepsilon > 0$ $\quad$ — $\varepsilon < 0$

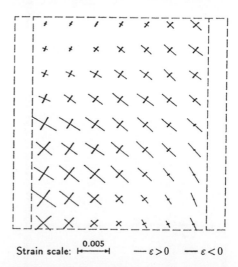

Strain scale: $\longmapsto$ 0.005 $\quad$ — $\varepsilon > 0$ $\quad$ — $\varepsilon < 0$

Figure 5: Strain field for test S1 before failure ($F_x = 686$ kN, $\delta_x = 20.07$ mm).

Figure 6: Strain field for test S2 before failure ($F_x = 918$ kN, $\delta_x = 9.30$ mm).

Figure 7: Test S1 after failure.

Figure 8: Test S2 after failure.

Figure 9: Test S5 after failure.

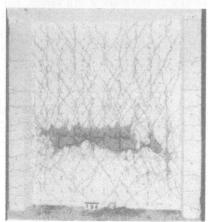

Figure 10: Test S7 after failure of wall.

Figure 11: Test S8 after failure of flange.

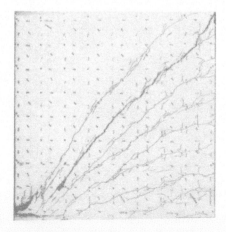

Figure 12: Test S9 after failure.

The following list presents observations and results particular for each test:

S1 At maximum load almost all vertical reinforcement outside the compression flange had reached the yield strain. The measurements with a baselength of 283 mm on the horizontal reinforcing bars never showed values above the yield limit. Horizontal cracking in the lower half of the compression flange outside could be observed just before failure.

S2 Steel strains over the yield limit were only measured at the final loadstep on the vertical reinforcement of the tension flange outside. Failure occurred without any previous notation and led to a total loss of bearing capacity.

S3 The strain measurements on the reinforcement proved yielding of the vertical bars in the tension flange and of the horizontal bars in the central part of the wall.

S4 The measurements on the vertical reinforcement of the lower right quarter of the wall showed values above the yield limit.

S5 The relation between shear force and horizontal displacement is shown in the left part of Figure 13 for the first and last cycle of each deformation level. Failure occurred during the seventh cycle on the fifth deformation level after a continuous shear force reduction of 28 %. Yielding strains were measured on the vertical reinforcement both for positive and negative shear force.

S6 More than half of the vertical and horizontal reinforcing bars came into yielding. This specimen was retested one year later with a changed direction of the applied shear force but without any repair of the damaged compression zone. Failure occurred at a maximum load $F_x = -645$ kN and a horizontal displacement $\delta_x = -40$ mm.

S7 The right side of Figure 13 presents the hysteretic behaviour of this test. During the load reversals on the fourth deformation level a progressive failure of the inner

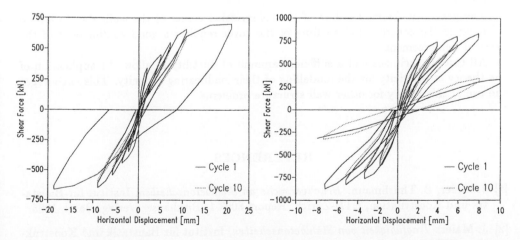

Figure 13: Hysteretic behaviour of the cyclic tests S5 (left) and S7 (right).

part of the wall was observed and the upper slab developed cracks. The shear force dropped down by 58 %. The final failure of the compression flange occurred during the first load reversal on the fifth deformation level. With the exception of one measurement on a horizontal bar in the latter failure zone of the wall, no yielding of the reinforcement was observed.

S8 Cracking started at the tension edge. The first crack in the concrete above the opening was observed at a shear force $F_x = 244$ kN. Horizontal cracks in the compression flange—one on the inside at the base and another on the outside at the level of the upper edge of the opening—announced the failure. The still undamaged rest of the wall allowed a continuous increase of the displacement leading to a slow reduction of the shear force. When the horizontal displacement of maximum load was surpassed by 75 %, the direction of the shear force was reversed. Now, with the opening in the tension zone, a maximum load of -656 kN was achieved.

S9 Because of the lack of horizontal reinforcement the failure surface could follow one of the diagonal cracks from the lower left to the upper right corner of the wall. Strains over the yield limit were measured on about 40 % of the vertical reinforcement.

S10 At maximum load all vertical steel bars of the hidden tension flange had started yielding. The strain measurements on the horizontal reinforcement reached a maximum value of $2.55 \cdot 10^{-3}$, which is still below the yield limit.

CONCLUSIONS

The test results showed a negligible influence of the horizontal reinforcement on the maximum shear force. However, the deformation capacity increased as a function of the horizontal reinforcing ratio. The tests with reversed shear force loading did not show a significant change in the behaviour when compared with the related monotonous tests. This allows the conclusion that the cyclic application of the shear force is not always necessary.

The generally observed failure of the compression corner of the walls due to lateral splitting of the concrete and spalling of the cover requires a good confinement of the vertical reinforcement.

All tested specimens had a sufficient amount of ductility to allow the application of the theory of plasticity for the modelling of their loadbearing capacity. This encourages the use of this theory for other wall structure problems.

REFERENCES

[1] J. Maier, B. Thürlimann: *Bruchversuche an Stahlbetonscheiben,* Institut für Baustatik und Konstruktion, ETH-Zürich, Report 8003-1, Birkhäuser Verlag, Basel, 1985.

[2] J. Maier: *Tragfähigkeit von Stahlbetonscheiben,* Institut für Baustatik und Konstruktion, ETH-Zürich, Report 169, Birkhäuser Verlag, Basel, 1988.

TESTS ON REINFORCED CONCRETE LOW-RISE SHEAR WALLS
UNDER CYCLIC LOADING

M.A. MANSUR, T. BALENDRA, AND S.C. H'NG
Department of Civil Engineering
National University of Singapore
Kent Ridge, Singapore 0511

ABSTRACT

Reported in this paper are the results of five low-rise shear walls tested under static cyclic loading to evaluate the effects of the type, orientation and disposition of reinforcement on their response during seismic excitation. Test results indicate that the use of welded wire mesh at 45-degree orientation is a good choice with respect to the requirements of strength, stiffness and ductility.

INTRODUCTION

In the construction of multistorey buildings it has become mandatory to use reinforced concrete walls primarily to control lateral deflections. Such walls, commonly known as shear walls, are cantilever structures fixed at the foundation. Since the height-to-width ratio of the walls for a tall building is relatively large, their behavior is governed by flexure which is readily predictable. But when the aspect ratio of such walls reduces to a value less than unity, such as those frequently occuring in low-rise buildings located in a seismic zone, shear deformation becomes predominent. Hence the usual flexural theory no longer applies.

A number of experimental investigations [1,2] had been carried out in the past on this type of walls, but no rational theory was avilable until the work by Hsu and Mo in 1985 [3]. They have employed the truss model concept coupled with a softened stress-strain relationship of concrete to predict successfully the complete structural response of low-rise shear walls. On the basis of the same theory, Mau and Hsu [4] have recently developed a method suitable for the direct design of such walls subject to monotonic loading.

The study reported here was aimed at assessing the performance of low-rise shear walls designed in accordance with Mau and Hsu's method under seismic excitation. It comprised testing five walls under cyclic loading.

The major parameters of the study were the type (welded wire fabric or expanded metal), orientation and distribution of reinforcement. In this paper, the results of these tests are presented and discussed in the light of seismic design philosophy.

<h2 style="text-align:center">EXPERIMENTAL PROGRAM</h2>

The five low-rise shearwalls, designated as W1-W5, were tested as isolated cantilevers under static cyclic loading. All specimens had identical dimensions and contained approximately the same amount of reinforcement. The type, orientation and dispersion of reinforcement in the wall were however varied.

Test Specimens
Fig.1 shows overall dimensions of the specimens. The height, width and thickness of a typical wall were 600 mm, 900 mm and 60 mm, respectively; the corresponding aspect ratio being 0.67. As shown in Fig. 1, all walls were provided with identical boundary elements. The design unultimate load was 360 kN.

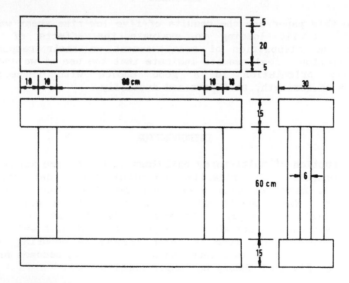

Figure 1.   Overall dimensions of the walls.

The reinforcement details are given in Table 1 and are shown in Fig. 2. Walls W1 to W4 were reinforced with welded wire mesh with square openings while expanded metal mesh was used in W5. The size of wires, opening size of the mesh and number of mesh layers were adjusted so as to achieve approximately the same amount of reinforcement for all the specimens. In walls W2, W4 and W5, the longitudinal direction of the mesh coincided with the longitudinal direction of the wall. But for the remaining two walls, these directions subtended an angle of 45 degrees, i.e., the wires were inclined at 45-deg. to the longitudinal direction of the wall.

TABLE 1
Details of walls

| Wall | Reinforcement | | | | | | f'_c |
	Type	Dia. of wire (mm)	Grid size (mm sq.)	Orientation (deg.)	Weight (kg/m^2)	No. of layers	(MPa)
W1	Welded wire mesh	6.00	150.0	45	8.89	2	33.7
W2		3.25	50.0	45	7.57	2	31.4
W3		3.25	50.0	0	7.57	2	31.3
W4		1.22	12.5	0	7.56	4	37.4
W5	Expanded metal			0	7.58	3	32.2

Each boundary element was reinforced with 8-10 mm diameter deformed bars with a sufficient number of ties as shown in Fig. 2. This amount of reinforcement was adequate to prevent flexural failure prior to shearing distress in the wall. Tension tests were performed on three representative samples for each of the 6 and 8 mm diameter bars. The tensile properties of mesh reinforcement were established in accordance with the recommendations of the ACI Committee 549 [5]. The properties of various types of reinforcement used in this program are summarized in Table 2.

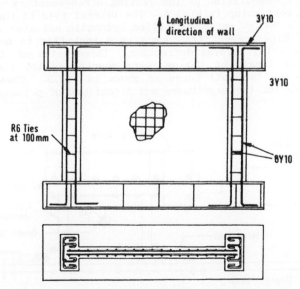

Figure 2. Reinforcement details of specimens

The specimens were cast in a plywood mold which was designed for the wall to be cast in the vertical direction. Ordinary portland cement, natural sand and crushed granite of 10 mm maximum aggregate size were used in the ratio 1:2.16:1.8 by weight for the concrete mix. The water-cement ratio used was 0.6. An admixture, Rapidard, was added at 2.27 litre per 50 kg of cement to accelerate the gain in strength of the concrete. Three 100 mm cubes and three 100 mm x 200 mm cylinders were cast for each wall as control specimens. The walls and control specimens were cured under wet hessian for seven days and kept under laboratory conditions thereafter. The concrete strengths at the end of each experiment are presented in Table 1.

TABLE 2
Properties of reinforcement

Bar/wire diameter	Mesh size	E_s	f_y	f_u
(mm)	(mm x mm)	(KPa)	(MPa)	(MPa)
10.00	isolated	200	460	495
6.00	150x150	206	327	455
3.25	50 x 50	200	429	454
1.22	12.5x12.5	178	359	382
1.00 mm-thick expanded metal		138	321	413

Test Setup and Procedure

The schematic representation of the testing arrangement is shown in Fig. 3 and the actual test setup in Fig. 4. The lateral load to the wall specimen was applied by a 500 kN servo-controlled hydraulic actuator through a stiff loading beam. Displacement tranducers were installed to monitor the base slip as well as total lateral deflection. Three pairs of 'Demec' targets, each with a gage length of 100 mm, were mounted in the form of a strain rossette on one of the wall faces as shown in Fig.3. These measurements enabled to establish the magnitudes and directions of principal strains.

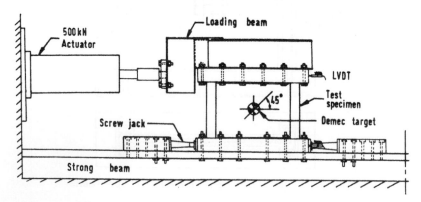

Figure 3. Test setup and instrumentation

Each specimen was subjected to a predetermined displacement history similar to that shown in Fig. 5. Initially, the increment in displacement amplitude used was small, but this was increased substantially at later stages of the loading history. At each displacement level four complete loading cycles were imposed. The specimens were tested to failure.

Figure 4. A view of the test assembly

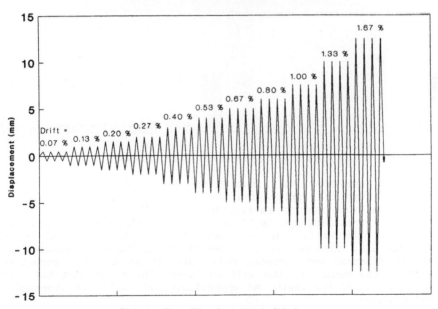

Figure 5. Displacement history

TEST RESULTS AND DISCUSSION

General Behavior

Cracking of the the walls occurred when a displacement amplitude of 0.5 mm was applied. The first crack developed was in a diagonal direction near the corner of the wall. As the cyclic loading continued, additional diagonal cracks were formed and the existing cracks started to propagate towards the two boundary elements. The system of well-developed diagonal cracks constituted a symmetric grid pattern on the wall surface as can be seen in Fig. 6. The cracks that traversed through the center of the wall usually developed into major cracks at later stages and finally led to the crushing of the concrete.

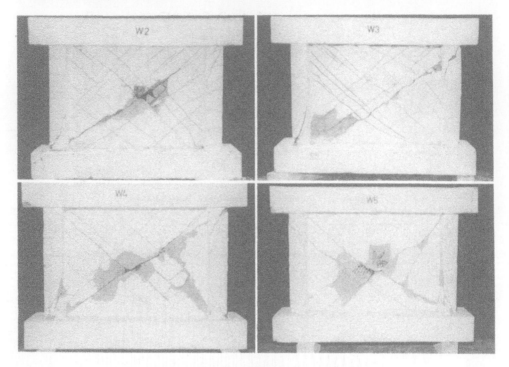

Figure 6. Specimens after failue.

For the specimen W1, unexpected and sudden faiure occurred immediatelty after the application of the loading cycles for the 10 mm deflection amplitude due to malfunctioning of the cotrolling tranducer. Failure of the remaining specimens was due to crushing of the concrete at the center of the wall. In specimens W4 and W5, traces of concrete crushing was also observed at the base of the boundary elements, and at failure the finemesh and expanded metal used in W4 and W5, respectively, fractured at the center of the wall and along the major diagonal cracks. For the four successful tests, no appreciable base slip was observed even at final collapse.

Strength
The lateral force-displacement hysteresis of the wall specimens tested in this program are shown in Fig. 7. The ultimate strength and its degradation expressed as a percentage of the ultimate value for each wall are presented in Table 3 along with the respective percentage of drift (lateral displacement at wall top x 100/height of wall). Because of malfunctioning of the displacement transducer, it was not possible to trace the degradation of strength for specimen W1.

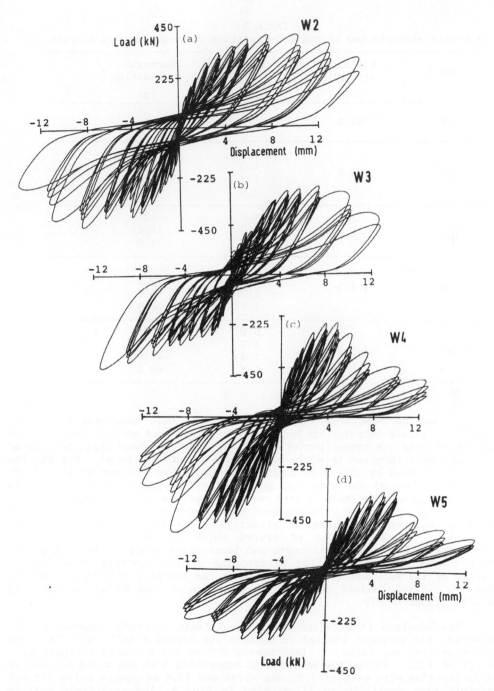

Figure 7. Load-displacement hysteresis

TABLE 3
Ultimate strength and strength degradation corresponding to a drift

Wall	Ultimate strength (kN)	Drift (%)	Strength degradation (%)
W1	397.4	1.4 Data is not available	0
W2	399.6	1.0 1.3 1.7 2.0	0 1.8 11.8 29.0
W3	378.8	1.0 1.3 1.7	0 12.6 34.0
W4	414.3	0.7 1.0 1.3 1.7	0 32.5 49.8 62.9
W5	345.9	0.8 1.1 1.4 1.7	0 34.5 49.8 60.5

It may be seen from Fig. 7 and Table 3 that specimens W2 and W3 exhibit a stable behavior up to a drift of 1%, beyond which strength start to deteriorate. A comparison of Figs. 7(a) and 7(b) shows that the rate of strength deterioration in W2 is more gradual than that in W3. For W2, the strength dropped by only 1.8% compared to 12.6% for specimen W3 at a drift of 1.3%. Although the type and amount of reinforcement were the same in both specimens, the mesh layers were orientated in different manners. Diagonal arrangement of reinforcement in W2 intercepted the cracks at right angles, and hence provided an effective means for the control of the initiation and propagation of cracks which may be attributed to the resulting higher ultimate strength and lower degradation of shear strength. Both specimens finally failed by crushing of the concrete close to the center of the wall and along the bottom of the boundary element as shown in Fig. 6. Final failure occurred in specimens W2 and W3 at 2% and 1.7% drifts, respectively.

The specimen W4 attained a higher ultimate strength compared to W3. However, its strength started to deteriorate beyond a drift of 0.7%. The deterioration was rapid; it lost nearly 50% of its ultimate strength at a drift of 1.3%. Unlike the welded mesh comprising 3.25 mmϕ wires as used in W-3, the fine wire mesh of 1.22 mmϕ wires and 12.5 mm square grids did not exhibit high ductility as illustrated in Fig. 8. The relatively brittle nature of the fine wire mesh may be attributed to the rapid loss of strength, and this was observed from the gradual fracturing of the wires in W4. The specimen failed at a drift of 1.7%.

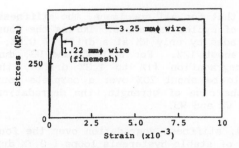

Figure 8. Stress-strain curves for welded wire mesh

In specimen W5 which was reinforced with expanded metal, the ultimate strength was reached at a drift of 0.8%. Beyond this level the deterioration of strength was rapid, losing 50% of its srength at a drift of 1.4%. This may be attributed to the scissors action of expanded metal in concrete subjected to compression. Moreover, expanded metal may be relatively brittle in nature as evidenced by fracturing of the mesh similar to that in specimen W4.

Stiffness

As shown in Fig. 7, the specimens were subjected to four complete loading cycles at each displacement level. The degradation of equivalent stiffness during the 2nd, 3rd and 4th cycles are depicted in Fig. 9 for different percentage of drift. In this presentation, equivalent stiffness at a particular loading cycle is defined as the gradient of the line joining the peaks of the load-displacement curves during forward and reverse loadings, and degradation is measured with reference to the first cycle and is expressed as a percentage.

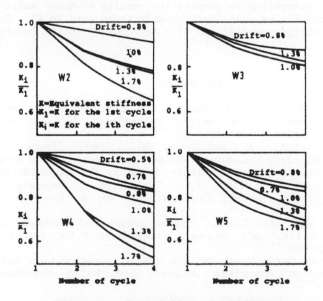

Figure 9. Degradation of stiffness

It may be seen that for all specimens, the stiffness degradation rate drops as the number of cycles increases. During the four complete cycles, stiffness of W2 degrades by only 9% at a drift of 0.8%; the corresponding value for W3 being about 12%. For both W2 and W3, during the phase of marginal strength degradation (1% to 1.3% drift), the degradation of stiffness is found to be about 20% over a complete four cycles. Beyond 1.3% drift, as in the case of strength, the degradation of stiffness is appreciable for both W2 and W3.

For specimen W4, stiffness degradation over the four cycles is about 16% during the phase of stable hysteresis loops (0.7% drift). Beyond this the stiffness degrades at a fater rate. For example, the degradation of stiffness is about 20% at a drift of 1%, whereas the corresponding value at 1.3% drift is about 40%. Specimen W5 suffered approximately the same order of stiffness degradation as W4.

CONCLUDING REMARKS

The results of a total of five pilot tests on low-rise shear walls are reported in this paper. Among these, test on wall W1 was partly unsuccessful due to halfway malfunctioning of the contolling transducer.

From the analysis and discussion of the remaining wall specimens, it becomes evident that the use of welded wire mesh at an orientation of 45 degrees to the longitudinal direction of the wall is a good proposition from the viewpoints of strength, stiffness and ductility requirements. Expanded metal or finer wire mesh are not the efficient form of reinforcement at least for low-rise shear walls.

It will be interesting to compare the results of shear walls reinforced with larger size of wire mesh, particularly those used in W1. Test data on such walls are being generated and will be reported in the future.

REFERENCES

1. Benjamin, J.R. and Williams, H.A., "The Behavior of One-Story Reinforced Concrete Shear Walls," Proceedings, ASCE, V. 83, No. ST3, May 1957, pp. 1254.1-1254.49.
2. Barda, F., Hanson, J.M., and Corley, W.G., "Shear Strength of Low-Rise Walls with Boundary Elements," Reinforced Structures in Seismic Zones, SP-53, American Concrete Institute, Detriot, 1977, pp. 149-202.
3. Hsu, Thomas, T.C., and Mo, Y.L., "Softening of Concrete in Low-Rise Shearwalls," ACI Journal, Proc. V. 82, No. 6, Nov.-Dec. 1985, pp. 883-889.
4. Mau, S.T., and Hsu, T.T.C., "Shear Design and Analysis of Low-Rise Structural Walls," ACI Journal, Proceedings V.83, No. 2, Mar.-Apr. 1986, pp.306-315.
5. ACI Committee 549, "Guide for the Design, Construction and Repair of Ferrocement," ACI Structural Journal, V.85, No. 3, May-June 1988, pp. 325-351.

105

HYSTERETIC SHEAR RESPONSE OF LOW-RISE WALLS

MURAT SAATCIOGLU
Associate Professor
Department of Civil Engineering
University of Ottawa
Ottawa, CANADA, K1N 6N5

ABSTRACT

The paper presents results of an experimental investigation on seismic resistance of low-rise shear walls. The experimental program includes tests of large scale wall specimens under slowly applied lateral load reversals. The emphasis is placed on shear response of walls, particularly with respect to different modes of shear behavior, and characteristics of hysteretic shear response observed during the tests. The results indicate that prevention of diagonal tension failure by the use of increased area of horizontal reinforcement may lead to shear sliding. Sliding reduces energy dissipation capacity of the walls, and may lead to premature failure prior to development of flexural or shear capacity. It is shown that walls reinforced with special sliding shear reinforcement develop higher lateral force resistance and energy dissipation characteristics. These walls however, are susceptible to web crushing caused by increased level of shear forces and reduced compressive strength of cracked concrete under cyclic loading.

INTRODUCTION

Reinforced concrete walls in earthquake resistant buildings provide strength and deformability against seismic forces. Deformability of walls in the inelastic range is an important aspect of seismic resistant design. It may be achieved relatively easily if the governing mode of deformation is flexure. While the dominance of flexural response is possible in multistory buildings, it may be difficult to achieve in low rise structures. High shear forces associated with low aspect ratios of low-rise walls may lead to reduced deformability and may result in premature diagonal tension, diagonal compression, or sliding shear failures.

The behavior of low rise shear walls was investigated experimentally by testing large scale wall specimens with aspect ratios of 1/2 and 1/4 under slowly applied lateral load reversals [1]. This paper provides a summary of experimental observations on three companion specimens, with an aspect ratio of 1/2, illustrating three typical modes of behavior in shear.

EXPERIMENTAL PROGRAM

Figure 1 illustrates overall geometric properties of the specimens tested in the experimental program. The specimens were prepared in stages to reflect the actual construction practice. Reinforcement cage for the foundation beam and the wall element was assembled first. Concrete for the foundation beam was cast next. A roughened concrete surface was provided along the construction joint with a relief depth of about 10 mm. Concrete for the wall element and the top beam was cast together at a later date. The foundation beam was heavily reinforced and bolted to the laboratory strong floor to provide full fixity during testing. The top beam, shown in the figure, was also heavily reinforced, and used to transfer the applied load to the wall element. Wall reinforcement arrangement consisted of uniformly distributed double rows of reinforcement in both vertical and horizontal directions. The end two rows of vertical reinforcement were tied by hoops to prevent premature buckling under compression. Table 1 provides a summary of specimen properties. Wall 1 was designed to have lower shear capacity than that corresponding to flexural strength. Walls 4 and 6 were designed to have higher shear capacities.

The loading program consisted of incrementally increasing lateral displacement cycles. Three cycles of displacement were applied at each deformation level, followed by a small cycle. The load was applied by a hydraulic jack, supported by a reaction frame. The specimens were well instrumented to measure deformations caused by shear, flexure, bar extension,

TABLE 1

Properties of Test Specimens

Specimen	f_c' (MPa)	Sliding Shear Reinf.	Vert. Reinf.		Horiz. Reinf.	
			f_y (MPa)	ρ (%)	f_y (MPa)	ρ (%)
Wall 1	25	NO	435	0.8	425	0.25
Wall 4	33	NO	480	0.8	480	0.80
Wall 6	35	YES	480	0.8	480	0.80

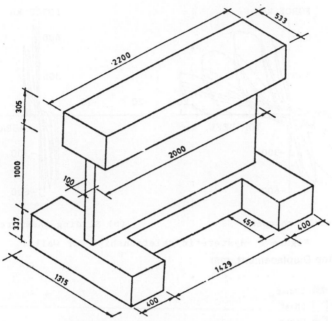

Figure 1. Specimen geometry

and sliding shear. Linear variable differential transducers (LVDT) and strain gauges were used to measure deformations.

OBSERVED BEHAVIOR AND TEST RESULTS

Wall 1 was reinforced with 11.3 mm vertical and 6.4 mm horizontal deformed bars. The first three deformation cycles were applied at 0.04 % drift ratio. The first flexural crack occured during this load stage along the construction joint. The following load stage consisted of three cycles at 0.2 % drift ratio. The first diagonal tension crack, and yielding of longitudinal reinforcement occured during this load stage. Diagonal cracks formed at approximately 35 degrees with the horizontal. Inclined cracks also formed between the upper corners and the opposite lower corners of the wall, crossing each other in the center. The computed flexural capacity, based on plane section analysis, was reached during the first cycle of deformation cycles at 0.4 % drift ratio. At this deformation level the flexural crack at the base became continuous along the construction joint. Widening of corner-to-corner inclined cracks was observed during the subsequent load cycle of 0.6 % drift ratio. The inclined cracks did not close completely upon load reversals. Widening of corner-to-corner cracks, and spalling and crushing of concrete at

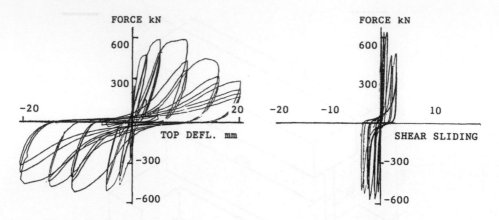

(a) Overall response   (b) Sliding shear response

Figure 2. Hysteretic relationships for Wall 1

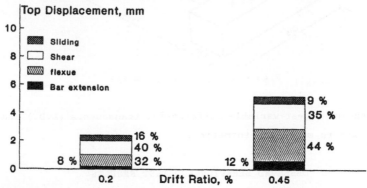

Figure 3. Components of horizontal displacement for Wall 1

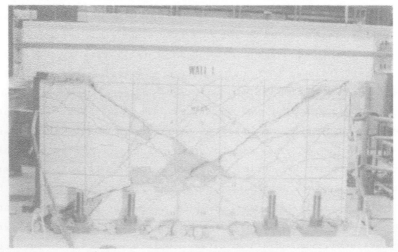

Figure 4. Wall 1 at the end of testing

the top corners and compression toes were observed.

Figure 2 shows hysteretic force deformation relationships recorded during the test. The overall response shown in Fig. 2(a) indicates a strength decay starting with a drift ratio of approximately 1.0 %. Pinching of hysteresis loops, typically observed in shear dominant response can be seen in the same figure. Fig. 2(b) depicts shear force-shear sliding hysteretic relationship. The figure indicates that sliding along the construction joint was insignificant. In fact some of the deformations included in horizontal sliding was due to inclined cracking that extended into the construction joint region where the LVDTs for sliding shear measurements were placed.

Other components of deformation due to flexure and extension of vertical reinforcement, were also measured. Fig. 3 illustrates the contribution of each deformation component to the overall response. It is clear from this figure that the governing mode of deformation in Wall 1 was shear. The specimen behaved in a ductile manner up to approximately 0.8 % drift ratio, and showed strength degradation under increasing deformation cycles. The failure was due to diagonal tension along corner-to-corner cracks. Figure 4 shows Specimen 1 at the end of testing.

Wall 4 was companion to Wall 1, and had similar properties. The only difference between the two walls was the use of 11.3 mm deformed bars as horizontal reinforcement, which resulted in increased reinforcement ratio of 0.8 % in Wall 4. This was done to investigate the wall behavior when diagonal tension failure was prevented or delayed. The fist three cycles of deformation applied to Wall 4 were at 0.1 % drift ratio. The first set of flexural and diagonal tension cracks occured during this load stage. The next three cycles were applied at approximately 0.35 % drift ratio. Additional diagonal cracking and widening of the base horizontal crack, up to 0.5 mm were observed. Sliding of wall along the construction joint started at this load stage. The wall could not develop the strength attained at the first cycle of this deformation level, during the subsequent two cycles. A significant drop in the load resistance was recorded as the wall slid back and forth to the imposed deformation level. The flexural crack along the construction joint became wider and the crack width was measured to be 1.5 mm during the cycles of 1.4 % drift ratio. Figure 5 illustrates the recorded force deformation hysteretic relationships. Excessive pinching of hysteresis loops is indicative of the observed shear sliding.

It is clear from Fig. 5 that a large portion of lateral deflection was due to shear sliding. The wall element could not develop its strength. The

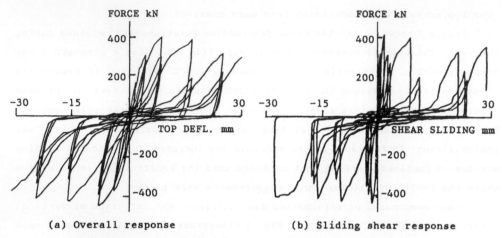

(a) Overall response   (b) Sliding shear response

Figure 5. Hysteretic relationships for Wall 4

Figure 6. Wall 4 at the end of testing

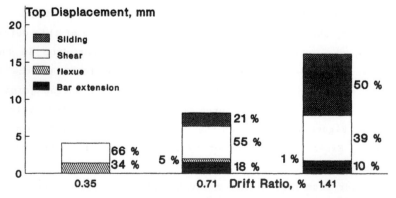

Figure 7. Components of horizontal displacement for Wall 4

maximum load resistance attained was 70 % of that computed on the basis of plane section analysis. The energy dissipation capacity was significantly hindered by sliding. Increased deformations in the subsequent cycles produced crushing of concrete around the vertical bars, caused by dowel action, followed by kinking of vertical reinforcement. The wall element did not suffer any further damage, while it behaved in the sliding shear mode. Fig. 6 shows the specimen at the end of testing. Figure 7 illustrates deformation components measured during the test.

Wall 6 was companion to Wall 4. The only difference between the two specimens was the presence of special sliding shear reinforcement in Wall 6. This reinforcement consisted of 11.3 mm deformed bars, placed in two rows between the vertical reinforcement, perpendicular to the construction joint. They were anchored into the foundation beam by extending at least the development length of the bar. The portion extending into the wall element was limited to 200 mm. Fig. 8 illustrates the sliding shear reinforcement during construction.

The behavior of Wall 6 was similar to that of Walls 1 and 4 up to deformation cycles of 0.35 % drift ratio. This time however, the sliding shear reinforcement controlled the flexural crack along the construction joint. This prevented sliding along the joint and enabled the wall to resist higher shear forces. Increased percentage of horizontal reinforcement, relative to Wall 1 prevented widening of the inclined cracks. Therefore the wall element sustained higher levels of shear force. The increase in shear force produced

Figure 8. Sliding Shear Reinforcement in Wall 6

a large number of well distributed inclined cracks. The wall did not show any sign of failure up to deformation cycles of 1.3 % drift ratio. The inclined web cracks passed over the tips of the sliding shear reinforcement, propagating towards the construction joint. The resultant crack pattern produced a sawtooth like appearance along the construction joint, where the concrete around the bars were held together with the foundation beam. The resultant concrete "shear keys" eliminated shear sliding completely. This led to well rounded hysteresis loops, as shown in Fig. 9(a), and improved energy absorption capacity of the wall. As the applied shear force level was increased, concentration of inclined cracks was observed near the tips of the sliding shear reinforcement. Crisscrossing of inclined cracks under reversed cyclic loading resulted in degradation of concrete compressive strength in this region. This led to crushing and spalling of concrete immediately above the sliding shear reinforcement, at about 1.7 % drift ratio (see Fig. 10). The wall eventually failed by web crushing of concrete. The maximum load resistance recorded was 86 % of that computed based on plane section analysis. Figures 11 shows the deformation components measured in Wall 6.

CHARACTERISTICS OF HYSTERETIC SHEAR RESPONSE

It is clear from the foregoing description of test results that shear behavior of low-rise walls can be dominated by diagonal tension, diagonal compression or sliding shear. When shear sliding is prevented, the hysteretic energy absorbed by the wall element improves. The primary force-deformation relationship under monotonic loading provides an envelope for the hysteresis loops until the onset of strength decay. Unloading and reloading curves indicate degradation of stiffness under cyclic loading. Upon unloading in one direction, reloading in the opposite direction meets with little initial resistance until the previously opened cracks are closed. The cracked surfaces along diagonal cracks slide against each other prior to closure. As the crack closure takes place, the resistance of wall element is gradually restored. This initial softening, followed by gradual built up of load resistance is reflected in the force deformation hysteretic relationship as "pinching" of loops. Pinching is relatively low in walls where the response is governed by diagonal compression. In this case the diagonal tension cracks develop as hairline cracks, well distributed over the web. The hysteresis loops are well rounded, and approach in appearance to those generated by flexure response. However, these curves are followed by a sudden strength loss, contrary to

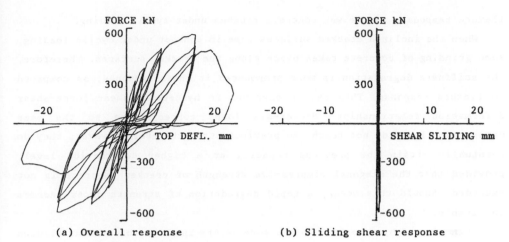

(a) Overall response   (b) Sliding shear response

Figure 9. Hysteretic relationships for Wall 6

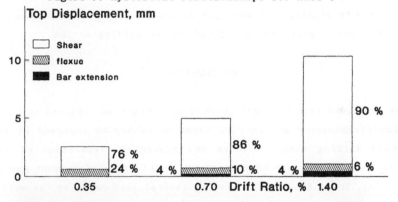

Figure 10. Components of horizontal displacement for Wall 6

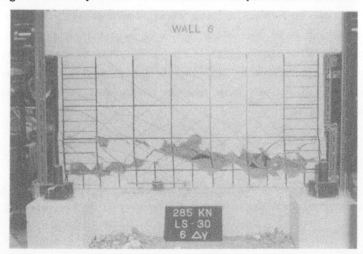

Figure 11. Wall 6 at the end of testing

flexure response, as the web concrete crushes under cyclic loading.

When the inclined cracked surfaces come in contact under cyclic loading, some grinding of concrete takes place along the cracked surfaces. Therefore, the stiffness degradation is more pronounced in shear response, as compared to flexure response. This can be observed in hysteretic shear force-shear deformation relationships. Hysteresis loops corresponding to the same deformation level do not reach the previous strength level. However, they do eventually attain the previous capacity at a higher deformation level, provided that the diagonal compressive strength of cracked concrete is not exceeded. Should this occur, a rapid degradation of strength decay becomes inevitable.

Pinching of hysteresis loops is more severe in shear force-shear sliding hysteretic relationship. In this case, the member becomes softer during reloading, due to sliding. The strength as well as energy dissipation capacity of the member are significantly reduced due to sliding action.

CONCLUSIONS

It may be concluded from the experimental investigation reported in this paper that seismic resistance of low-rise shear walls may be improved if diagonal tension and sliding shear failures are prevented. Prevention of diagonal tension failure by adequate horizontal and vertical reinforcement may trigger shear sliding. Shear sliding reduces the lateral load capacity, as well as the capacity to dissipate earthquake induced energy. Shear sliding can be prevented by the use of special sliding shear reinforcement. This may lead to brittle web crushing under increased levels of deformation cycles.

Prevention of diagonal tension and sliding shear failures improved the deformability of walls tested in this investigation approximately from 0.8 % to 1.6 % drift ratio. However, any further improvement in deformability was limited by web crushing.

REFERENCES

1. Saatcioglu, M, Wasiewicz, Z.F, Pilette, C.F., and Wiradinata, S., "Tests of Low-Rise Shear Walls Under Simulated Seismic Loading," Research Report, Department of Civil Engineering, University of Ottawa, Ottawa, Canada, 1990.

PREDICTION OF HYSTERETIC LOOPS FOR R.C. SHEAR
WALLS WITH 0.65 TO 1.90 SHEAR-SPAN TO WALL-WIDTH RATIOS

MAW-SHYONG SHEU
Prof., Department of Architecture
National Cheng-Kung University
Tainan, Taiwan, Rep. of China

GWANG-SHING HUANG
Graduate Student, Dept. of Architecture
National Cheng-Kung University
Tainan, Taiwan, Rep. of China

ABSTRACT

This paper proposed the empirical equations for the prediction of load-deflection curves for one-story and two-story R.C. shear walls under static monotonic or reversed cyclic loads plus axial compressive loads. The width of wall is 70 cm; the thickness of wall is 7 cm; the height of wall for each story is either 50 cm or 75 cm; the cross-section of boundary columns is 17 cm x 15 cm. The shear-span to wall-width ratios are from 0.65 to 1.90. The P-Δ curves under monotonic loading are predicted first. The hysteretic rules are based on the deteriorating envelopes. The deterioration of residual ultimate capacity of wall is a function of energy dissipation and the maximum lateral deflection of wall in the previous loading cycles. All the predicted load-deflection curves are compared to the experimental results with reasonable accuracy.

INTRODUCTION

Consider a R.C. frame with shear walls as shown in Fig.1(a). The shear walls are identical for every story. If we are going to investigate the interactions between frame and shear walls, we might condense the walls as big columns. Then the moment diagram of wall is shown in Fig.1(b). It is clear that shear-spans for walls are different from story to story. Thus, the ultimate capacity and changes of stiffness under reversed cyclic loadings might also be different from story to story. This paper tries to investigate the ultimate capacity and hysteretic rules for one-story and two-story shear walls through experimental tests. The shear-span to wall-width ratios for the specimens are 0.65, 0.90 and 1.90 for each group.

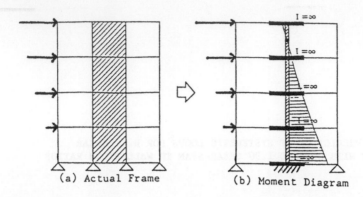

(a) Actual Frame   (b) Moment Diagram

Figure 1. Bending Moment and Shear Span of Shear Walls

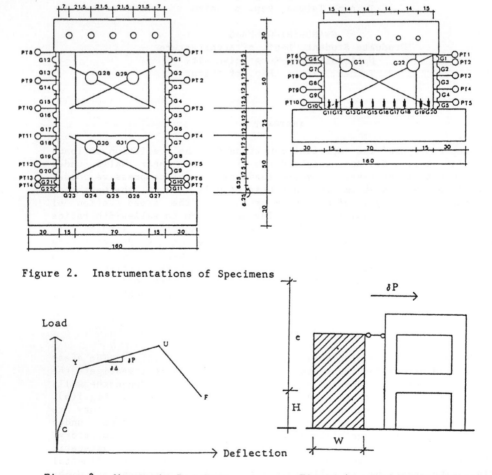

Figure 2. Instrumentations of Specimens

Figure 3. Monotonic P-Δ Curve   Figure 4. Analytical Models

EXPERIMENTAL TEST

There are 15 specimens tested in this paper, 9 of them are one-story walls and 6 of them are two-story walls. The parameters for investigation are listed in TABLE 1. The instrumentation is shown in Fig.2. The specimens are loaded horizontally through 5 high-tension bolts by one hydraulic jack pushing and another jack pulling. The total lateral deflections of wall are measured by linear LVDT'S or potentiometers. The lateral deflections due to bending are calculated by the clip-on gages on both sides of wall. The lateral deflections due to shear deformation are calculated by substructing the bending deflections from total deflections.

TABLE 1   Parameters of Specimens

Specimen	Dimension of Wall (Boundary Column) (cm)	Shear Span wall width	Axial Comp. Loads (t)	Horiz. Loads	ρ_v (%)	ρ_h (%)	f_y (kg/cm²)	f'_c (kg/cm²)
SWB-6A	70x50x7 (15x17)	0.65	0	cyclic	0.85	0.81	5005	273
SWB-7A	70x50x7 (15x17)	0.65	0	cyclic	0.85	0.81	5005	307
SWB-8A	70x50x7 (15x17)	0.65	0	cyclic	0.85	0.81	5005	225
SWB-17A	70x50x7 (15x17)	0.65	0	cyclic	0.85	0.81	5005	229
SWB-1B	70x75x7 (15x17)	0.90	0	mono	1.52	1.45	4716	315
SWB-2B	70x75x7 (15x17)	0.90	0	cyclic	1.52	1.45	4716	335
SWB-3B	70x75x7 (15x17)	0.90	0	mono	1.52	1.45	4716	285
SWB-4B	70x75x7 (15x17)	0.90	0	cyclic	1.52	1.45	4716	303
SWB-5B	70x75x7 (15x17)	0.90	0	mono	1.52	1.45	4716	380
SWB-1C	2-70x50x7 (15x17)	1.90	0	cyclic	0.85	0.81	5005	235
SWB-2C	2-70x50x7 (15x17)	1.90	0	cyclic	0.85	0.81	5005	248
SWB-3C	2-70x50x7 (15x17)	1.90	0	cyclic	0.85	0.81	5005	240
SWB-1D	2-70x50x7 (15x17)	1.90	18.5	cyclic	0.85	0.81	5005	295
SWB-2D	2-70x50x7 (15x17)	1.90	18.5	cyclic	0.85	0.81	5005	300
SWB-3D	2-70x50x7 (15x17)	1.90	18.5	cyclic	0.85	0.81	5005	233

PREDICTION OF LOAD-DEFLECTION
CURVES UNDER MONOTONIC LOADINGS

The theoretical load-deflection curves for R.C. shear walls without boundary columns, using constitutive laws of concrete and rebar, were discussed in Huang[*1,*2]. However, such analytical solutions are very much a waste of computer time. This paper presents the empirical predictions of load-deflection curves under monotonic and reversed cyclic loadings. The cracking load of wall is calculated by Equ.1 or Equ.2, whichever is less:

$$P_{fc} = 2(1.99 \sqrt{f'_c} + N/A_g)I_g/(H_sW) \qquad (1)$$

$$V_C = 0.53(1 + 0.0357N/A_g)(1.14 - 0.23H_s/W) \sqrt{f'_c} A_g \qquad (2)$$

where P_{fc} and V_C are the flexural and diagonal cracking loads; N is axial compressive force to the wall; A_g is the gross cross-sectional area; H_S is shear span of wall; W is width of wall including boundary columns. Metric units are used in Equ.1 and 2.

The yield load is calculated by dividing the yield moment by shear span, that is $P_y = M_y/H_S$. Where M_y is calculated from the balance of cross-sectional force, when the extreme vertical rebars of tensile boundary column reach the beginning of strain hardening.

The ultimate load P_u is picked up the smaller from ultimate flexural capacity, $P_{fu} = M_u/H_S$, and ultimate shear capacity, V_u. For calculation of M_u, the ultimate strain of concrete at bottom of boundary columns is calculated by the equation proposed by Mattok[*3]:

$$\varepsilon_u = 0.003 + 0.02b/H_S + 0.2\rho_S \qquad (3)$$

where b is the width of boundary column and ρ_S is percentage of hoops. The stress-strain curve of concrete is employed by Kent and Park[*4]:

$$f_c = f_c' \, (1,000\,\varepsilon_c - 250,000\,\varepsilon_c^2) \qquad (4)$$

when $\varepsilon_c < 0.002$, and

$$f_c = f_c' \, [1 - 0.5(\, \varepsilon_c - 0.002)/(\, \varepsilon_{50u} + \varepsilon_{50h} - 0.002)] \qquad (5)$$

For the concrete outside the core of boundary columns, ε_{50h} is zero in Equ.5. In balancing of cross-sectional forces, strain is assumed to be linear distributed and the strain hardening of rebars are considered.

For calculation of V_u, the concept of ACI Code is used. But the detailed for V_C and V_S are modified as:

$$V_u = V_C + V_S$$

that is $\qquad V_u = V_C + 0.33(A_{sh} + A_{sb})f_y + 0.18(W/H)^{1.1} A_{sv} f_y \qquad (6)$

where V_C is calculated by Equ.2, $A_{sh}\,(A_{sv})$ is the total cross-sectional area of horizontal (vertical) rebars in wall, A_{sb} is the total cross-sectional area of hoops in boundary column within $45°$ inclined crack, H is the clear height of wall, W is the total width of wall and columns.

The total lateral deflection Δ_T is separated into bending deflection Δ_B and shear deformation Δ_S experimentally and analytically. The load-deflection curves under monotonic load, P-Δ_B and P-Δ_S, are assumed to be quadrilinear as shown in Fig.3. The analytical models for determination of each segment deflection of the quadrilinear are shown in Fig.4. Let γ and λ be the reduction functions for tangential flexural stiffness, EI, and tangential shear rigidity, GA. Let $G_c = E_c/3$, then

$$\delta P = \gamma[6EI_{wt}/(2H + 3e)H^2 + 24EI_{ct}/H^3]\delta\Delta_B \qquad (7)$$

and $\qquad\qquad\qquad \delta P = \gamma[E_c A_{gt}/3H]\delta\Delta_S \qquad (8)$

where
$$\gamma = A[1 + D(N/A_g) + E(N/A_g)^2][1 + F\sqrt{f_c'}\,][1 + G\rho_v]$$
$$/[1 + B(H_s/W) + C(H_s/W)^2] \qquad (9)$$

and
$$\lambda = A'[1 + D'(N/A_g) + E'(N/A_g)^2][1 + F'\sqrt{f_c'}\,][1 + G'\rho_h]$$
$$/[1 + B'(H_s/W) + C'(H_s/W)^2] \qquad (10)$$

$$\Delta_T = \Delta_B + \Delta_S \qquad (11)$$

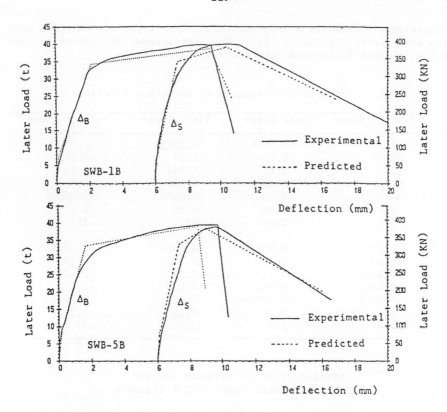

Figure 5. Comparison of Monotonic P-Δ Curves

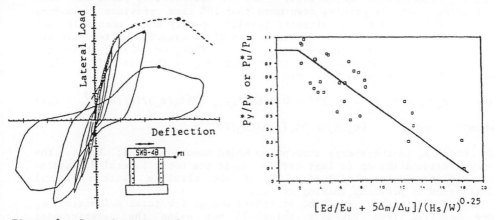

Figure 6. Deterioration of
Residual Capacity

Figure 7. Trend of Deteriorated Capacities

The correlation constants A, B, C, ..., A', B', C'... etc. for each stage are listed in TABLE 2. The comparison of predicted and experimental load deflection curves are shown in Fig.5.

TABLE 2 Correlation Constants of Reduction Functions

	Elastic Stage	Crack Stage	Yield Stage	Failure Stage
A	2.35	-0.0734	0.00259	-1.03
B	17.10	0.10	-2.36	-3.52
C	8.97	0.12	0.548	0.23
D	0.094	0.035	0.026	0.01
E	-0.002	-0.00057	-0.00061	-0.00006
F	0.2	0.06	0.06	0.1
G	526	-699	-527	-135
A'	0.0362	0.0188	0.0024	-0.00088
B'	0.43	0.479	-0.18	-1.35
C'	-0.0703	-0.113	0.27	0.455
D'	0.096	0.042	0.079	-0.016
E'	-0.0024	-0.00089	-0.0024	0.00135
F'	0.2	0.2	0.18	0.2
G'	46.6	45.7	22.9	-37.6

DETERIORATION OF ENVELOPE AND THE HYSTERETIC RULES UNDER CYCLIC LOADINGS

Fig.6 shows the comparison of P-Δ_T curves of SWB-4B, under cyclic loads, to the corresponding curves of SWB-3B, under monotonic load. It is obvious the residual ultimate capacity, after certain cycles of reversed loads, deteriorates quite lots. And the intersecting point for each reloading path is getting down more than 10% from previous unloading peak. Let P_u (P_y) be the ultimate (yield) capacity under monotonic load and P_u^*(P_y^*) be the residual ultimate (yield) capacity after certain loading cycles. The deterioration ratio, P_u^*/P_u or P_y^*/P_y, from the experimental test of this paper is shown in Fig.7. That is $P_u^*/P_u = P_y^*/P_y$ = 1 for horizontal axis less than 2,

and
$$P_u^*/P_u = P_y^*/P_y = 1.1125 - 0.0563(E_d/E_u + 5\Delta_m/\Delta_u)/(H_S/W)^{0.25} \qquad (12)$$

when
$$(E_d/E_u + 5\Delta_m/\Delta_u)/(H_S/W)^{0.25} > 2.0$$

In which E_u is the energy absorption under monotonic load; E_d is the energy dissipation up to last cycle; Δ_u is the total lateral deflection at ultimate monotonic load; Δ_m is the absolute maximum lateral deflection in previous loading cycles.

The hysteretic rules of stiffness change for total deflection Δ_T are modified from Takeda's Rules[*5] but using the deteriorating envelopes.

In cracking stage (Fig.8a): unloading path remains vertical at beginning 10% of peak unloading force. Then③ parallels to OB (the midline of OC and OY) until U_b, 10% of peak unloading force. ④ directs to yield point on reversed side.

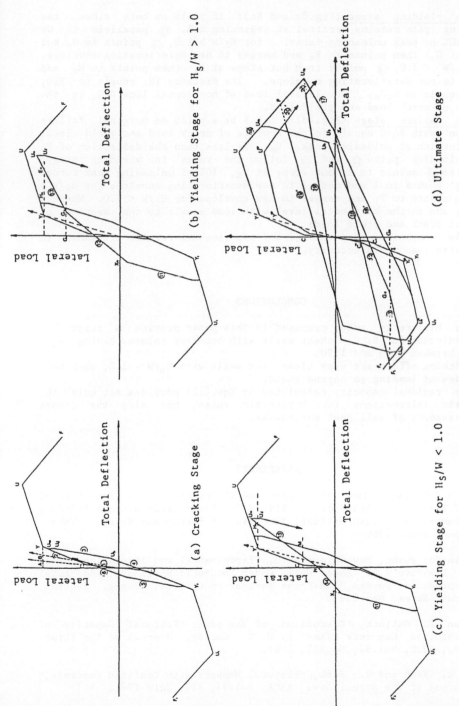

(a) Cracking Stage

(b) Yielding Stage for $H_S/W > 1.0$

(c) Yielding Stage for $H_S/W < 1.0$

(d) Ultimate Stage

Figure 8. Hysteretic Rules under Cyclic Loadings

In yielding stage (Fig.8b and 8c): If yield on both sides, the unloading path remains vertical at beginning 10%. ⑪ parallels to OY until 10% of peak unloading force. for $H_S/W > 1.0$, ⑫ points to H_n but stops at G_n; then points to E_n and merges to new deteriorating envelope. For $H_S/W < 1.0$, ⑭ points to M but stops at N; then points to H_n and merges to new deteriorating envelope. The distance YH_n equals to X_nO; H_nE_n equals to G_oG_n. The lateral load of horizontal line G_oG_n is the average of crack load and yield load.

In failure stage (Fig.8d): Let F be a point on original failure envelope with load equals to the average of crack load and yield load. If deflection of unloading peak, U_m, is less than the deflection of F, the unloading paths ⑫ and ⑬ follow the rules in yielding stage. Otherwise ⑰ points to Y_1 but stops at U_b, 10% of unloading peak force. Then ⑱' points to I and merges to new deteriorating envelope for $H_S/W > 1.0$; ⑱ points to F_1 and merges to new envelope for $H_S/W < 1.0$. Where I and F_1 are on the horizontal level with load equals to the average of original crack and yield loads.

The comparison of predicted and experimental results are shown in Fig.9 with reasonable accuracy.

CONCLUSIONS

1. The hysteretic rules proposed in this paper provide a reasonable prediction for R. C. shear walls with boundary columns having H_S/W in between 0.65 and 1.90.
2. Pinching effect are very clear, for walls with $H_S/W < 1.0$, when both sides of loading go beyond yield.
3. The residual capacity calculated by Equ.(12) provides not only the basic informations for hysteretic rules, but also the damage assessment of wall post earthquake.

REFERENCES

1. Chin-Chi Huang, Maw-Shyong Sheu and Shyong-Ming Guo, "Experimental and Theoretical Study of Low-Rise R.C. Shear Walls without Boundary Columns," 7th Japan National Symposium on Earthquake Engrg., Tokyo, Japan, Dec., 1986.

2. Chin-Chi Huang, Maw-Shyong Sheu, "Experimental and Theoretical Study of Low-Rise R.C. Shear Walls under Monotonic Horizontal and Axial Compression Forces," US-Korea Seminar on Critical Engrg. Systems, Seoul, Korea, May 1987.

3. Alan H. Mattock, Discussion of the paper "Rational Capacity of Reinforced Concrete Beams" by W. G. Corley, Journal of the Struc. Div., ASCE, Vol.93, No.ST2, 1967.

4. D. C. Kent and R. Park, "Flexural Members with Confined Concrete," Journal of the Struc. Div., ASCE, Vol.97, ST7, July 1971.

5. T. Takeda, M. A. Sozen and N. W. Nielsen, "Reinforced Concrete Response to Simulated Earthquakes," Journal of Struc. Div., ASCE, Vol.96, ST12, Dec.1970.

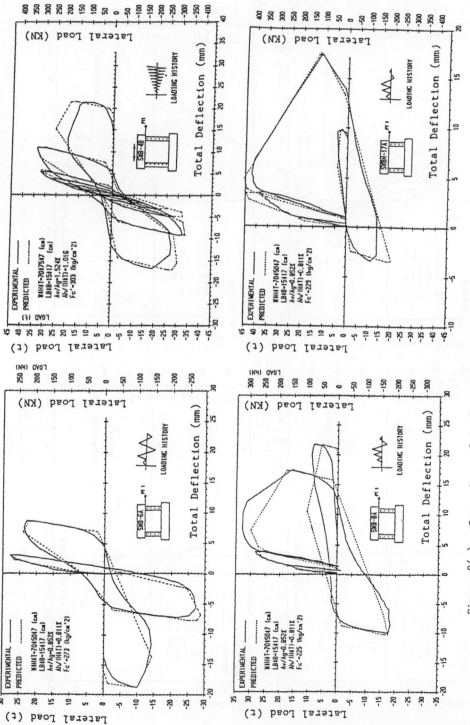

Figure 9(a). Comparison of Cyclic Load-Deflection Curves (For $H_S/W < 1.0$)

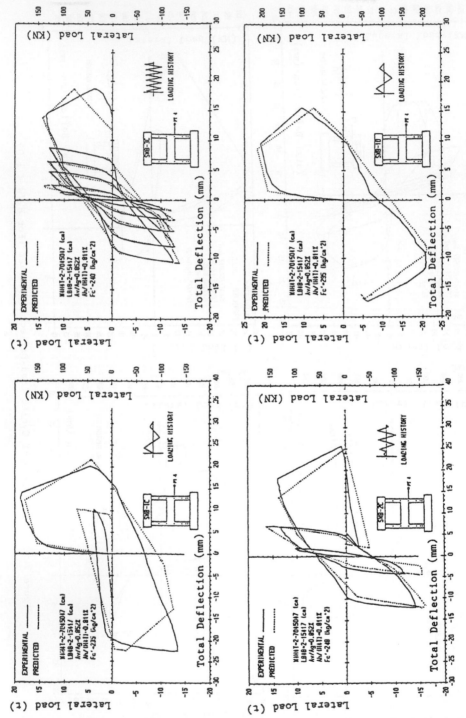

Figure 9(b). Comparison of Cyclic Load-Deflection Curves (For $H_S/W > 1.0$)

FAILURE MODES OF RC TALL SHEAR WALLS

EHUA FANG
Dept. of Civil Engineering
Tsinghua University, Beijing, China

ABSTRACT

This paper classifies the failure modes for moderate and tall shear walls---
flexure, flexural-shear, out-of-plane and shear failures. Parametric study is
given and discussed based on the experimental data. The differences and
similarities in failure patterns of shear walls with and without openings are
also discussed.

INTRODUCTION

Experimental study of RC shear walls around the world in the last decades
demonstrated that the seismic behavior of tall shear walls is much better
than that of squat walls because of their high ductility. However, being
susceptible to shear deformation, the effect of shear on the tall shear wall
cannot be ignored. Some brittle-like failure, such as flexure-shear failure
and out-of-plane failure, may occur and reduce the ductility. A variety of
failure modes were observed in experiments with shear wall specimens
designed and tested by the author for various aims. This paper will
classify the failure modes of tall shear walls, and discuss their primary
parameters based on experimental results from tests done by the author
(listed in Table 1) and other researchers.

FAILURE MODES AND PARAMETRIC STUDY

For moderate and tall shear walls, it is possible to design them as a
ductile wall with plastic hinges. However, the brittle shear failure before
yielding and the less-ductile failure after yielding of main reinforcement
have been observed. There are many parameters affecting the failure
mode of shear walls. According to the study, the main parameters may be

selected as: (1) shear-to-span ratio, (2) the proportion between shear and flexural strength of wall panel, (3) the level of nominal shear stress, (4) the shape of cross section and the detailing of walls.

The first three parameters are organized in three figures. Figures 1 and 2 are V_u/V_s vs. τ/f_c and V_s/V_f vs. m, respectively. Among them, V_u is the measured shear force during test, V_s and V_f are the shear force associated with calculated shear and flexural strength, respectively. All the calculations use the formulas specified in the Chinese code, but based on measured strength of material. "m" is shear-to-span ratio (M/Vl_w) at wall bottom, and τ/f_c is the nominal shear stress divided by compressive strength of concrete. Fig. 3 is reproduced from Fig. 1, but uses index $\tau/\sqrt{f_c}$ instead of τ/f_c.

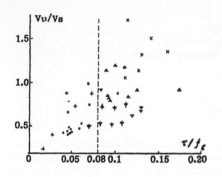

Figure 1. V_u/V_s vs. τ/f_c

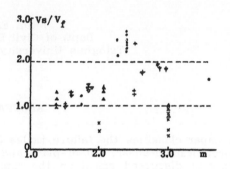

Figure 2. V_s/V_f vs. m

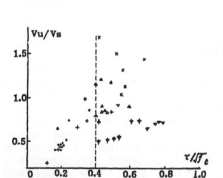

Figure 3. V_u/V_s vs. $\tau/\sqrt{f_c}$

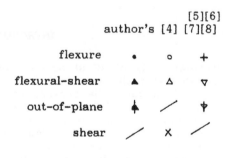

In addition to the data presented in Table 1, some other data, quoted from references [4][5][6][7][8] recalculated by the author, are also dotted in these figures using different symbols. Concrete strength of specimens being counted in figures range from 12 to 48 MPa. Unfortunately, many test results reported in other references cannot be taken due to lack of necessary data or insufficient clarity of the failure patterns.

The modes of failure in moderate and tall shear walls are classified and discussed below.

TABLE 1

Specimen	m	Concrete f_c (MPA)	Reinforcement			Axial force N (kN)	Calculated			Measured				Failure mode
			ρ_c (%)	ρ_v (%)	ρ_h (%)		V_f (kN)	V_s (kN)	$\frac{V_s}{V_f}$	V_u (kN)	τ/f_c	$\tau/\sqrt{f_c}$	$\frac{V_u}{V_s}$	
S-1A	1.79	12.28	0.17	0.28	0.17	80	34.9	44.3	1.28	34.3	0.070	0.25	0.76	F
S-1B	1.79	12.49	0.17	0.28	0.17	160	54.1	58.3	1.08	51.4	0.102	0.36	0.88	F
S-1C	1.38	14.56	0.27	0.418	0.25	80	65.8	77.4	1.17	66.8	0.114	0.44	0.86	F-S
S-1D	1.38	18.36	0.27	0.418	0.25	80	65.8	85.3	1.29	68.7	0.093	0.40	0.81	F-S
S-2	2.26	12.56	0.27	0.40	0.25	80	32.1	69.7	2.17	33.5	0.059	0.21	0.48	Frac.
S-3	2.26	14.04	0.27	0.38	0.25	80	31.1	78.5	2.52	31.8	0.046	0.17	0.41	Frac.
S-4A	1.39	12.82	0.17	0.28	0.17	80	53.2	54.7	1.02	50.9	0.124	0.44	0.93	F-S
S-4B	1.39	12.35	0.17	0.28	0.17	160	75.8	74.5	0.98	68.6	0.173	0.60	0.92	F-S
S-5	2.37	15.19	0.37	0.41	0.25	83	30.9	74.4	2.40	33.4	0.049	0.19	0.45	F
S-6(1)	2.37	13.9	0.37	0.41	0.25	88	31.9	72.3	2.26	38.1	0.061	0.23	0.53	F
S-6(2)	2.37	17.2	0.37	0.41	0.25	88	31.9	78.8	2.47	37.4	0.048	0.20	0.47	F
S-7	2.37	18.5	0.37	0.41	0.25	83	30.9	80.3	2.61	37.5	0.045	0.19	0.46	F
S-8	2.37	17.2	0.37	0.43	0.25	80	31.3	73.2	2.33	32.7	0.048	0.20	0.45	F
S-9(1)	2.37	19.17	0.37	0.43	0.25	80	31.3	76.5	2.45	37.0	0.048	0.21	0.48	F
S-9(2)	2.37	21.8	0.37	0.43	0.25	80	31.3	81.9	2.62	35.1	0.040	0.19	0.43	F
SW-1	2.04	24.84	5.84	0.46	0.612	400	264.4	371.1	1.40	277.3	0.085	0.42	0.74	Out-of-Pl.
SW-2	2.04	24.84	6.5	0.46	0.612	400	293.4	341.4	1.16	294.4	0.090	0.45	0.86	F-S
SW-3	2.04	24.84	3.12	0.46	0.612	400	278.3	370.8	1.33	306.5	0.093	0.47	0.83	F-S

Flexural Failure

This failure mode is characterized by fracture of tensile reinforcement bars or concrete crushing and buckling of bars in the compression zone after forming a plastic hinge in the wall. Usually, it has a large value of ductility ratio with only horizontal cracks and few if any inclined crackings. It can be found in Fig. 1-3 that most of the shear walls, which failed by flexure, have great shear strength, i.e. $V_s/V_f > 1.0$, and all have low nominal shear stress, i.e. $\tau/f_c \leq 0.08$ or $\tau/\sqrt{f_c} \leq 0.4$.

Flexural–Shear Failure

If a shear wall fails by shear along a main inclined cracking or web crushing after the yielding of flexural reinforcement bars, it can be called flexural-shear failure. Strictly speaking, it can be called shear deformation failure, because its strength is determined by the flexural resistant capacity, but due to rapidly developed shear deformation after yielding, failure comes before the flexural deformation capacity is exhausted. Thus, ductility ratio decreases to a certain extent. Most of such walls, represented by triangles in Fig. 1-3, have $V_s/V_f > 1$ (Fig. 2), but higher nominal shear stress, $\tau/f_c > 0.08$ (Fig. 1) or $\tau/\sqrt{f_c} > 0.4$ (Fig. 3).

There are two different cases in this category:

(1) Shear sliding failure   When quite a large part of shear strength of a tall shear wall depends on the shear reinforcement, it may have a relatively high value of V_s/V_f, but probably possesses higher nominal shear stress, then it will likely fail due to shear sliding (see Fig. 4a), because excessive shear reinforcement cannot prevent walls from shear deformation. It can only delay the crush of web concrete to a certain extent.

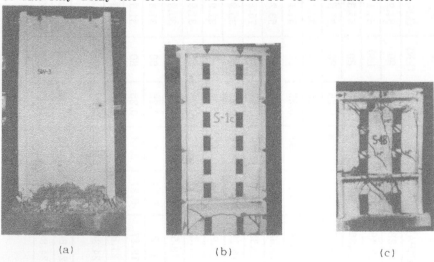

(a)   (b)   (c)

Figure 4.   Flexural-shear failure

(2) Shear along an inclined crack   When a moderate or tall shear wall has smaller shear-to-span ratio, say less than 1.5, it most likely will fail soon after the yielding of wall by forming a significant inclined crack rather than the crushing of the web concrete (see Fig. 4b and c). Evidently, its plastic deformation will be relatively small as compared with that of a wall which failed due to shear sliding. An explanation should be given here that, in some cases, for example, some of the tall shear walls in a structure will undertake the large shear forces transferred from other column-supported shear walls [3], then the shear-to-span ratio of their first story may be smaller than 1.5. The specimen S-1C was simulating this case, in which a concentrated horizontal force acted at the first story in addition to a reversed triangular distributed loading, thus the shear-to span ratio of the first story became 1.39.

Out-of-Plane Failure
This mode often occurs in the walls with rather thin (equal to or less than $l_w/10$) rectangular cross section. Crack distribution and the yielding of reinforcement are about the same as that of the shear sliding condition. However, when the web concrete begins to crush by crossing cracks, or when the residual width of cracks in tensile side becomes large, the wall will lose its strength suddenly by out-of-plane failure during its recompression under reversed loading [5]. Not only it reduces the plastic deformation, but also it could possibly lead to the slab toppling down. The latter is the real disadvantage of such failure. In order to draw attention, this one is separated from the flexural-shear failure mode in this paper and they are symbolized by "▲" or "▼" in Fig. 1-3. It seems that the column-like reinforcement (heavy flexural reinforcement with stirrups at the ends of wall) could not prevent walls from out-of-plane failure. The example listed in Table 1 is SW-1 (see Fig. 5). However, another specimen SW-2 has the same cross section as SW-1 but, reinforced with I steel at two ends of the section, did not lose its strength until the web concrete crushed seriously (see Fig. 6).

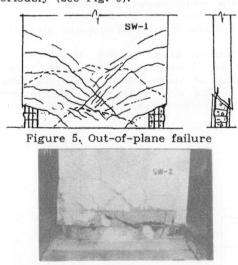

Figure 5. Out-of-plane failure

Figure 6. I steel reinforced wall specimen   Figure 7. Shear failure

Shear Failure
When too much flexural reinforcement is put in the tall shear walls, before they yield the shear failure may occur owing to the insufficiency of shear strength. The symbols of "×" in Fig. 1-3 represent those shear walls which failed due to shear. It can be seen from these figures that the shear

strength of each of them is smaller than its shear force associated with flexural strength, i.e. $V_s/V_f < 1$. Even though some specimens have very large shear-to-span ratio (m=3.0), shear failure still may occur without any plastic hinge developed (see Fig. 6) [4]. Of course, such brittle failure mode is undesirable. In fact, it can be avoided as long as a strong shear-weak flexure design, i.e. $V_s/V_f > 1$, is realized.

FAILURE PATTERNS OF SHEAR WALLS WITH OPENINGS

The four categories of failure mode described above cannot include the failure patterns which happened in the shear wall with openings. They are much more complicated than those in cantilever walls, and depend on, first, the proportioning of piers and coupling beams, and second, the relative strength of the coupling beams [1].

Strong Pier-Weak Beam Type

If the coupling beams have relative small stiffness and/or less reinforcement (compared with the need for elastic bending moment), the coupled walls will have the plastic hinges developed in the coupling beams first, and lose the loading capacity until the failure of piers, which may be one of the four modes mentioned above. Several specimens listed in Table 1, such as SW-1A, 1B, 5, 6 and 8, 9 belong to this type. Fig. 8 shows two examples, S-8 and 9. They are the same in all respects except for the reinforcement in coupling beams. The percentage of reinforcement in beams of S-9 is 0.31%, which just meets the requirement of elastic bending moment. In S-8, it is 0.24%. The coupling beams started to yield earlier in S-8 than in S-9. Fig. 8 indicates that S-8 has lower stiffness and strength than S-9.

It should be noticed that the coupling beams of S-8 failed due to cross diagonal crackings (see Fig. 8b) because of the large shear deformation developed after the yielding of those beams. When coupling beams yield early, the adjacent pier at one side would not uplift under lateral load, thus the drifts of the beams become large. Similar phenomena

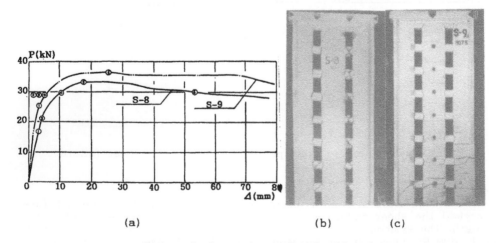

(a)   (b)   (c)

Figure 8. Comparison of S-8 and S-9

have been found in other tests. It indicates that the capability to resist shear deformation of the coupling beams is comparatively important in the strong pier-weak beam type coupled walls.

Cantilever-Like Walls

If the coupling beams are strong enough in stiffness and/or strength, then the coupling wall may be characterized as a cantilever wall, i.e. one of the above mentioned four types of failure may occur in a certain critical cross section of the whole wall, while only very light cracks can be discovered in coupling beams. Fig. 9 shows the comparison of load-deflection curves in three walls listed in Table 1. S-1A and 2 have different openings, and only in S-1A did the coupling beams yield. In Fig. 9, it is shown that, having higher stiffness and smaller ductility, S-2 is just like the S-3--a cantilever wall without openings. It was seen during the tests that the crackings at the bottom of S-2 and 3 were much larger than those in S-1A, and there was fracture of reinforcement at the edges of both.

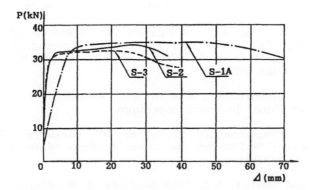

Figure 9. Comparison of S-1A, S-2, S-3

For seismic structures, usually strong pier-weak beam coupled walls have better seismic behavior than cantilever walls, because the former have larger ductility and energy dissipation capacity caused by more than one plastic hinges. Of course, the ductile coupling beams are expected and they are able to be designed in such walls [2].

SUMMARY

This paper classifies and discusses the variety of failure modes of shear walls with or without openings.

The shear (strength) failure can be avoided by designing the strong shear-weak flexural walls. The pure flexural failure will occur only when the shear wall has rather low nominal shear stress, $\tau/f_c \leq 0.08$ or $\tau/\sqrt{f_c} \leq 0.4$. Therefore, the flexural-shear failure is a common mode for moderate and tall shear walls. To avoid a shear-to-span ratio of less than 1.5 in a tall shear wall is beneficial for ductility. Finally, attention should be paid to prevent walls from out-of-plane failure.

Bar-belled and flanged walls are always better than walls with rectangular cross section. When rectangular cross section has to be used, it is a good practice to put some section steels at the edges of the walls (only within the plastic hinge zone). It is also possible to alter the detailing of the walls, mainly in the plastic hinge zone, to further improve their seismic behavior.

For seismic structures, usually strong pier-weak beam coupled shear walls have better behavior than cantilever walls. Ductile coupling beams are needed.

REFERENCES

1. Fang, Ehua and Lee, Guo Wei "Investigation of the Behavior of the RC Shear Wall With Openings" Seismic Behavior of RC Structures. Technical Reports TR-3 by Research Laboratory of Earthquake and Blast Resistant Engineering, Tsinghua University. 1981 (in Chinese).

2. Fang, Ehua "Experimental Investigation and Seismic Design of Coupling Beams in The Tall Shear Wall Structures" Proceedings of the 4th International Conference On Tall Buildings. Vol.2, Hong Kong and Shanghai, Apr/May 1988.

3. Fang, Ehua "Construction, Design and Research of RC Shear Wall Structures in China" in this Proceedings.

4. Hao, Ruikun "Experimental Study of RC Cantilever Shear Walls" Technical Report No.7701 by the China Academy of Building Research. Jan. 1978 (in Chinese).

5. Paulay, T. and Goodsir, W.J. "The Ductility of Structural Walls" Seismic Design of Reinforced Concrete Structures, June 1988.

133

6. Cardenas, A.E. and Magura, D.D. "Strength of High-Rise Shear Walls-- Rectangular Cross Sections" Response of Multi-story Concrete Structures to Lateral Forces, SP-36, 1973.

7. Wang, T.Y. and Bertero, V.V., Popov, E.P. "Hysteretic Behavior of Reinforced Concrete Framed Walls" Report No. EERC 75-23, Univ. of California, Berkeley, Dec. 1975.

8. Vallenas, J.M. and Bertero, V.V., Popov,E.P. "Hysteretic Behavior of Reinforced Concrete Structural Walls" Report No. UCB/EERC 79-20, Univ. of California, Berkeley, Aug. 1979.

9. Wood, Sharon L. "Minimum Tensile Reinforcement Requirements in Walls" ACI Structural Journal Sept.-Oct. 1989.

10. Fang, Ehua and Zhou, Yunlong "Nonlinear Deformation of RC Shear Walls and The Influence of Cross Section Shape and Reinforcement Types" Proceedings of International Concrete Conference 90, Tehran, IRAN, May, 1990.

EXPERIMENTAL STUDY OF STRENGTH AND DEFORMABILITY
OF REINFORCED-CONCRETE WALLS AND THEIR JOINTS
UNDER ALTERNATING LOADS

T.Zh. ZHUNUSOV
Dr. of Sc.(Tech), Professor, Director of KazNIISSA, Alma-Ata, Kazakh SSR

V.A. SHAPILOV
Cand. of Sc. (Tech), Head of laboratory, KazNIISSA

A.A. PAK
Scientific worker, KazNIISSA

ABSTRACT

The dynamic response of a 12-storied building with monolithic bearing RC walls was obtained during shaker-excited vibration test. The data had been used for more accurate definition of design-load values in main building elements. The tests of RC shear wall models and their joints with different types of reinforcement under the action of permanent and reversal loads were carried out. The elements strength and deformability had been estimated.

BUILDING DYNAMIC TESTS

The tests had been performed in order to receive the data which can be used in building analysis and to obtain the dynamic characteristics of multistoried building under the action of loads imitating the seismic ones. The shaker-excited vibration test of a 12-storied building with RC monolithic bearing walls had been performed in Chimkent city. The building internal walls, the thickness of which was 16 santimeters(sm), were made of common concrete (compression strength 7.7 Mpa) and the external walls were made of ceramsiteconcrete (compression strength 4.0 Mpa) and had the thickness of 36 sm. The floor slabs with the thickness of 16 sm were made of common concrete too. The external and internal walls joints had reinforcing lateral bars and tongues in internal walls.

The internal wall crossings were reinforced by spatial reinforcing frames. Wall panels were reinforced by separate diagonal bars. There were frames in the wall spandrels strengthened by diagonal reinforcing bars.

Resonance-vibration excitation was fulfilled by vibromachine V-3 that had been installed on the roof slab of the investigated building. The test has been carried out at few stages. The accelerations and displacements of each building floor and displacements which were induced by base slab turning and shearing were registered.

The investigated building had a complicated shape in the plan, one unusual for seismic regions. As the experimental data processing showed, that because of the building shape, the vibrations were complex and included two transverse,longitudinal and torsional modes (fig.1). The value of top building displacement at the last test stage under the action of maximum exciting force was 13 mm. The shear forces at the level of base slab achieved the value 7131 KN. At that stage the initial vibration period value decreased by 17%. The damping ratio increased by 28%. The test data processing demonstrated that dynamic characteristics change was caused by building construction damages. Construction observation during loading and its inspection after testing revealed vertical cracks of 0.3 mm wide arise in during dynamical loading. The remaining crack width was less 0.05 mm.

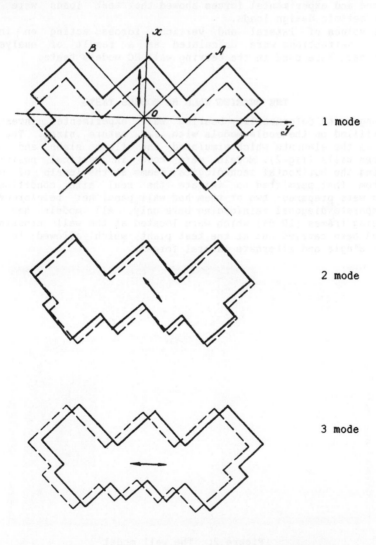

Figure 1   The modes in plan of building vibrations

The analysis of building structure stress-strain state was fulfilled with the use of the experimental data obtained. The spatial multi-degree-of-freedom design model of building was taken into account in the spatial performance of building structures during the seismic loading. The bending and rigid wall characteristics of designed model were prescribed with the use of experimental data obtained. The comparison of calculated and experimental X-axis's vibration-period values determined that they differed from each other by 23%. The calculated Y-axis's transverse and torsional vibration-period values coincided with experimental ones. The difference between calculated and experimental-mode shape-coefficient values did not exceed 4%.

The improving design models had been used in building analysis under the action of seismic loads intensity 8 degrees. The comparison of calculated and experimental forces showed that test loads were amounted 76% from seismic design loads.

The values of lateral and vertical forces acting on individual building constructions were calculated as a result of analysis. The obtained data were used in the bearing wall RC models tests.

THE BEARING WALL RC MODELS TESTS

The strength and deformability bearing wall experimental investigations were fulfilled on the scale models with 1/3 nature sizes. The testing models had the elements which simulated the floor slabs and parts of transverse walls (fig.2). Models were made in vertical position with organizing the horizontal technological seams at the levels of wall top and bottom. That permitted to imitate the real site conditions. Four specimen were prepared: two of them had wall panel net reinforcement and two - separate diagonal reinforcing bars only. All models had vertical reinforcing frames (10 d4) which were located at the wall crossings. The tests had been carried out at the test plant which allowed to put two vertical single and alternate lateral forces.

Figure 2.   The wall model

The model loading had been started with the applying of vertical loads with excentricity followed by lateral loading applied step by step. The applied point of vertical load had been changed with the change of lateral loading direction.

Four RC wall models tests were fullfilled with the use of the methods applied. Testing results obtained are shown in table 1.

TABLE 1
Wall models testing results

Model No	Compression strength (MPa)	Panel reinforcement	Crushing load (kN)	Displacement, (mm)	Damping coefficient
M1	22.5	nets	240.0	7.0	1.326
M2	18.4	diagonal	260.0	5.95	1.093
M3	22.1	diagonal	240.0	3.5	1.144
M4	28.4	nets	240.0	3.75	1.043

From Table 1 it is obvious that the type of wall panel reinforcement has influence at the values of collapsing forces and displacement. It is evident from crack patterns that the type of reinforcement defines walls cracks characteristics (figures 3,4). The collapse of models with the reinforcing nets was more plastic. However the comparison of damping ratios obtained for four models didn't confirmed the increasing of plasticity of models with reinforcing nets.

As it was found from the tests the horizontal technological seams influenced greatly on the deformability of walls during the action of lateral forces. For all tested models exact boundary of elastic performance was not determined because horizontal technological seams

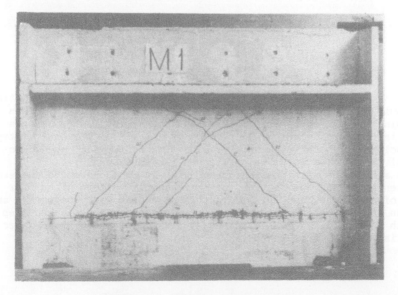

Figure 3. The M1 crack pattern

opening took a place from the beginning of the second step of tests and that was the cause of nonlinear nature of dependence "restoring force-displacement" (fig.5).

The model collapse begun after inclined cracking occurred in a wall panel and a part wall displacing along inclined crack with following rupture between transverse and longitudinal walls and along a horizontal joints between the floor slab and the wall.

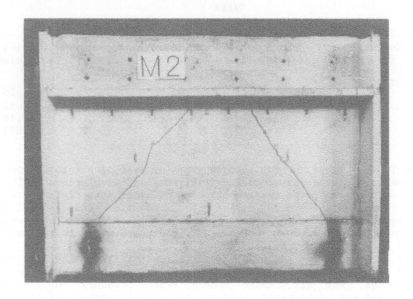

Figure 4.   The M2 crack pattern

THE VERTICAL WALL JOINTS TESTS

The three types of models of joints were subjected to investigations. These models had been in made 1/3 natural scale. The wall elements had tongue joints with trapeziform (1 type) and triangular (2 type) forms and linear jointing surface which was covered by steel wicker net (3type). Three specimens of each types were prepared for test. All specimens had horizontal reinforcing bars, which crossed joints planes.

In accordance with the test procedure it was foreseen determine shear strength under the action of permanently acting load for one specimen of each type, and for others this parameter was estimated under the action of alternating loads. The load was applied to specimen step by step with value 0.2 from expected ultimated force. The specimen characteristics are shown in table 2.

TABLE 2
The specimens characteristics

Type of specimens	Type jointing surface	Concrete type	Compression strength (MPa)	Modulus of elasticity (kPa)	Reinforcement
1-1	tongue	CC	--	--	
		LC			
1-2	trapeziform	CC	31.0	24.11	12 ⌀2
		LC	23.12	15.26	
1-3		CC	23.80	23.57	
		LC	20.60	15.16	
2-1	tongue	CC	21.40	20.41	
		LC	21.30	13.47	
2-2	trianugular	CC	28.31	29.10	7 ⌀2
		LC	18.0	11.74	
2-3		CC	--	--	
		LC			
3-1		CC	20.95	23.99	
		LC	20.90	15.29	
3-2	linear	CC	27.43	24.55	7 ⌀2
		LC	22.35	11.37	
3-3		CC	24.40	24.58	
		LC	16.86	11.95	

CC   -   common concrete
LC   -   ceramsiteconcrete

All the tested specimens collapsed as a result of the cracking which occurred and the extension of crack along the whole element's length, independent of loading type. In tongue joints, cracking and collapse occurred along ceramsiteconcrete tongues; in specimens of the third type - along jointing surface between light and common concrete.

The test data obtained for joints of three types under action of permanently acting loads are shown on figure 6. The highest value of crushing load was gotten for the third type model. The first type joint, was the most deformable. During the loading of the third type joint no deformations were registered in the range of 76% of crushing load. Then, under the next increase of loads, the cracking along the shear plane was followed by specimen collapse.

The main curves of the test hysteresis loops obtained are shown in fig. 7. In specimens 2-2 and 3-3, there were the special thin seams which were made during the concreting of joints. That permitted an estimating the influence of shrinkage cracks on strength and deformability of types of joints investigated.

The experimental data revealed that the joints of the second and third types had the highest values of strength and deformability. The shrinkage cracks resulted in a significant increase in joint deformability. In this case, the strength deformability of joints depends on the amount of joint reinforcement and plastic characteristics of reinforcing bars.

140

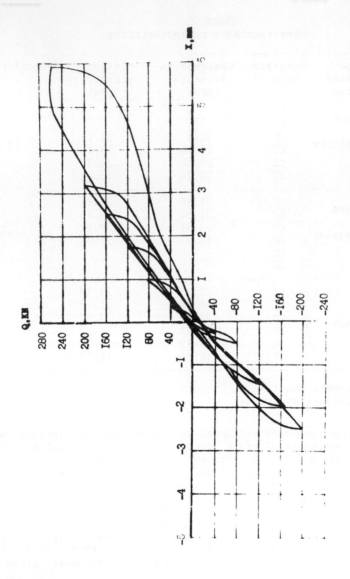

Figure 5. "Shearforce – Displacement" dependence under action of alternative loads

TABLE 3
The test results of model wall joints

Type of specimens	Cracking			Crushing shear force (kN)
	shear force (kN)	shear (mm)	rigidity (kN/mm)	
Permanently acting loads				
1-1	120	0.54	222	140
2-1	130	0.24	542	170
3-1	180	0.08	2250	250
Alternating loads				
1-2	120	0.33	364	160
2-2	120	0.27	144	180
2-2	80	0.65	123	150
3-2	165	0.167	988	200
3-3	100	0.50	200	175

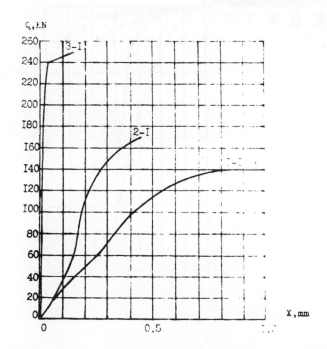

Figure 6. "Shearforce — Displacements" dependence under action of monotonic loads

142

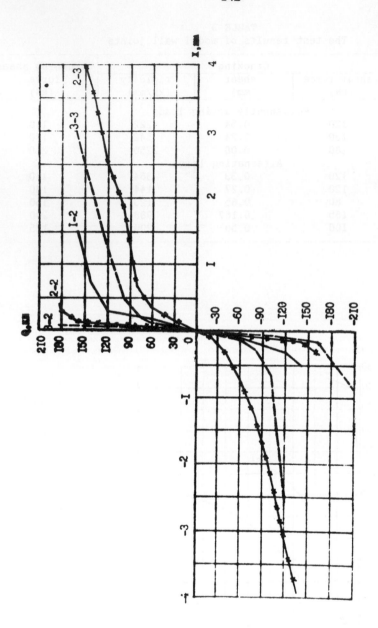

Figure 7. "Shearforce - Displacement" dependence under action of alternative loads

ON THE BEHAVIOUR OF CONCRETE COMPOSITE STIFFENING DIAPHRAGMS WITH SEISMIC-TYPE ALTERNATING LOADS

T.Zh. ZHUNUSOV
Dr. of Sc. (Tech), Director of KazNIISSA, Alma-Ata, Kazakh SSR

Yu.G. SHAKHNOVICH
Cand. of Sc. (Tech), Head of Laboratory, KazNIISSA

T.A. DZHAZYBAYEV
Engineer, Head of the Experimental Building Design Department,
KazNIISSA

ABSTRACT

Designs of multi-storey frame buildings with composite stiffening diaphragms to be used in regions of high seismic activity have been developed at the KazNIISSA (research institute of earthquake-resistive construction and architecture) of the Kasakh SSR Gosstroy. An increased reliability of buildings is achieved due to avoiding brittle failure of columns and diaphragms and disadjusting the frame from resonance state in vibrations.

Taking into account a higher dissipation of vibration energy, the braced frame with composite stiffening diaphragms is a system of active earthquake protection for structures.

RESULTS OF RESEARCH WORK

A study of plane models of a braced frame was carried out with application of repeated alternating loads of seismic type.

Four types of specimens were tested: fig.1a-specimen 1-a complete stiffening diaphragm; specimens 2 and 3-composite diaphragms with rigid connection of the upper and lower girders; specimen 4-composite diaphragms with the upper girder connected by means of keys that have vertical (3,6 mm) and horizontal (5,0 mm) clearance.

The presence of clearances ensures a certain engagement lag of the diaphragm after free displacement of the frame with horizontal loading. Long-term vertical loads of 317 kN were applied to columns of experimental specimens.

The test results showed the performance of models at loading stages including the limiting state, which revealed the most effective variants of reinforcement of composite diaphragms and the methods of connecting them with girders of the frame.

Specimen 1 failed in shear of the diagonal section passing through the upper part of the column and the diaphragm (fig.1b) to form many diagonal cracks in both directions.

The limiting state of specimen 2 was characterized by concrete bearing on the bearing surface of composite diaphragms, followed by inclined cracks and concrete crushing along the lateral faces. At the same time the diaphragm failure by the diagonal section was accompanied by bulging of the compressive reinforcement of the columns and accommodation of plastic deformations in the tensile reinforcement.

The limiting state of specimen 3 was characterized by the appearance in the angular surface of composite diaphragms of inclined cracks from diagonal tension-compression forces and by the development of diagonal cracks. The opening of diagonal cracks in diaphragms was accompanied by irreversible deformations of tensile bar diagonals, as well as plastic deformations of the tensile reinforcement in the joint zone of columns.

In specimen 4 the failure of composite diaphragms took place in the lower fixing-in-surface by normal sections and was accompanied by plastic deformations of inclined reinforcement bars. The failure of diaphragms was preceded by the appearance, first in the intermediate and then in the lower surfaces, of inclined cracks the intensity of which grew as the horizontal loads were increased. The exclusion of composite stiffening diaphragms from operation led to development of plastic deformations of the reinforcement in the bearing sections of columns and the girder. The envelope forms of the "load-displacement" relationship for specimens 2, 3 and 4 are different from the standard specimen 1 (fig.2). The largest plastic deformations were found in specimens 3 and 4 where the envelope curves are of a similar form. The envelope for specimen 4 is displaced on the initial area where composite diaphragms were not engaged in operation. The limiting displacements of the models were: for the specimen 1 - Δ_1=6,0 mm, for the specimen 2 - 12,0 mm ($2\Delta_1$), for the specimen 3-16 mm ($2,7\Delta_1$) and for the specimen 4 - 21,00 mm ($3,5\Delta_1$). The energy capacity of the specimen characterizing the volume of work spent for a specimen failure is determined by the area of diagram deformation and is respectively A_1=240kNcm, A_2=1,4A_1; A_3=2,3A_1 and A_4=2,6A_1.

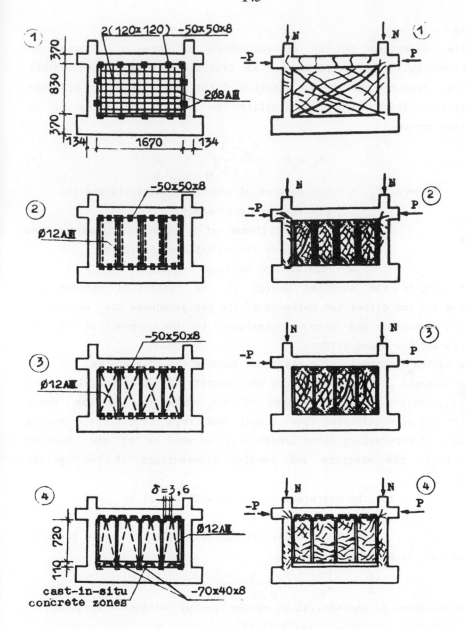

a) Model specimens of frames with stiffening diaphragms

b) Cracking and failure schemes of frame-diaphragm models

Figure 1.

Analysis

Composite diaphragms in the design scheme of bracing frames of multi-storey buildings are presented as bracing systems of equivalent stiffness. From the equilibrium equation of the node A (fig.3a) and the conditions of deformation compatibility the bracing stiffness of an equivalent system:

$$E_b . A = \frac{C_g . 1}{2 . \sin^2 \alpha} \tag{1}$$

where: E_b, A - the modulus of concrete elasticity and the bracing sectional area;

$C_g = P/\Delta_r$ - the stiffness of a composite diaphragm (the horizontal force-displacement ratio);

α - the bracing gradient angle, $\sin \alpha = b / 2l$

According to the accepted design of the connection between the diaphragm and the girder two variants of the design scheme can be had - without a clearance and with a clearance in the connection of the diaphragm and the upper girder.

The system state without a clearance between the diaphragms and the outline elements is characterized by the concrete key bearing along the upper (λ_1) and the lower (λ_2) faces of the diaphragm, by the shear deformability of reinforcing free length and keys (λ_3), the tensile compliance of reinforcing force length (λ_4), as well as by the bearing surface (λ_5), the shearing and bending deformations of the proper diaphragm (λ_6).

For example, for the diaphragm in fig.3b with S=0,21 m;

$E_b = 2,75.10^7 kN/m^2$; $\eta_o = h/b = 2$

$\Delta_r = \Sigma_6^1 \lambda_i = (0,054+0,038+0,04+0,245+0,058+0,545).10^{-4} = 0,98.10^{-5} m/kN$

Then the diaphragm stiffness

$$c_g^0 = \frac{1}{\Delta_r} = \frac{1}{0.98.10^{-5}} = 102.10^3 kN/m$$

and the stiffness of the equivalent system bracing (without clearances)

$$E_b . A = \frac{102.2,47.10^3}{2.0,243^2} = 5,19.10^5 kN$$

Taking into account the scale factor $\alpha = 1/3$ and the similarity criteria in the transition from a real construction to a model the model stiffness was obtained to be $c_g^0 = 30,6.10^3 kN/m$, which differs from the value determined in the tests of an individual diaphragm, fig.4, C = 31,8.10 kN/m by 4%.

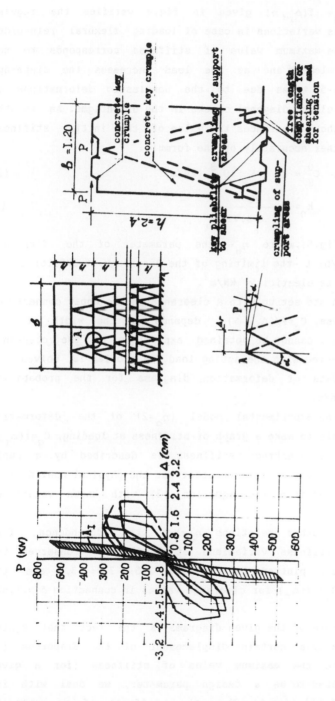

Figure 2. Deformation diagrams $\Delta = f(P_\Gamma)$ of frame diaphragms with alternating static loading

a) an equivalent bracing system b) tie elements of a composite diaphragm with a girder

Figure 3.

The dependence $C_g = f(\Delta_r, \eta)$ given in fig.4 verifies the regular character of stiffness variations in case of loading flexural reinforced concrete elements; the maximum value of stiffness corresponds to the initial state of the element and as the load increases the diaphragm stiffness lowers by 3-5 times due to the nonelastic deformations of concrete and the normal and diagonal cracks in the diaphragms. As to the diaphragms set up without clearances the value of the initial stiffness can be determined rather accurately by the formula:

$$c_g^0 = k_\eta \cdot t \cdot E_b \cdot 10^{-3} \quad (kN/m) \qquad (2)$$

Here
$$k_\eta = \frac{28}{\eta_o^3 + 0,75 \cdot \eta_o} \qquad (3)$$

or in the diagram (fig.5), where η_o - the parameter of the diaphragm relative height $\eta_o = h/b$; t -the limiting of the diaphragm, $t = S/0,16$; E_b - the modulus of concrete elasticity, kN/m^2.

In case diaphragm are set up with a clearance in the upper connection, the diaphragm stiffness, $C_g = P/(\Delta_i + \delta)$, depends on the width of the clearance. The design , cannot be obtained explicitly as the clearance size (δ) is not connected with the acting load. The problem is solved by applying empirical data of deformation diagrams for the probability parameter values $\eta_o = h/b$.

On the basis of the experimental model ($\eta_o = 2$) of the deformation diagram it was possible to make a graph of stiffness at loading, $C_g = f(\Delta_r)$. It is meant that the diaphragm stiffness is described by a cubic parabola, whereas the factor η that characterizes the physical nonlinearity of an element is defined from the equation (3) with known values of stiffness.

Taking that for other relations $\eta_o = h/b$ the dependence form characterizing the stiffness variations at loading is conserved the respective curves are plotted for $\eta_o = 1,5$; $2,5$; $3,0$; $3,5$ and the deformation diagram $C_g = f(\Delta_r)$ for clearance values in connection $\delta = 1,80$; $3,60$ and $5,40$ mm.

The specific feature of the given diagrams is the fact that system stiffness corresponds to a certain displacement of the diaphragm (an altered state). Since the maximum value of stiffness (for a given clearance) is considered to be a design parameter, we deal with its variations, $C_g/C_g^0 = f(\delta)$ (fig.6). Then the stiffness of a composite

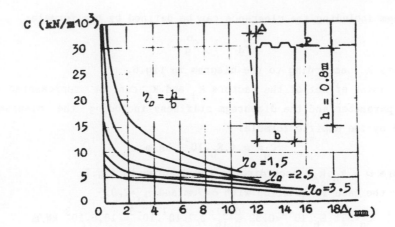

Figure 4. Stiffness of diaphragm models excluding the vertical clearence

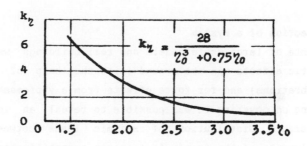

Figure 5. Variations of the initial stiffness of a composite diaphragm

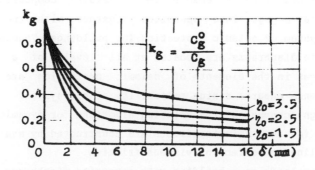

Figure 6. Effect of the vertical clearence on the
composite diaphragms stiffness

diaphragm including the clearance can be defined by the formula;

$$C_g = k_\delta . C_g^o \qquad (4)$$

where k_δ- according to the diagram in fig.6.

The joint effect of the factors k_η and k_δ can be represented as a single parameter and the diaphragm stiffness including the clearance is defined by the empiric formula;

$$C_g = \nu.t.E_b.10^{-3} \text{ kN/m} \qquad (5)$$

where $\nu = k_\eta.k_\delta$- according to the diagram in fig.7.

For the case in question with δ=7mm and η_o=2,0;

$$C_g = \nu.t.E_b.10^{-3} = 0,52 \frac{0,21}{0,16} 2,7.10^7.10^{-3} = 18,8.10^3 \text{ kN/m}$$

The bracing stiffness of an equivalent system by formula(1) $E_b.A = 1,11.10^5$.kN, i.e.,it is lower by 4,7 times as compared to a system without clearances.

Dynamic reaction of a system

Dynamic tests of large-scale models and real buildings were carried out for kinematic effects via the ground (with the help of a generator of seismic vibrations) and for force effects from a vibromashine installed on the building on covering.It was possible to reveal an increased dumping and a favourable distribution of seismic loads between the vertical carrying elements of the frame. As a result,a stabilization of the system reaction (floor displacement) was achieved in case of external force and kinematic effects (fig.8).

The obtained results enable to consider composite stiffening diaphragms that engage into operation together with the frame as a structural system of seismic protection for buildings. Thanks to the fact that composite diaphragms are located at all levels of a building the specific forces in the elements of seismic protection are lowered and their recurrence ensures a high reliability of the system.

The design schemes and the solution algorithm for dynamic load design of buildings for real seismic effects are constructed by analogy with the system modelling the effect of viscous friction.

The design loads on a building with composite diaphragms are lowered by two times as compared to the constructive scheme accepted as an analogy.

151

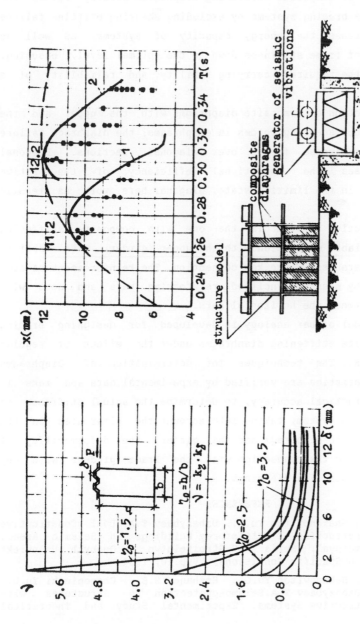

Figure 7. Stiffness of composite diaphragms including the vertical clearance.

Figure 8. Dynamic reaction of an experimental structure $Q_s = 45,5$ kN (1), $Q_s = 91.6$ kN (2).

CONCLUSION

The investigations carried out enabled the following to be established:

1. The application of composite stiffening diaphragms greatly raises the efficiency of frame-bracing systems by excluding shearing brittle failure of columns and increases the energy capacity of systems, as well as allowing lowering of frame stiffness down to the optimal level, ensuring, at the same time, the required carrying ability and reliability of a building.

2. The stiff connection of composite diaphragms with the upper and the lower girders induces shearing forces in diaphragms; the diaphragm failure is induced by transverse forces over diagonal sections. Diagonal reinforcement raises the operational efficiency of the active reinforcement and, in the limiting state, diagonal bars work as tensile bracing.

3. The most effective system is the one with composite stiffening diaphragms with pliable connection with the upper girder, with vertical and horizontal clearances in the key connection. The reinforcement of such diaphragm is done by means of inclined reinforcing bars and nets with narrower spaces between the horizontal bars.

4. An effective model of bar analogy is developed for designing bracing frames with composite stiffening diaphragms under the effect of regular and seismic loads. The techniques for determination of diaphragms stiffening characteristics are verified by experimental data and make it possible, within practical accuracy, to determine the actual stiffness of an equivalent bracing system, taking into account the clearances in the upper connection of the diaphragm and the girder, the deformability of diaphragm elements and the connection with the frame girders with any dimensions of the sides of a diaphragm.

REFERENCES

1   Zhunusov T.Zh., Shakhnovitch Yu.G., Dzhazybaev T.A. Dual Constructive Systems for Construction of Multistorey Buildings in Seismic Areas. "Express Informatsya", Seria XIV-Stroitelstvo v Osobykh Usloviakh (Construction in Specific Conditions), Vyp.10, 1984.

2. Zhunusov T.Zh., Sakhnovitch Yu.G., Erzhanov S.E., Cherepinski Yu.D., Gorovits J.G.,Dzhazybaev T.A.Seismo-protection to Structures with Different Constructive Systems. Experimental Study and Theoretical Analysis.

3. Shakhnovitch Yu.G., Dzhazybaev T.A. On the bar analogy in investigations of bracing frames with stiffening composite diaphragms. Transactions of the Kazpromstroyniiproekt institute "Investigations of Seismic Stability of Structures", issues 16-17, 1990.

THE BEHAVIOR OF REINFORCED CONCRETE NUCLEAR CONTAINMENT STRUCTURES UNDER EARTHQUAKE LOADING

RICHARD N. WHITE, James A. Friend Family Professor of Engineering
and
PETER GERGELY, Professor of Structures
School of Civil and Environmental Engineering, Cornell University
Hollister Hall, Ithaca, NY 14853, USA

ABSTRACT

The behavior of cracked reinforced concrete containment structures subjected to seismic loading is explored. After a brief review of experiments on reinforced concrete panels under combined in-plane biaxial tension and cyclic shear loads, the effects of reduced shear stiffness on containment stresses and response is presented, using results from analyses conducted at Cornell University and at Sargent & Lundy and CBI Na-Con, Inc. The primary emphasis in this paper is on the how reduced, degrading concrete shear stiffness affects the overall behavior of reinforced concrete containment structures.

INTRODUCTION

Reinforced concrete containment vessels experience cracking as a result of the structural acceptance test pressure. Accident conditions may lead to much higher levels of internal pressure, which would increase the existing crack widths and produce additional cracking in both the hoop and meridional directions. Seismic ground motions induce shearing stresses which can produce inclined cracking in the structure. The combination of accident pressure and seismic forces may be of interest because an earthquake conceivably could initiate accident pressure conditions, with subsequent shocks acting on the pressurized containment.

Several primary questions arise in analyzing these cracked structures under earthquake loadings -- (1) what is the shear stiffness of the cracked concrete for selected combinations of internal pressure and seismic loading?, (2) how does the stiffness degrade under repeated cyclic loads, (3) how nonlinear is the resulting dynamic response of the containment?, and (4) what is the simplest type of analysis needed to adequately predict response?

There are two other issues that deserve attention. Substantial loss of in-plane shearing stiffness might result in significant redistribution of stresses in the containment vessel. Some of the shear carried by tangential shearing action (in the direction of ground motion) may be redistributed and carried by radial shearing action in regions of the wall 90 degrees removed from the region of peak tangential shear, thus resulting in potential overstress in areas where a linear elastic analysis would predict small shearing stresses. Secondly, if the concrete should experience a severe decrease in stiffness, the internal steel liner in the containment would support loads that were neglected in design, and questions of liner integrity may arise. If the concrete shear stiffness reduces to, say, 10% of the value for that of uncracked concrete, then 1/2 inch of steel becomes equivalent to 50 inches of concrete in terms of resisting shear forces.

Strength considerations are not discussed in this paper; the interested reader is referred to Ref. [1] based on extensive studies by the Design Subgroup in ASCE-ACI Committee 359.

TYPICAL CONTAINMENT STRUCTURE GEOMETRY

The geometry and reinforcement details for a typical reinforced concrete containment building will be defined for a specific power plant -- the Clinton Power Station -- to be used later in an analysis example. This containment (Fig. 1a) houses a Boiling Water Reactor of the Mark III type and is founded on soil. It consists of a right circular cylinder (3 ft thick and I.D. = 124 ft) with a hemispherical domed roof (2.5 ft thick) and a flat basemat (9.7 ft thick). The height above the top of the basemat is 215 ft. The containment wall is reinforced with two layers of hoop and meridional reinforcement and one layer of diagonal rebars (see Fig.1b). The dome is reinforced in two directions. An internal steel liner provides protection against leakage.

TESTS ON RC ELEMENTS UNDER COMBINED TENSION AND SHEAR

Substantial experimental evidence exists on the behavior of reinforced concrete panels subjected to various combinations of biaxial tension and in-plane cyclic shear. The brief summary given here will be restricted to the set of results obtained from experiments on more than 50 six-inch thick panels designed to simulate a section of a typical containment wall carrying combined pressurization and seismic shear [2]. The Cornell test specimen (Fig. 2) contained orthogonal reinforcement stressed in tension. Shearing action was induced independently by applying compression forces to one set of diagonally opposed thickened corner regions and tensile forces along the other diagonal.

Phase I specimens were orthogonally reinforced with two layers of #6 (19 mm) bars in each direction, providing reinforcing ratios of 0.0122 and 0.0244, respectively. Phase II specimens were similar except that bar size was reduced to #4 (13 mm), with resulting reinforcing ratios of 0.0055 and 0.0110. Some Phase II specimens also had #3 (9mm) diagonal reinforcement to simulate the seismic shear bars used in containments. All specimens in Phases I and II had equal levels of tensile stress applied to the orthogonal reinforcement that protruded from the edges of the specimen (Fig. 1); diagonal steel was not stressed with external forces. Phase III specimens were similar but with axial stress in only one direction.

Every specimen was precracked by stressing the orthogonal reinforcement to 0.6fy. After removing the tensile loads, a selected level of tension (either 0, 0.3fy, 0.6fy, or 0.9fy) was applied to each bar and held constant during the subsequent application of shear loading. Either monotonic or fully reversing cyclic shear loads were applied -- cyclic shear typically started with 10 cycles at 125 psi, followed by 10 at 175 psi, 10 at 225 psi, etc., up to failure. A summary of strength and stiffness results is presented in [2], with details provided in [3] and two other NRC NUREG reports, CR-2049 and CR-2788, listed in [2].

Typical behavior is given in Fig. 3, with the response of a Phase II specimen with 0.6fy biaxial tension plotted for the first and second cycle of shear loading at each level of shear force. The initial loading part of the first cycle was always nearly linear. Subsequent cycles showed increasing softness at low shear stresses, with this "free slip" terminating at about 50 psi shear stress. Shear modulus ratios (R = measured shear modulus divided by uncracked concrete shear modulus) in this range of behavior ranged from R = 0.05 down to R = 0.001 for the complete group of specimens. At shear stresses above about 50 psi, shear modulus increased sharply but was still far below that of uncracked concrete,with average tangent modulus ratios falling in the 0.04 to 0.12 range in most specimens (outlier values of R ranged from 0.02 to 0.4). When only a few cycles of load were applied, shear stiffness for stresses above 50 psi was essentially the same as for monotonically loaded specimens; extensive cycling increased the "free slip" and reduced overall stiffness.

The cyclic shear loading reduced strength levels by about 15% below that of the monotonically loaded specimens. Reduced stiffness effects were most pronounced at higher tension levels of 0.6fy and 0.9fy. Increasing levels of tension reduced both shear strength and shear stiffness, with the effects on stiffness becoming significant when shear stress levels (above about 200 psi) produced extensive shear cracking. Using a lower percentage of orthogonal steel had a modest effect on stiffness at low to moderate shear stress levels; the effect was much higher at shear stress levels producing shear cracking. The relatively small amounts (0.20%) of diagonal reinforcement used in some of the Phase II specimens provided significant enhancement of both strength and stiffness.

Results from a parallel investigation on larger specimens at the Portland Cement Association Laboratories [4] showed effective tangent shearing modulus ratios as low as 0.05, and model structures tested in Japan exhibited similarly low values of effective shear stiffness.

NONLINEAR 3-D FINITE ELEMENT APPROACH

A nonlinear 3-dimensional finite element analysis was carried out by Conley [5,6] at Cornell University to quantity the influence of nonlinear cyclic shear stiffness of cracked concrete. Specific issues addressed include: (1) peak values of containment displacements, (2) redistribution of shear and bending stresses, (3) redistribution of tangential and radial shear stresses, and (4) values of principal stresses in the steel liner.

Shear stiffness modeling

The nonlinearities exhibited in the panel tests described above were captured in an approximate fashion by considering behavior after the first few cycles of loading. A bilinear idealization of the cyclic shear stress-strain history was introduced (Fig. 4), with an initial shear modulus ratio R = 0.017 and a subsequent value of 0.07. These values, which were used for the first four load cases considered (Table 1), were chosen on the basis of expected maximum tangential seismic shear stresses of about 400 psi in the containment wall, and with reinforcement stressed to 0.6fy from equivalent internal pressurization effects. However, the analysis results indicated that the peak shear stresses were only about 200 psi, and the bilinear shear modulus ratios were then increased to 0.027 and 0.12 for load cases 5-7 (Table 1).

TABLE 1 - Containment Analysis Stiffness Assignments, Cases 1-7

		G(X-Z)		G(X-Y, Y-Z)
Case	Comments	Zone 1	Zone 2	Zones 1 & 2
1	Normal linear	0.5	0.5	0.5
2	Bilinear, Zone 2	0.14	0.017, 0.07	0.5
3	Bilinear, Zones 1 and 2	0.017, 0.07	0.017, 0.07	0.5
4	Linear approximation, Zone 2	0.14	0.05	0.5
5	Stiffened bilinear, Zone 2	0.14	0.027, 0.12	0.5
6	Reduced X-Y & Y-Z moduli	0.14	0.027, 0.12	0.14
7	Include 3/8 in. steel liner	0.14	0.027, 0.12	0.5

The constitutive model used for the concrete permitted two independent values of shear moduli -- X-Z and (X-Y, Y-Z), with the X-Z value being either bilinear or linear, and the (X-Y, Y-Z) value being strictly linear. Table 1 shows the various combinations of shear modulus ratio R used in the several analyses of the containment.

Another approach to modeling shear stiffness softening was used by Miyashita and Sozen [7] -- they modeled aggregate interlock and doweling of the reinforcement at cracks and reported adequate levels of prediction of both test specimen response and model containment displacements under shearing loads.

Finite element modeling and dynamic analysis assumptions

20-node isoparametric hexahedron elements were used to approximate the containment shell geometry, as shown in Fig. 5. The hemispherical dome was replaced with a stiff ring of elements to simplify the analysis. The value of ring stiffness was determined by comparative analysis of a ring and a pinched sphere, with adjustment of the ring dimensions and assigned modulus (70 times larger than the rest of the cylinder) such that the top ring of elements properly approximated the effects of the hemisphere in restraining deformations of the cylinder.

The assignment of different values of shear modulus to each element of the cylinder would be best controlled within the computer program, but in the interests of simplification, two zones were defined apriori and used throughout the analysis. Zone 1 covered from 0 to 45 degrees and from 135 to 180 degrees, and Zone 2 covered the 90 degree segment centered about the neutral axis of bending, from 45 to 135 degrees (Fig. 5). Zone 2 is associated primarily with in-plane shear behavior; thus using the shear stiffness from the panel tests was appropriate. Zone 1 is associated primarily with shear normal to the wall; hence a shear modulus (R = 0.14) was determined from single crack experiments done earlier at Cornell.

A static loading that simulated the inertia loads was employed instead of doing a full dynamic analysis. A displaced shape obtained in an earlier approximate analysis of a lumped mass model subjected to a 0.4g maximum ground acceleration was used to determine load distribution over the height of the vessel. The total shear load was determined on the basis on maximum elastic (sinusoidally distributed) shear stress of 200 psi.

Results

Seven cases (Table 1) were analyzed, beginning with the commonly used approach of a linear elastic analysis with an effective shear modulus ratio R = 0.5. Subsequent cases were: (2) bilinear shear stiffness in Zone 2, using experimental results for a realistic model of a cracked containment; (3) bilinear in both Zones 1 and 2 to test for sensitivity to variations of the X-Z modulus in Zone 1; (4) linear approximation of stiffness in both Zones 1 and 2 to determine if an equivalent linear analysis would suffice; (5) Case 2 but with stiffened bilinear values in Zone 2 to reflect the fact that peak shear stresses were less than originally expected; (6) Case 5 but with a reduced value for (X-Y, Y-Z) stiffness, and (7) Case 5 modified to include a 3/8 inch thick steel liner in the containment.

The effects of using the bilinear reduced shear stiffness in Zone 2 may be seen by comparing plots of **tangential shear** stresses in the first (bottom) layer of elements for Cases 1, 2, 5, and 6 (see Fig. 6). The maximum value of tangential shear (Case 2) nearly doubles in comparison with Case 1, and its location in the cylinder shifts from 90 degrees to about 35 degrees when the shear softening is considered. **Radial shear** stresses also increase sharply, with maximum values increasing from about 20 psi for Case 1 to about 65 psi for Case 2. Comparisons of first layer element **bending** stresses for load cases 1, 2, 5, and 6 are shown in Fig. 7; there is a marked redistribution of bending stresses towards the extreme fibers (0 degrees). Case 2 maximum extreme fiber bending stress is about 3 times that of Case 1. The reduced shear stiffness leads to a dramatic increase in **displacements,** with the following springline values: 0.47, 2.93, 1.75, and 2.07 in. for Cases 1, 2, 5, and 6, respectively.

If the liner is neglected in the analysis, then Case 5 is the most realistic representation of behavior, with the following results: (a) maximum tangential shear stress is at about 35-40 degrees and is substantially larger than the peak linear elastic value at 90 degrees, (b) maximum radial shear stress is about twice the linear case value, and is about 20% of the peak tangential shear stress, (c) maximum bending stress is about 2.4 times the linear case result, and (d) displacement at the springline is about 3 times the linear displacement, but is still less than 2 inches for this medium intensity earthquake loading.

When the liner is included (Case 7), all concrete stresses are reduced. Cases 5 and 7 shear stresses are compared in Fig. 8. Case 7 peak values as a percentage of Case 5 results are: tangential shear stress, 72%; radial shear stress, 80%; bending stress 85%, and displacement at springline, 72%. The maximum principal stress in steel liner was 11,000 psi.

The equivalent linear model (Case 4) provided accurate displacement prediction and slightly overestimated the peak shear and bending stresses, but stress distributions were not accurately reproduced in comparison to the bilinear model results.

Commentary - Suggested improvements needed in the analysis include refined definition of Zones 1 and 2, improved directional assignment of reduced shear values as a function of calculated shear stress, better accounting for diagonal reinforcement effects, variable E to reflect tension cracking effects, and a full dynamic analysis capability with appropriate modeling of damping. More attention needs to be given to analysis of the region at the base of the wall to better define the possible critical combined effects of axial, bending, and shear forces, along with interaction effects with the base-mat.

"PRACTICAL" DYNAMIC ANALYSIS OF CONTAINMENTS

The Clinton containment described above was analyzed recently by Sargent & Lundy and CBI Na-Con, Inc. as part of their analytical study [8] of the seismic threat to containment integrity. The approach and selected results are summarized here to provide some insight into how consulting engineers approach the cracked containment analysis problem. Analyses were done for an initial main earthquake and three load combinations that included the combined effects of a severe accident (internal pressure and elevated temperature) initiated by the main shock and an aftershock that occurred at various times during the progression of the severe accident.

Two levels of analysis were done for the Clinton containment:
1. A seismic beam model (hereafter referred to as "simplified analysis") that accounted for soil-structure interaction and for basemat uplift. Nonlinear moment-curvature relations were used and the reduction in wall shearing stiffness was accounted for in an approximate manner

as described below. Cracked-section, element-level analyses were done for critical sections that accounted for yielding of the liner and reinforcing bars.

2. A three-dimensional quasi-static finite element model that accounted for the effects of concrete cracking and yielding of the liner and reinforcing bars at the structural system level. The program ADINA was used to determine shell force distributions and to examine how peak liner and reinforcing strains are influenced by cracking and yielding.

The seismic analyses were done with horizontal and vertical acceleration time histories derived to have their response spectra enclose the Nuclear Regulatory Guide 1.60 response spectra for a specified maximum horizontal ground acceleration.

Reduction in Wall Shear Stiffness Produced by Cracking

The 3-D analysis results confirmed that containment response is sharply affected by the value of effective shear stiffness used in the analysis, in contrast to the relative insensitivity to values of moment of inertia and normal area for the containment beam model. This is true because containments deform primarily in the shear mode under seismic excitation.

The authors of [8] propose an effective shear modulus ratio R as based on experimental studies reported in Refs. 3, 4, and 9. They conclude that the ratio is a strong function of normal stress: 0.05 for concrete in shear plus tension, 0.10 for concrete in pure shear, and a function of compressive stress when compression is less than 1400 psi, with no reduction when compression exceeds 1400 psi. These shear modulus ratios are translated into an effective shear stiffness for the analytical beam model by calculating normal stresses resulting from the combination of 2.5(Safe Shutdown Earthquake) + DL and then assigning shear modulus ratios to twelve 15-degree sectors covering one half of a containment circumference, as follows: R = 0.05, 0.05, 0.05, 0.05, 0.10, 0.10, 0.14, 0.16, 0.19, 0.26, 0.26, and 0.26. The assigned values on the central six elements, where most of the tangential shear is resisted, are then averaged to get a final ratio of R = 0.12 for the Clinton containment. Note that this same approach provides an R value of 0.42 for the Zion prestressed concrete containment.

Results of Analyses, with peak acceleration value of 1.0g:

1. The ADINA model showed considerable cracking in the basemat and lower portion of the containment wall up to 42 ft. above the basemat.
2. Values of lateral displacement at springline were:
   - 1.76 in. with simplified analysis and a shear stiffness ratio R = 0.12
   - 1.46 in., ADINA "cracked" analysis
3. Circumferential redistribution of stresses from cracking effects produced a large shift in the neutral axis of the circular containment section, leading to compressive strains about 20% higher than predicted by the beam model. But effects are not critical in containment behavior.
   The total in-plane shear force is plotted in Fig. 9a and shows that cracking produces a major

decrease near the edge of the cracked zone. Fig. 9b compares in-plane shear force in the liner plate as predicted by ADINA and the simplified analysis.

4. Through-thickness redistributions caused by cracking produces (in general) values of peak rebar strain, transverse shear in the wall, and liner principal strains that are conservative in comparison to beam model results, except when in-plane shear is significant.

5. Beam model results predicted critical strains in the basemat liner that were much higher than in the ADINA results, where only one element in this region reached yield.

CONCLUDING REMARKS

The sharply reduced shearing stiffness of cracked reinforced concrete requires special analysis efforts to predict adequately such important response parameters as concrete stresses, steel liner stresses, and displacements of reinforced concrete containments subjected to earthquake loading and pressurization. Simply using a decreased linear shear stiffness may not adequately capture the very substantial softening effects that can occur under certain conditions of biaxial tension and repeated cyclic shearing action. It is suggested that with increasing levels of combined seismic shear and biaxial tension, the associated degraded shear stiffness of the concrete may lead to redistribution effects and potentially large displacements that might play an important role in potential containment failure modes. A more satisfactory solution to this problem is within relatively easy reach, it is believed, with a proper combination of current understanding of cyclic shear effects and advanced finite element analysis capabilities.

REFERENCES

1. Oesterle, R.G., Design Provisions for Tangential Shear in Containment Walls, NUREG/CR-5209, U.S. Nuclear Regulatory Commission, Aug 1988.
2. Conley, C.H., White, R.N., Hilmy, S., and Gergely, P., Design Considerations for Concrete Nuclear Containment Structures Subjected to Simultaneous Pressure and Seismic Shear, Nuclear Engineering and Design, Vol. 69, No. 2, May 1982, pp. 241- 252.
3. Perdikaris, P.C., White, R.N., and Gergely, P., Strength and Stiffness of Tensioned Reinforced Concrete Panels Subjected to Membrane Shear, Two-Way Reinforcing, NUREG/CR-1602, July 1980.
4. Oesterle, R.G. and Russel, H.G., Shear Transfer in Large Scale Reinforced Concrete Containments, Report No. 1, NUREG/CR-1374, U.S. Nuclear Regulatory Commission, April 1980.
5. Conley, C.H., and White, R.N., Predictions of Displacements and Stresses in Cracked Reinforced Concrete Containment Vessels Subjected to Seismic Loads, Nuclear Engineering and Design 79(1984).
6. Conley, C.H., White, R.N., and Gergely, P., Analysis of Reinforced Concrete Containment Vessels with Nonlinear Shearing Stiffness, NUREG/CR-3255, April 1983.
7. Miyashita, T., and Sozen, M.A., Nonlinear Analysis of Reinforced-Concrete Containment Vessel Using Shear Transfer Stiffness of Cracked Elements, Proceedings, International Conference on Structural Mechanics in Reactor Technology, 1981.
8. Amin, M., Agrawal, P.K., and Ahl, T.J., An Analytical Study of Seismic Threat to Containment Integrity, NUREG/CR-5098, U.S. Nuclear Regulatory Commission, July 1989.
9. Vecchio, F., and Collins, M.P., The Response of Reinforced Concrete to In-Plane Shear and Normal Stresses, Report No. 82-03, Dept. of Civil Engineering, Univ. of Toronto, March 1982.

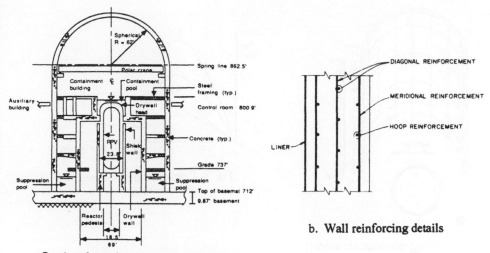

a. Section through containment building

b. Wall reinforcing details

Fig. 1 Clinton Power Station

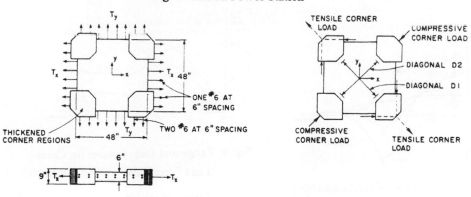

Fig. 2 Cornell biaxially tensioned shear panel

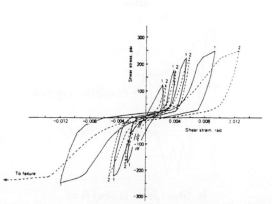

Fig. 3 Typical shear stress-shear strain plot

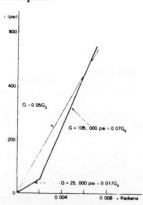

Fig. 4 Typical shear stress vs. shear strain idealization for analysis

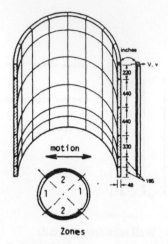

Fig. 5 Analytical containment model;
definition of Zones 1 and 2 for
stiffness assignments

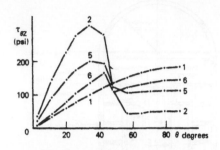

Fig. 6 Comparison of tangential
shear stresses

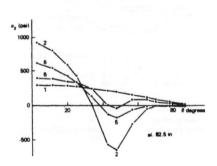

Fig. 7 Comparison of bending stresses

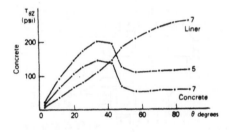

Fig. 8 Tangential shear stresses for Cases
5 and 7 (with and without liner plate)

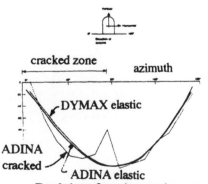

a. Total shear force in containment wall

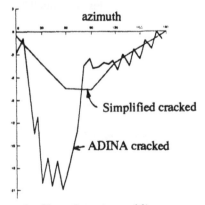

b. Shear force in steel liner

Fig. 9 Circumferential variation of in-plane shear

REPORTS ON TESTS OF NUCLEAR PRESTRESSED CONCRETE CONTAINMENT VESSELS

TOSHIKAZU TAKEDA   TSUNEO YAMAGUCHI   KAZUHIRO NAGANUMA
Technical Research Institute, Obayashi Corporation
4-640, Shimokiyoto, Kiyose-shi, Tokyo 204, Japan

ABSTRACT

This paper discribes the results of tests and analyses of a model of a prestressed concrete containment vessel (PCCV) subjected to lateral loads during earthquake. The test results revealed sliding shear collapse at the lower parts of the cylindrical wall. Therefore, in order to examine the overall behavior and the failure mechanisms of concrete, in-plane shear loading tests of reinforced concrete (RC) panels were conducted. From those test results, the constitutive law for concrete was proposed. Nonlinear finite element analyses of the PCCV model were done and the results were compared with the test results. It was found that the proposed constitutive law for concrete reproduced the test results with reasonable accuracy.

INTRODUCTIONS

PCCVs adopted in Japan have their reinforcement quantities determined based on shear force during earthquake. In lateral loading test[1] of a 1/8-scale PCCV model, while the strength became high due to the large amount of reinforcement, on the other hand, concrete failure appeared in the zone where in-plane shear was predominant.

When carrying out a nonlinear analysis of a RC structure, the strength of which is decided by failure of concrete, the constitutive law of concrete in the in-plane shear field governs the analytical results. Therefore, in-plane shear loading tests [2][3] were conducted on 32 RCpanel specimens, and the constitutive law of concrete was derived from those test results.

This paper describes an outline of the RC panel tests and the results of lateral loading test and simulation anaylyses of the PCCV model.

IN-PLANE SHEAR LOADING TESTS OF REINFORCED CONCRETE PANELS

Test Method
The loading apparatus is shown in Fig.1. It is possible to apply pure shear or combined shear and compression. It is devised so that deformation of each side of the specimen will not be restrained and that determination of

strength due to progress of local failure will be prevented.

The dimensions of a specimen were 120× 120 × 20cm with isotropic double reinforcement using deformed bars longitudinally and transversely, the spacing between bars being 20 cm.   The parameters of the tests were concrete strength and reinforcement ratio for pure shear loading tests, and axial compressive force level for combined shear and compression loading tests. All of the tests were conducted up to failure by monotonic loading.   In combined shear and compression loading tests, application of shear force was done while maintaining axial compressive forces.

Outline of Test Results

Representative relationships between shear stress (v) and shear strain (γ) are shown in Fig.2   Fig.2(a) compares differences in reinforcement ratios.

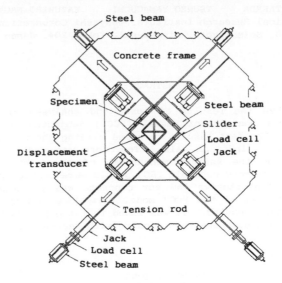

FIGURE 1. Loading apparatus for RC panel tests

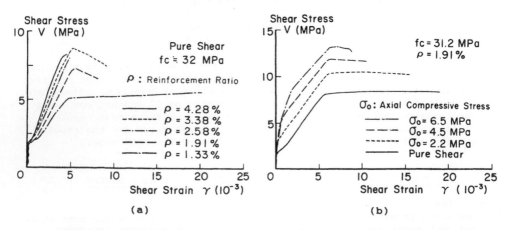

FIGURE 2. Relationships between shear stress and shear strain

Stiffness after cracking becomes greater with increase in reinforcement ratio but since the failure mode shifts from reinforcement yielding to concrete failure, a ceiling is seen in the rise in maximum load. Fig.2(b) compares differences in axial compressive stress level. Cracking load becomes lager with increase in compressive stress levels, and there is a property of v-γ curves moving upward parallel to each other.

The relationship between maximum shear stress(v_u) and cylinder strength (f_c) is shown in Fig.3 for the specimens with strength determined by concrete crushing. With the results of pure shear loading as objects and assuming v_u to be expressed in the form of $a \cdot f_c^b$ and determining optimal coefficients a and b, the following equation is obtained.

$$v_u = 0.95 f_c^{0.66} \text{ (MPa)} \tag{1}$$

The v_u of biaxial compression and shear loading specimens exceed the v_u of pure shear loading ones. Fig.4 shows the relationships betwen v_u and the

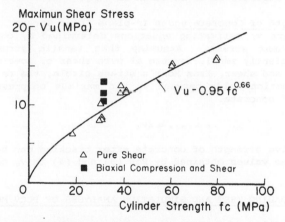

FIGURE 3. Relationship between maximum shear stress and cylinder strength

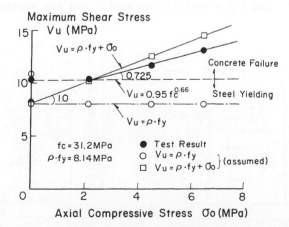

FIGURE 4. Relationship between maximum shear stress and applied biaxial compressive stress

acting biaxial compressive stress(σ_o). The data in case of assuming v_u to be equal to $\rho \cdot f_y$ and ($\rho \cdot f_y + \sigma_o$) are also shown in the figure, where, ρ is reinforcement ratio and f_y is yield strength of reinforcing bar. It can be seen from this figure that axial compressive forces change reinforcement yielding mode into concrete failure mode, further, increase the strength.

Based on these results, the shear strength estimation equations in case there is compressive axial force are hypothesized as follows;

$$v_u = \rho \cdot f_y + \sigma_o \text{ (MPa) (when } \rho \cdot f_y + \sigma_o < 0.95f_c^{0.66}) \qquad (2)$$

$$v_u = 0.95f_c^{0.66} + 0.725 \cdot \{\sigma_o - (0.95f_c^{0.66} - \rho \cdot f_y)\} \text{ (MPa)} \qquad (3)$$
(when $\rho \cdot f_y < 0.95f_c^{0.66}$, and moreover, $\rho \cdot f_y + \sigma_o \geqq 0.95f_c^{0.66}$)

$$v_u = 0.95f_c^{0.66} + 0.725\sigma_o \text{ (MPa) (when } \rho \cdot f_y \geqq 0.95f_c^{0.66}) \qquad (4)$$

In the above equations, the smaller of the longitudinal and transverse compressive axial stresses is to be taken as σ_o.

Compressive Strength of Concrete under In-plane Shear Stress

Eqs.(3) and (4) are v_u estimating equations determined by concrete failure under in-plane shear stress. Assuming that tensile stress of cracked concrete is sufficiently small, in case of pure shear or combined biaxially equal compression and shear, from Mohr's stress circle, the relationship of the following equation holds between v_u and maximum compressive principal stress($_c\sigma_{cmax}$) of concrete.

$$_c\sigma_{cmax} \fallingdotseq 2v_u \qquad (5)$$

The compressive strength of concrete after cracking can be calculated by substituting the values obtained by Eqs.(3) and (4) in v_u of Eq.(5).

LATERAL LOADING TEST AND SIMULATION ANALYSES OF PCCV MODEL

Outline of PCCV Model

An outline of the PCCV model is shown in Fig.5. The structure is composed of a hemispherical dome and a cylindrical shell, and has vertical buttresses

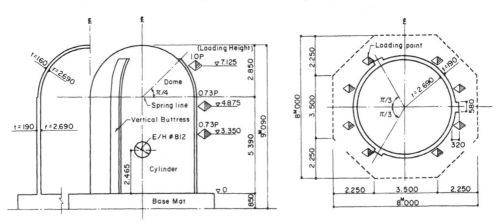

FIGURE 5. Outline of PCCV model

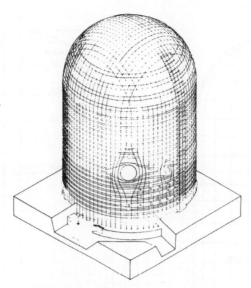

FIGURE 6. Tendon arrangement

provided for fixing of hoop tendons at three places on the circumference. Hoop tendons and inverted U tendons are arranged as shown in Fig.6.

Modeling of PCCV Model

The PCCV model has numerous openings provided, and especially, in the surroundings of a large opening, increase in wall thickness and the existence of opening reinforcement cannot be ignored. Therefore, it was decided to consider the large opening in analysis, and in modeling of the opening, the round opening was made a rectangular opening of equal area, and the increase in wall thickness and opening reinforcement were considered. Regarding the base mat, it was replaced by an equivalent spring where the rotating deformations at the base of the PCCV become equal. As for buttresses, since it was considered that their influences on strength were small, they were ignored in modeling.

Finite element mesh and reinforcement ratios are shown in Fig.7. The mesh has 416 layered shell elements. Each element consists of 10 concrete layers and reinforcment layers. The reinforcing bars and tendons were replaced by equivalent reinforcement layers having stiffness in one way only.

Analytical Method

Analyses were performed for the three cases indicated in Table 1. The uniaxial stress-strain relationships of concrete and reinforcement used in the analyses are shown in Figs.8 to 9. For tensile strength (f_{cr}) of concrete, the calculation equation for diagonal cracking strength of ACI[4] indicated below was used.

$$f_{cr} = 0.33 \sqrt{f_c} \text{ (MPa)} \tag{6}$$

With regard to the tension stiffening characteristics, as shown in Fig.8, analyses using the hypotheses of the two cases of tension cut-off and the model proposed by Vecchio and Collins were performed, and the degrees of

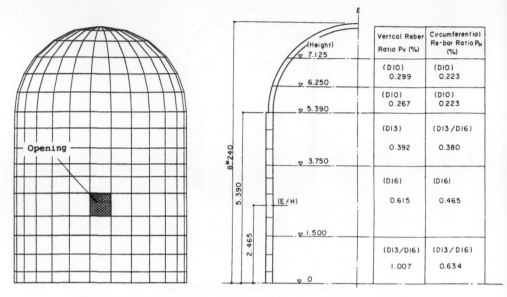

FIGURE 7. Finite element mesh and reinforcement ratios

TABLE 1
Analysis parameters

	Case 1	Case 2	Case 3
Compressive strength of cracked concrete	Cylinder strength	Reduced strength derived from panel tests	
Tension stiffening	Proposed model by Vecchio and Collins		Tension cut-off

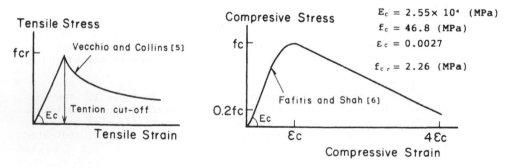

FIGURE 8. Stress-strain relationships of concrete

their influences were grasped. For compressive strength of concrete after cracking, the two cases of maintaining f_c and of reduction by the before-mentioned model based on the results of panel experiments were set up.

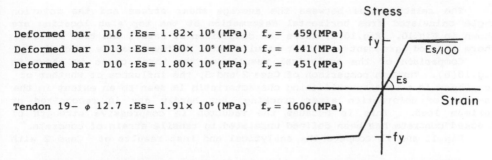

Deformed bar   D16 :Es= 1.82×10^5(MPa) f_y = 459(MPa)

Deformed bar   D13 :Es= 1.80×10^5(MPa) f_y = 441(MPa)

Deformed bar   D10 :Es= 1.80×10^5(MPa) f_y = 451(MPa)

Tendon 19- ϕ 12.7 :Es= 1.91×10^5(MPa) f_y = 1606(MPa)

FIGURE 9. Stress-strain relationship of reinforcing bars and tendons

Prestress loads were given as equivalent external pressures and horizontal forces as nodal loads at the individual loading point locations. The prestresses transferred were 5.90 MPa in the vertical direction and 7.75 MPa in the circumferential direction.

The incremental load method was employed in analyses. Firstly, after prestressing load was applied, gradually increasing horizontal force was applied. The incremental loads at the various steps were 0.2 MPa up to average shear stress at the bottom part 5.0 MPa, 0.1 MPa up to 5.5 MPa, and subsequently, 0.05 MPa up to failure.

Analytical Results

Comparisons of the various loads and deformation amomg test and analytical results are shown in Table 2. The maximum load in analyses was taken to be the point at which stiffness became too small to continue the analysis. Cracking and reinforcing bar yielding load corresponded well, approximately, for all cases. The maximum load, with the exception of Case 1, corresponded well with the test value. In case 1, since the reduction of compressive strength of cracked concrete is not considered, the maximum load exceeds the test value by approximately 30%.

TABLE 2
Comparisons of test results and analytical results

	Test	Case 1	Case 2	Case 3
Flexural cracking load[1]	0.76 (1.00)	0.80 (0.95)	0.80 (0.95)	0.80 (0.95)
Shear cracking load[1]	2.62 (1.00)	2.80 (0.94)	2.80 (0.94)	2.80 (0.94)
Yielding load[1] of reinforcement	4.23 (1.00)	4.40 (0.96)	4.40 (0.96)	4.20 (1.01)
Ultimate strength[1]	5.91 (1.00)	7.65 (0.77)	6.10 (0.97)	6.05 (0.98)
Displacement[2] at ultimate strength	4.15 (1.00)	12.94 (0.32)	4.96 (0.84)	5.88 (0.71)

*1: Average shear stress at bottom (MPa)
*2: Horizontal displacement of top slab (cm)
(   ): Ratio of test result to analytical result

The relationships between the average shear stress and the rotation angle calculated from horizontal deformation at the top slab location are shown in Fig.10. Fig.10(a) shows the comparison of Case 2 and test results. There is good agreement with test results until maximum load is reached.

Comparison of the analytical results of the three cases is shown in Fig.10(b). Through comparison of Case 2 and 3, the influence of whether or not there is a tension stiffening characteristic is seen to an extent in the amount of deformation after cracking, but hardly any is recognized at maximum load. This is because the reducton in compressive strength of cracked concrete has been defined unrelated to tensile strain of concrete.

Fig.11 and 12 compare the analytical and test results of Case 2 with

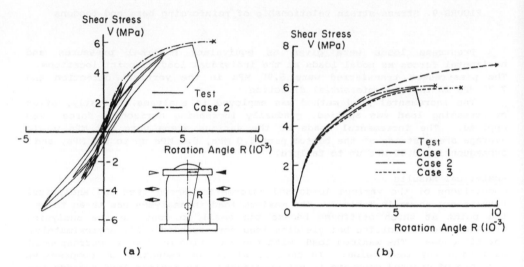

FIGURE 10. Comparisons of relationships between shear stresses and rotation angles

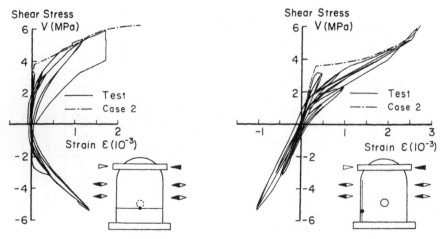

FIGURE 11. Comparisons of relationships between shear stresses and strains of reinforcing bar

171

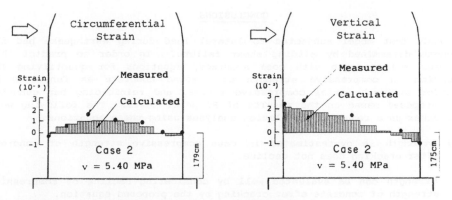

FIGURE 12. Comparisons of strain distributions of reinforcing bars

FIGURE 13. Comparisons of cracking and failure distributions

regard to the average shear stress-reinforcing bar strain relationships at representative locations and strain distributions of reinforcing bars. It can be seen from these that correspondence between analytical and test results is good concerning strain properties of reinforcing bars also.

Fig.13 compares the distributions of cracking and concrete failure at maximum load of Case 2 with the test results. The distribution of cracking obtained from the analysis correspond approximately with the test results. In the test, sliding shear failure occurred near the middle between the opening and the bottom part. Compressive failure of concrete was prominent at the bottom part in the analysis.

CONCLUSIONS

The PCCV test model subjected to lateral load during earthquake had its strength determined by sliding shear failure. In order to predict this behavior and strength with good accuracy, equations for quantifying the reduction in compressive strength of cracked concrete as functions of cylinder strength, axial compressive stress, and reinforcing bar quantity were proposed based on the results of RC panel tests. The following were made clear as a result of simulation analyses using these equations:

1) Strength is overestimated in case compressive strength of concrete after cracking does not decline.

2) Strength can be evaluated well by considering decline of compressive strength of concrete after cracking by the proposed equation.

3) Differences in the tension stiffening characteristic of concrete have hardly any effect on the strength.

REFERENCES

1. Ogaki,Y., Kobayashi,M., Takeda,T., Yamaguchi,T. and Yoshioka,K., Horizontal Loading Tests on Large-scale Model og Prestressed Concrete Containment Veseel, Proceedings of the 6th SMiRT, J4/2, 1981.

2. Yamaguchi,T., Koike,K., Naganuma,K. and Takeda,T, Pure Shear Loadiing Test on Reinforced Concrete Panels, Annual Meeting of Architectural Institute of Japan,(in Japanese), 1988.

3. Yamaguchi,T. and Naganuma,K., Mechanical Properties of RC Panels with Axial and In-plane Shear Forces, Annual Meeting of Architectural Institute of Japan,(in Japanese), 1989.

4. ACI Comittee 318, Building Code Requirements for Reinforced Concrete (ACI 318-77),American Concrete Institute,1977

5. Vecchio,F.J.and Collins,M.P.: The Modified Compression-Field Theory for Reinforced Concrete Elements Subjected to Shear,ACI Journal,Proceedings Vol.83,No.2,Nov.-Dec.,1986, pp.219-231.

6. Fafitis,A.and Shah,S.P.:Lateral Reinforcement for High-Strength Concrete Columns, Publ.ACI,No.SP-87,1985, pp.213-232.

A UNIFIED EVALUATION OF RESTORING FORCE
CHARACTERISTICS OF REACTOR BUILDINGS

HIROYUKI AOYAMA
Professor of Structural Engineering
Department of Architecture, University of Tokyo
Bunkyo-ku, Tokyo 113, Japan

ABSTRACT

Shear stress vs. shear strain relationship and bending moment vs. curvature
relationship of reinforced concrete shear walls were formulated by two sets
of equations for skeleton curves and two sets of hysteretic rules, to be
used in nonlinear earthquake response analysis of reactor buildings. The
proposed evaluation method of restoring force characteristics of reinforced
concrete shear walls was developed on the basis of review of available test
data, and hence it was shown that calculated skeleton and hysteresis curves
coincided with measured ones with practically sufficient accuracy.

INTRODUCTION

In the structural design of nuclear reactor buildings, nonlinear dynamic
response analysis is carried out for the assessment of seismic safety as
well as the evaluation of the response of equipment. For the uniformity of
the analytical techniques involved, a unified evaluation method of
restoring force characteristics was proposed to, and subsequently approved
by, the Special Examination Committee on Seismic Design for Nuclear Power
Plants of the Japan Electric Association, chaired by the writer,
established under the auspices of the Resources and Energy Agency of the
Ministry of International Trade and Industry. This proposal was the result
of the studies carried out by representatives of Tokyo Electric Power
Company and five major construction companies [1], sponsored by the ten
electric power companies in Japan. Major portion of the work was based on
the long-ranged effort by Dr. Y. Inada [2].

Currently prevalent method of dynamic response analysis employs
modeling of reactor buildings into one or several vertical cantilevers,
capable of undergoing inelastic shear and flexural deformation. Thus it
would suffice for an analytical model to be provided with shear stress vs.
shear strain relationship and bending moment vs. curvature relationship of
a vertical cantilever consisting mainly of reinforced concrete (R/C) shear
walls.

In recent years, a number of model tests were carried out of R/C shear walls, with the configuration simulating those in reactor buildings, subjected to a reversal of horizontal loading. They provided a large stock of test data of restoring force characteristics. The proposed evaluation method was derived from the systematic review of these test data.

OUTLINE OF TEST DATA

A survey was made of existing test data of R/C model shear walls reflecting structural characteristics of either BWR or PWR buildings. Test data of 103 specimens were collected; 22 square box walls, 26 cylindrical walls, 19 cylindrical walls with prestress and/or internal pressure, 3 truncated conical walls, 9 octagonal tube walls, and 24 I-shaped section walls simulating one web of a box wall.

As to the dimensions of specimens, the center-to-center distance between flanges was found to be between 100 cm and 200 cm, while the wall thickness was between 5 cm and 10 cm. This corresponds to 1/10 to 1/30 scale of actual dimension of shear walls in reactor buildings. The shear span ratio M/VD, which in most cases corresponds to height-width ratio H/D, was between 0.5 and 1.5.

The compressive strength of concrete of major portion of specimens was between 24 MPa and 26 MPa, while specimens simulating prestressed concrete containment vessel (PCCV) were made of concrete with compressive strength greater than 40 MPa. Reinforcement was arranged to simulate prototype reactor structures in terms of reinforcement ratio, which was in most cases lower than 1.2%.

Out of 103 specimens, 67 were utilized for the discussion of skeleton curves, and 48 were further used for the discussion of hysteresis rules, while 23 specimens were referred to only for the discussion of ultimate strength.

SKELETON CURVE FOR SHEAR

The skeleton curve for shear stress vs. shear strain relationship is idealized into a trilinear curve with three control points as shown in Fig. 1. The first control point (τ_1, γ_1) corresponds to the formation of a shear crack. The third control point (τ_3, γ_3) corresponds to the ultimate shear strength. The second control point (τ_2, γ_2) does not represent any specific phenomenon, but it was selected in order to simulate the trend of $\tau-\gamma$ curve.

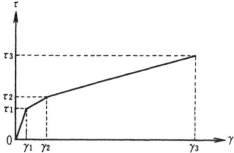

Figure 1. Skeleton Curve for $\tau-\gamma$ Relationship

Three control point are defined by the following equations.

$$\tau_1 = \sqrt{\sqrt{\sigma_B/10}\ (\sqrt{\sigma_B/10} + \sigma_V)} \tag{1}$$

$$\gamma_1 = \tau_1/G \tag{2}$$

$$\tau_2 = 1.35\,\tau_1 \tag{3}$$

$$\gamma_2 = 3\,\gamma_1 \tag{4}$$

$$\tau_3 = \{1 - \tau_s/(4.5\sqrt{\sigma_B/10})\}\,\tau_0 + \tau_s$$

$$\text{for}\ \ \tau_s \leq 4.5\sqrt{\sigma_B/10} \tag{5-1}$$

$$\tau_3 = 4.5\sqrt{\sigma_B/10} \quad \text{for}\ \ \tau_s > 4.5\sqrt{\sigma_B/10} \tag{5-2}$$

$$\gamma_3 = 4.0 \times 10^{-3} \tag{6}$$

where

$$\tau_0 = (3 - 1.8M/VD)\sqrt{\sigma_B/10} \quad \text{for}\ \ M/VD \leq 1 \tag{7-1}$$

$$\tau_0 = 1.2\sqrt{\sigma_B/10} \qquad\qquad \text{for}\ \ M/VD > 1 \tag{7-2}$$

$$\tau_s = (\rho_V + \rho_H)\,\sigma_y/2 + (\sigma_V + \sigma_H)/2 \tag{8}$$

σ_B: compressive strength of concrete (MPa)
G : shear modulus of concrete (MPa)
M/VD : shear span ratio where M and V are bending moment and
       shear at critical section and D is the center–to–center
       distance of flanges
σ_V: axial stress in vertical direction (MPa)
σ_H: axial stress in horizontal direction (MPa)
ρ_V: reinforcement ratio in vertical direction
ρ_H: reinforcement ratio in horizontal direction
σ_y: yield stress of reinforcing bar (MPa)

Table 1 summarizes the comparison of measured values in the tests and calculated values using above equations. Measured shear stress is an average value, determined by the shear force divided by the effective web area of the wall section. Measured shear strain is also an average value, determined in the following way. First the flexural deformation was determined by integrating measured curvatures at each horizontal segment of the wall. Then the shear deformation was determined by subtracting the flexural deformation from the measured total deformation, and it was then divided by the wall height.

The equation (1) corresponds to the average shear stress at the initiation of diagonal crack in the central portion of walls subjected to in-plane shear stress [2]. As shown in Table 1, the average of the ratios of measured vs. calculated values is 0.98, indicating that the approximation by eq. (1) is satisfactory. Fig. 2 shows that the effect of concrete strength is sufficiently taken care of in the equation.

The equation (2) assumes that the shear stress vs. strain relationship is elastic up to the formation of diagonal crack. Table 1 indicates that actual stiffness at the first control point is slightly less than the shear modulus G. Nevertheless eq. (2) is used for simplicity.

TABLE 1. Ratios of Measured vs. Calculated
Values at Control Points of τ-γ Relationship

Items	Range of specimens	Number of specimens	Average	Standard deviation
τ_1	All specimens	57	0.98	0.21
	Except①,②,③	51	0.98	0.18
$G = \dfrac{\tau_1}{\gamma_1}$	All specimens	49	0.93	0.16
	Except①,②,③	42	0.94	0.16
τ_2	All specimens	58	0.99	0.17
	Except④	51	1.00	0.17
τ_3	All specimens	86	1.04	0.19
	Except④	75	1.01	0.15
γ_3	All specimens	48	1.74	0.76
	Except④	37	1.80	0.82

NB : ① box walls subjected to diagonal loading
② specimens made of mortar
③ PCCV
④ specimens with additional flexural reinforcement in flange
walls

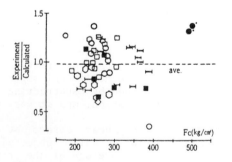

Figure 2. Comparison of Experimental vs. Calculated τ_1 values

Experimental τ-γ relationship beyond cracking shows a convex shape,
or in other words, gradual reduction of stiffness. To approximate this
trend, eq. (4) was assumed for the strain at the second control point, and
the corresponding shear stress τ_2 was found from the test data. The ratios
of measured τ_2 to τ_1 was plotted against concrete strength, as shown in

Fig. 3. The average of the ratio, 1.35, was used in eq. (3), and it gives a good agreement as indicated in Table 1.

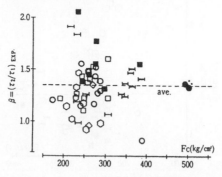

Figure 3. Ratio of τ_2 vs. τ_1 Values Obtained from Test Data

The equation (5-1) for the ultimate shear stress was derived taking into account the stress carried by reinforcement as well as the stress carried by concrete [3]. The stress carried by reinforcement and axial stress (due to external load and/or prestress) is expressed by τ_s, and the stress carried by concrete τ_0 is linearly reduced as τ_s is increased. Eq. (5-2) gives an upper limit of τ_s. As shown in Table 1 and Fig. 4, the evaluation of these equations was fairly favorable; the ratios of measured vs. calculated values had an average of 1.04. Moreover, some of the I-shaped section walls had an exceptionally large amount of vertical reinforcement in flange walls in order to surely produce shear failure in the experiment. This is a condition not found in any real reactor buildings. When these specimens were excluded, the average of ratios of measured vs. calculated ultimate shear stress was improved to 1.01, and the standard deviation was also reduced.

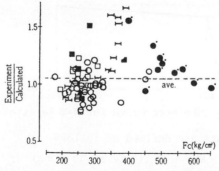

Figure 4. Comparison of Experimental vs. Calculated τ_3 Values

As to the strain associated with the ultimate shear stress, it was found that the experimental τ-γ relationship of cylindrical walls and octagonal tube walls had relatively large deformation capacity as compared to box walls and I-shaped section walls. However, it was considered appropriate, at least at present, to use commonly adopted ultimate shear strain of eq. (6) in this study, because of large fluctuation in observed values and insufficient amount of test data for separate evaluation between shear walls of different configuration. The degree of approximation is shown in Table 1 as well as in Fig. 5.

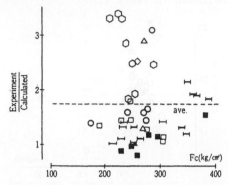

Figure 5. Comparison of Experimental vs. Calculated γ_3 values

SKELETON CURVE FOR BENDING

The skeleton curve for bending moment vs. curvature relationship is idealized into another trilinear curve with three control points as shown in Fig. 6. The first control point (M_1, ϕ_1) corresponds to the formation of a flexural crack. The second control point (M_2, ϕ_2) corresponds to the state where the most tensile reinforcement reaches the yield point. The third control point (M_3, ϕ_3) corresponds to the ultimate flexural failure dictated by crushing of the most compressive concrete.

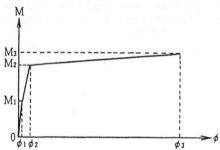

Figure 6. Skeleton Curve for M-ϕ Relationship

Three control points are defined as follows.

$$M_1 = Z_e \left(1.2\sqrt{\sigma_B/10} + \sigma_V\right) \tag{9}$$

$$\phi_1 = M_1/(E\,I_e) \tag{10}$$

$$M_2 = M_y \tag{11}$$

$$\phi_2 = \phi_y \tag{12}$$

$$M_3 = M_u \tag{13}$$

$$\phi_3 = 0.004/x_{nu} \leqq 20\phi_2 \tag{14}$$

where
Ze : section modulus including reinforcement (m^3)
Ie : second moment of section including reinforcement (m^4)
E : Young's modulus of concrete (MPa)
My : bending moment when the most tensile reinforcement reaches the yield point, calculated by assuming linear strain distribution
ϕy : curvature associated with My
Mu : ultimate moment calculated by assuming full plastic stress distribution, taking concrete stress of $0.85\sigma_B$ and steel stress of σ_y
xnu: distance from extreme compression fiber to the neutral axis at full plastic moment Mu.

To compare experimental values of three control points with the calculation from above equations, bending moment M at the wall base and flexural rotation R were used, where the latter was defined as the flexural deformation determined as the integral of measured curvatures, divided by the wall height. The observed M-R relationship was further idealized into a trilinear skeleton curve to obtain experimental control points. Table 2 summarizes the comparison of measured vs. calculated values.

TABLE 2. Ratios of Measured vs. Calculated
Values at Control Points of M-R Relationship

	Items	Number of specimens	Average	Standard deviation
M_1 :	first flexural crack	68	0.87	0.29
	first control point	50	1.10	0.25
M/R:	initial stiffness	41	0.93	0.35
M_2 :	second control point	36	1.07	0.17
R_2 :	second control point	36	1.44	0.54
M_3 :	ultimate moment	18	1.09	0.15

The equation (9) assumes that the tensile stress in the concrete at the first flexural crack is expressed as $1.2\sqrt{\sigma_B/10}$ [2]. In Table 2, M_1 was compared in two ways. First, the calculated values were compared with the moment at the observed first flexural cracking, and second, they were compared with the moment at the first control point of experimentally determined trilinear curve. Figs. 7 and 8 also shows the comparison with respect to concrete strength. The average of ratios for first flexural cracking was 0.87, indicating that eq. (9) somewhat overestimates the cracking moment. The average for the first control point was 1.10, which means that eq. (9) slightly underestimates the turning point in the observed M-R relationship.

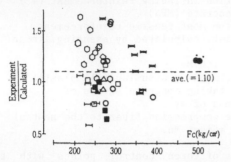

Figure 7. Comparison of Experimental vs. Calculated M_1 Values
(initiation of flexural cracking)

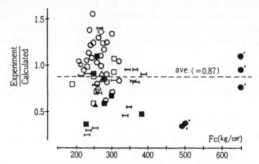

Figure 8. Comtarison of Experimental vs. Calculated M_1 Values
(approximation by trilinear skeleton curve)

The equation (10) assumes a linear elastic stiffness up to the moment M_1. Measured initial stiffness was much smaller than the calculated values, indicating some additional deformation would have occurred at the wall base. Hence a rotation due to slip and elongation of tensile reinforcing bars in the base [4] was calculated and added. This modification will be justified when one considers that the effect of bar slip and elongation in the base becomes significant when the specimen is small, while it becomes negligible in case of prototype wall structures. Even with this modification, measured initial stiffness was slightly lower than the calculated one, as shown in Table 2, and was associated with a large fluctuation.

The second control point from eqs. (11) and (12) corresponds to the yielding of most tensile reinforcement. When the calculated moment M_2 was compared with that at the second control point of experimentally determined trilinear curve, it was found that average of ratios was 1.07. This is also shown in Fig. 9.

To compare the rotation at the second control point R_2, calculated curvature distribution along the wall height was integrated. Further the rotation due to slip and elongation of tensile reinforcing bars in the base was added. However, as shown in Table 2 and Fig. 10, the comparison was not so favorable; the average of 1.44 indicates that experimental deformation is considerably greater at the second control point.

181

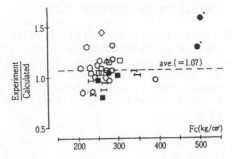

Figure 9. Comparison of Experimental vs. Calculated M_2 Values

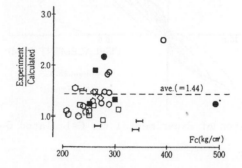

Figure 10. Comparison of Experimental vs. Calculated R_2 Values
(rotation at the second control point)

The third control point corresponds to the ultimate flexural failure by the crushing of the extreme compression fiber concrete, but eq. (13) assumes that the ultimate moment M_3 can be approximated by the full plastic theory taking stress of concrete and steel to be $0.85\sigma_B$ and y, respectively. The comparison with observed ultimate moment in Table 2 shows that the average of ratios is 1.09, indicating that even the full plastic stress distribution was a conservative assumption. The strength enhancement is presumably due to strain hardening of reinforcing bars.

The ultimate curvature in eq. (14) is obtained by assuming that the strain of concrete at crushing is 0.004 and that location of neutral axis is approximated by the full plastic theory. In case of walls with small axial stress and reinforcement ratio, the neutral axis often moves into the compression flange of the section, making the neutral axis depth xnu very small. However, this is unrealistic for actual reactor buildings. Therefore an arbitrary upper limit of ultimate curvature was set to be twenty times the yield curvature. No attempt was made to compare eq. (14) with test results, because relatively few specimens reached the flexural ultimate state; most specimens failed in shear in earlier loading stage. Furthermore, flexural deformation constitutes relatively small portion of total deformation, and hence it does not play an important role in the evaluation of restoring force characteristics.

COMPARISON OF SKELETON CURVES

Figure 11 illustrates shear force vs. total displacement relationship measured in the experiment and corresponding skeleton curves calculated as the total of shear and flexural deformations in the preceding sections. The calculated skeleton curves coincided with the measured ones with sufficient accuracy. Consequently, the skeleton curves of the restoring force characteristics of shear walls in reactor buildings can be evaluated by the proposed method of analysis.

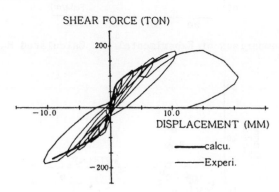

Figure 11. Comparison of Experimental vs. Calculated Q-δ Relationship

HYSTERESIS MODELS

The hysteresis due to reversal of loading was idealized by simple models that are frequently used in dynamic response analysis. It was considered in the adoption of these models that the practical applicability was more important than the accuracy in simulating the observed behavior of test specimens.

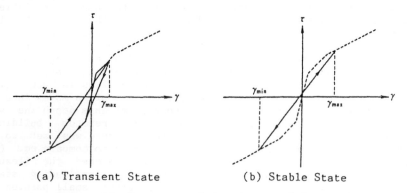

(a) Transient State   (b) Stable State

Figure 12. Hysteresis Rules for τ-γ Relationship

For the shear stress vs. shear strain (τ-γ) relationship, a peak-oriented hysteresis model was adopted. As shown in Fig. 12 (a), an unloading path from a maximum (or minimum) value point beyond the first control point aims straight at the minimum (or maximum) value point on the opposite side. If the peak value on the opposite side dose not yet exceed the first control point, the unloading path aims at the first control point on the opposite side.

As long as the response remains within the previous peaks in the positive and negative directions, the stable loop shown in Fig. 12 (b) is used. A characteristic point of this stable loop is that it does not have any hysteretic area. This is an attempt to underestimate energy dissipation in shear deformation, which is originally small, to start with.

As to the bending moment vs. curvature (M–Φ) relationship, another peak-oriented hysteresis model was adopted, which is a combination of ordinary peak-oriented model and degrading trilinear model.

As long as the peak value in either direction does not exceed the second control point, the hysteresis rules are identical to the peak-oriented model of τ–γ relation, as shown in Fig. 13 (a). The stable loop in this stage does not have any area, as shown in Fig. 13 (b). This assumption is in accordance with the intended underestimation of energy dissipation prior to yielding of reinforcing steel.

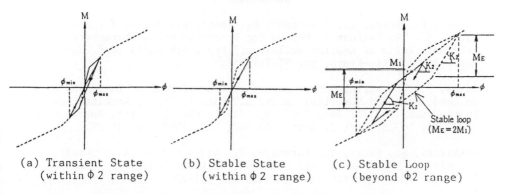

(a) Transient State   (b) Stable State   (c) Stable Loop
  (within Φ2 range)   (within Φ2 range)   (beyond Φ2 range)

Figure 13. Hysteresis Rules for M–Φ Relationship

If the response exceeds the second control point, unloading path forms a stable loop, shown in Fig. 13 (c), in such a way that the path is directed towards the peak ever reached in the opposite direction. If the opposite peak does not yet reach the second control point, the unloading path aims at the second control point on the opposite side.

The shape of the stable loop is a parallelogram whose area corresponds to the viscous damping determined by the following equation.

$$h_e = 0.15 \frac{\phi_{max} - \phi_2}{\phi_3 - \phi_2} \tag{15}$$

where h_e : equivalent viscous damping of stable loop
        ϕ_{max} : absolute value of maximum curvature.

The break point of the parallelogram is determined by the moment difference M_E, which is taken twice the moment M_1 from eq. (9), as illustrated in Fig. 13 (c). Together with the equivalent viscous damping of eq. (15), one can determine slopes K_1 and K_2 of the parallelogram.

CONCLUSION

For the uniformity in nonlinear earthquake response analysis of nuclear buildings, a unified evaluation method of restoring force characteristics of reinforced concrete shear walls was proposed. It consists of shear stress vs. shear strain relationship and bending moment vs. curvature relationship, in the form of nonlinear skeleton curves and sets of hysteresis rules. The proposed method was developed on the basis of extensive review of existing test data of reinforced concrete shear walls in Japan. Reliability of equations to specify each control point on the skeleton curves was examined by comparing calculated and experimental values. It was concluded that restoring force characteristics of shear walls evaluated by the proposed method would represent the actual dynamic behavior of nuclear buildings with practically sufficient accuracy.

REFERENCES

1. Tanaka, H., Imoto, K., Yoshizaki, S., Emori, K., Inada, Y., and Nanba, H., "An Evaluation Method for Restoring Force Characteristics of R/C Shear Walls of Reactor Buildings," Proc. 9th World Conf. Earthquake Eng., 1988, Tokyo-Kyoto, pp. VI-747-752.

2. Inada, Y., "Restoring Force Characteristics of Reactor Buildings Based on Load Tests and Numerical Analysis," Part 1, Trans. Architectural Inst. Japan, No.371, Jan. 1987, pp. 61-71; Part 2, ditto, No.378, Aug. 1987, pp. 16-26; Part 3, ditto, No.382, Dec. 1987, pp. 19-29.

3. Yoshizaki, S., Ezaki, T., Korenaga, T., and Sotomura, K., "Shear Strength of Shear Walls with Numerous Small Openings," 5th Annual Meeting, Japan Concrete Inst., 1983, pp. 201-204.

4. Otani, S., "Inelastic Analysis of R/C Frame Structures," J. of Structural Div., ASCE, July 1974, pp. 1433-1449.

THEME II:

THEORETICAL STUDIES OF MEMBRANE SHEAR BEHAVIOR

STATE-OF-THE-ART OF MEMBRANE SHEAR BEHAVIOR - EUROPEAN WORK

PETER MARTI
Professor of Structural Engineering
ETH Hönggerberg, 8093 Zürich, Switzerland

ABSTRACT

European contributions to the development of truss model, compression field and limit analysis approaches are reviewed and a brief account is given of the application of such methods in practice.

INTRODUCTION

This paper provides an overview on European developments of truss model, compression field and limit analysis approaches to reinforced concrete membrane shear behavior. The presentation is largely based on two more comprehensive publications [17, 18] which contain further details and references and, in particular, relate the European work to the significant theoretical and experimental contributions made by North American engineers during the past two decades.

THEORETICAL DEVELOPMENTS

Truss Models and Compression Field Approach

The idea of using truss models for following the flow of internal forces in reinforced concrete structures dates back to the end of the 19th century [27]. Over the first two decades of this century, Mörsch [19] greatly advanced these concepts and introduced the classical 45-degree truss model which has been adopted by most codes of practice as the basis of their shear and trosion design provisions.

From tests on beams subjected to flexure and shear Mörsch [19] knew that with decreasing amounts of transverse reinforcement diagonal cracks in the web become flatter and that their inclination to the beam axis may be considerably less than 45 degrees.

However, he thought it would be practically impossible to compute the inclination of these cracks, and he recommended to continue using the simple and conservative 45-degree truss model.

By analysing a truss model consisting of linearly elastic members and neglecting the concrete tensile strength Kupfer [12] solved Mörsch's problem and found an equation for the inclination of the diagonal cracks. Baumann [1] derived a similar equation for the crack direction in orthogonally reinforced concrete membrane elements, i.e.,

$$\tan^2\theta\rho_x(1+\rho_y n)+\tan\theta\rho_x\frac{\sigma_y}{\tau_{xy}}=\frac{\sigma_x}{\tau_{xy}}\ \rho_y\cot\theta+(1+\rho_x n)\rho_y\cot^2\theta \tag{1}$$

where θ = inclination of diagonal cracks to x-axis; ρ_x and ρ_y = reinforcing ratios in x- and y-direction; $n = E_s/E_c$ = modular ratio; and σ_x, σ_y and τ_{xy} = applied stress components. It is noteworthy that Eq. (1) can be obtained from a simple generalization of Kupfer's equation by substituting $(\sigma_x+\tau_{xy}\cot\theta)/(\rho_x E_s)$ and $(\sigma_y+\tau_{xy}\tan\theta)/(\rho_y E_s)$ for the average axial strain at mid-depth of the web and the average stirrup strain, respectively.

Using a space truss model similar to Kupfer's approach Thürlimann and Lüchinger [35] determined the deformations of beams subjected to torsion and flexure. Pré [25] and Teutsch and Kordina [34], who generalized this work for prestressed girders and combined actions, based their analysis on Collins' compression field approach [17]. Finally, Potucek [24] used a finite element model based on Baumann's work to analyse the webs of reinforced concrete girders in flexure and shear.

Kupfer's and Baumann's investigations [12, 1] and related work [35, 24] constitute a linear compression field approach as opposed to Collins' more general non-linear compression field approach [18]. In essence, the compression field approach idealizes the cracked concrete as a material with coinciding principal stress and strain axes which are free to adapt their direction as required by the applied loads [17].

Kupfer et al. [13] pointed out that in general, the principal stress and strain axes do not coincide and that they also deviate from the crack direction. For members subjected to shear they considered a uniform diagonal compression field in the web and derived an equation for its inclination using a simplified aggregate interlock relationship based on Walraven's research [40]. Kirmair and Mang [10] generalized this investigation for members subjected to shear and axial load. Dei Poli et al. [5, 6] adopted a similar approach as Kupfer et al. [13] but used a different aggregate interlock relationship.

Rather than determining the compression field direction from an aggregate interlock consideration, Reineck and Hardjasaputra [26] assumed that the cracks open perpendicularly to the compression field direction. Thus, the crack direction and the compression field direction are characteristics of the strain field due to crack opening and the associated principal directions bisect the angles formed by the cracks and the compression struts. Taking the additional compressive strain in the concrete struts into account one concludes that the average principal compressive strain direction is somewhat closer to the strut direction than the crack direction and an analysis similar to that by Kupfer et al. [13] can be performed.

The research summarized above [13, 10, 5, 6, 26] has brought valuable insight into the behavior of cracked reinforced concrete membranes. It supplements limit analysis - based dimensioning methods and it provides a new interpretation of the "concrete contribution" to shear resistance introduced in various codes of practice. This is similar to Collins' modified compression field approach [18] which accounts for tensile stresses in the concrete between the cracks. However, the European work in this area has been limited to uniform stress fields; i.e., B-regions in Schlaich et al.'s terminology [28-30], and, in contrast to the modified compression field approach, potential concrete crushing failures have not been addressed in much detail. Notable exceptions include the experimental investigations by Kollegger and Mehlhorn [11] and by Schlaich and Schäfer [31]. Compared to predictions based on North American research these tests seem to indicate a less pronounced softening of the concrete in compression due to simultaneous transverse tensile action.

Limit Analysis

Limit analysis procedures have mainly been promoted by two groups of researchers around Thürlimann in Zürich and Nielsen in Copenhagen. With a colloquium organized by IABSE in Copenhagen a first state-of-the-art report was produced [9] and later, Thürlimann et al. [39] and Nielsen [23] presented comprehensive monographs.

Based on the classical theory of plasticity, Nielsen [22] established a general theory of plane stress in reinforced concrete. He discussed a yield criterion for isotropically reinforced concrete membrane elements and presented various complete limit analysis solutions for deep beams.

Using the same notation as for Eq. (1) and assuming that failure is governed by yielding of the reinforcement, Nielsen's yield criterion leads to the requirements

$$\rho_x f_y \geq \sigma_x + k\tau_{xy} \quad , \quad \rho_y f_y \geq \sigma_y + \tau_{xy}/k \tag{2}$$

where f_y = yield strength of the reinforcement and $k = \cot\theta$. Eq. (2) is frequently used for the dimensioning of the reinforcement, e.g., in conjunction with a linearly elastic finite element analysis of the uncracked structure resulting in a statically admissible stress field with components σ_x, σ_y and τ_{xy}.

Eq. (2) is only valid if the concrete does not crush, i.e., if the requirement

$$(\rho_x + \rho_y)f_y - (\sigma_x + \sigma_y) \leq f_c \qquad (3)$$

is satisfied, where f_c = effective concrete compressive strength. If Eq. (3) is not satisfied then at least one reinforcement remains elastic and the concrete crushes. In particular, for $\sigma_x = \sigma_y = 0$, $\rho_y f_y \leq f_c/2$ and $(\rho_x + \rho_y)f_y > f_c$ one gets Braestrup's web crushing criterion

$$\tau_{xy}^2 = \rho_y f_y (f_c - \rho_y f_y) \qquad (4)$$

describing a failure in pure shear governed by crushing of the diagonally compressed concrete and yielding of the weak reinforcement in y-direction while the strong reinforcement in x-direction remains elastic [2].

Müller [21] treated the yield criterion for arbitrarily reinforced membrane elements and presented a comprehensive discussion on the stress and deformation conditions in reinforced concrete walls for all possible yield regimes. Yield criteria including the concrete tensile strength were also derived [16].

The application of limit analysis methods to torsion problems was beneficial for the later treatment of shear problems. Lampert and Thürlimann [14] treated beams in torsion and flexure using a variable angle space truss approach. Thürlimann et al. [36] extended this work to cover warping torsion as well as combined flexure, shear and torsion, leading to a supplement to the Swiss Code [33] which largely influenced the 1978 CEB-FIP Model Code [3, 37, 38].

Starting from the skew bending approach for members subjected to torsion Elfgren et al. [7] also treated the interaction of torsion, bending and shear in concrete beams. Müller [20, 21] supplemented the space truss approach by describing compatible failure mechanisms and discussed the relationship between his complete solutions and earlier lower-bound (space truss) and upper-bound (skew bending) solutions.

Maier [15] applied both lower- and upper-bound methods to the ultimate strength analysis of shear walls and compared the theoretical predictions with various large-scale experiments. Schlaich [32] developed an interactive computer program for the design of reinforced concrete shear walls based on the truss model approach. A similar computer program using discontinuous stress fields composed of homogeneously stressed triangular areas was developed by Hajdin [8].

PRACTICAL APPLICATIONS

Dimensioning Based on Elastic Analysis

Usually, designs are based on a linearly elastic analysis of the uncracked structures. The resulting sectional forces and moments are then used in dimensioning the reinforcement and checking for both serviceability and ultimate limit states. Redistributions of internal forces are only rarely taken into account.

Finite element programs are now quite commonly used to perform an elastic structural analysis. Typically, the necessary reinforcement is determined for ultimate limit state conditions, e.g., by applying Eq. (2), and concrete dimensions are checked by applying Eq. (3). Finally, serviceability limit states are checked, e.g., by applying Eq. (1) and determining likely deformations according to a linear compression field approach [17] combined with some conventional crack width computation.

In applying the described procedure a certain amount of engineering judgement is required. In particular, average stresses over sufficiently large areas rather than local peak stresses should be considered to arrive at rational reinforcement layouts and to avoid unnecessary conservatism.

Design Based on Strut and Tie Models

Practising engineers have commonly applied strut and tie models for the dimensioning and detailing of reinforced concrete structures. However, as there was little official support, many engineers had some doubts about the justification of their methods. This has changed in recent years. For example, the current draft CEB-FIP Model Code [4] acknowledges the potential of strut and tie models and gives detailed rules regarding their application.

The application of strut and tie models was strongly promoted by Schlaich et al. [28, 30]. They introduced the term "D-region" for regions in the vicinity of geometric, static or material discontinuities or singularities while for regular regions, with gradually changing parameters, they used the term "B-region". They pointed out that, whereas rather refined methods are available for treating B-regions, there is no equivalent for D-regions. Hence, to arrive at a rational dimensioning and detailing, they proposed to consistently apply strut and tie models for B-regions as well as D-regions. As a guideline for the development of these models they recommended to visualise the internal force flow according to a linearly elastic analysis and to orient the struts and ties such that they would roughly correspond to the resultants of the elastically determined stresses. Thus, the same model can be used for serviceability and ultimate limit states considerations.

Strut and tie models permit to visualize the force flow and to proportion a member. While they indicate the necessary amount, the correct position and the required detailing of the main reinforcement, such models should only be applied in conjunction with a well distributed minimum reinforcement in all those regions of a member where no main reinforcement is required. In order to prevent brittle failures and to enable the desired redistribution of the internal forces in the cracked concrete, minimum reinforcement must be capable of replacing a significant fraction of the concrete tensile strength that is overcome locally upon cracking.

Based on this discussion it is recommended to provide a sufficient minimum reinforcement after having selected the cross-sectional dimensions. The minimum reinforcement enhances the ductility and it may significantly contribute to the ultimate resistance. A rather simple equilibrium solution is usually sufficient for determining the additional main reinforcement needed to transfer the applied loads. Consideration of elastically determined stresses is not essentially superior compared with other equilibrium analyses since there are always residual stress states which cannot reliably be determined in practice and which are typically ignored in any structural analysis. However, as linearly elastic analyses are readily available it is quite reasonable to use them as a guide in developing strut and tie models. Further hints regarding practical applications are given elsewhere [18].

CONCLUSIONS

European work on membrane shear behavior has mainly concentrated on the response to static loading.

Strut and tie models as well as linearly elastic analyses combined with linear compression field and static limit analysis approaches are in common use.

Non-linear analyses, compression field approaches considering aggregate interlock, and limit analysis-based CAD-programs so far serve only as research tools.

Recent developments have favoured the application of static limit analysis methods while perhaps, kinematic methods have been thrust too much into the background [18].

There is a need for better understanding of bond and development of reinforcement, and more guidance is required for the selection of minimum reinforcement [18].

193

REFERENCES

[1] Baumann, T., "Zur Frage der Netzbewehrung von Flächentragwerken (On the Problem of Net Reinforcement of Surface Structures)", Bauingenieur, Vol. 47, No. 10, Oct. 1972, pp. 367-377.

[2] Braestrup, M.W., "Plastic Analysis of Shear in Reinforced Concrete", Magazine of Concrete Research, Vol. 26, No. 89, Dec. 1974, pp. 221-228.

[3] Comité Euro-International du Béton, "CEB-FIP Model Code for Concrete Structures", Third Edition, Paris, 1978, 348 pp.

[4] Comité Euro-International du Béton, "CEB-FIP Model Code 1990", First Draft, Bulletins d'Information, No. 195 and No. 196, Lausanne, March 1990.

[5] Dei Poli, S., Gambarova, P.G., and Karakoç, C., "Aggregate Interlock Role in R.C. Thin-webbed Beams in Shear", Journal of Structural Engineering, ASCE, Vol. 113, No. 1, Jan. 1987, pp. 1-19.

[6] Dei Poli, S., Di Prisco, M., and Gambarova, P.G., "Stress Field in Web of R.C. Thin-webbed Beams Failing in Shear", Journal of Structural Engineering, ASCE, Vol. 116, No. 9, Sep. 1990, pp. 2496-2515.

[7] Elfgren, L., Karlsson, I., and Losberg, A., "Torsion-Bending Shear Interaction for Concrete Beams", Proceedings, ASCE, Vol. 100, No. ST8, Aug. 1974, pp. 1657-1676.

[8] Hajdin, R., "Computerunterstützte Berechnung von Stahlbetonscheiben mit Spannungsfeldern (Computer-aided Analysis of Reinforced Concrete Walls Using Stress Fields)", Institute of Structural Engineering, ETH Zürich, Report No. 175, 1990, 115 pp.

[9] IABSE, "Colloquium on Plasticity in Reinforced Concrete", Copenhagen 1979, International Association for Bridge and Structural Engineering, Introductory Report, Vol. 28, 1978, 172 pp; and Final Report, Vol. 29, 1979, 360 pp.

[10] Kirmair, H., and Mang, R., "Das Tragverhalten der Schubzone schlanker Stahlbeton- und Spannbetonträger bei Biegung und Längskraft (Response of the Shear Zone of Slender Reinforced and Prestressed Concrete Girders Subjected to Flexure and Axial Load)", Bauingenieur, Vol. 62, 1987, pp. 165-170.

[11] Kolleger, J., and Mehlhorn, G., "Material Model for Cracked Reinforced Concrete", IABSE Colloquium Delft 1987, International Association for Bridge and Structural Engineering, Report, Vol. 54, 1987, pp. 63-74.

[12] Kupfer, H., "Erweiterung der Mörsch'schen Fachwerkanalogie mit Hilfe des Prinzips vom Minimum der Formänderungsarbeit (Generalization of Mörsch's Truss Analogy Using the Principle of Minimum Strain Energy)", Comité Euro-International du Béton, Bulletin d'Information, No. 40, Paris, Jan. 1964, pp. 44-57.

[13] Kupfer, H., Mang, R., and Karavesyroglou, M., "Bruchzustand der Schubzone von Stahlbeton- und Spannbetonträgern - Eine Analyse unter Berücksichtigung der Rissverzahnung (Ultimate Limit State of the Shear Zone of Reinforced and Prestressed Concrete Girders - An Analysis Taking the Aggregate Interlock into account)", Bauingenieur, Vol. 58, 1983, pp. 143-149.

194

[14] Lampert, P., and Thürlimann, B., "Ultimate Strength and Design of Reinforced Concrete Beams in Torsion and Bending", International Association for Bridge and Structural Engineering, Publications, Vol. 31-I, 1971, pp. 107-131.

[15] Maier, J., "Tragfähigkeit von Stahlbetonscheiben (Ultimate Strength of Reinforced Concrete Walls)", Institute of Structural Engineering, ETH Zürich, Report No. 169, 1988, 93 pp.

[16] Marti, P., and Thürlimann, B., "Fliessbedingung für Stahlbeton mit Berücksichtigung der Betonzugfestigkeit (Yield Criterion for Reinforced Concrete Considering the Concrete Tensile Strength)", Beton- und Stahlbetonbau, Vol. 72, 1977, pp. 7-12.

[17] Marti, P., "Strength and Deformations of Reinforced Concrete Members Under Torsion and Combined Actions", Comité Euro-International du Béton, Bulletin d'Information, No. 146, Jan. 1982, pp. 97-138.

[18] Marti, P., "Dimensioning and Detailing", IABSE Colloquium on Structural Concrete, Stuttgart, 10-12 April 1991, International Association for Bridge and Structural Engineering, Zürich, in press.

[19] Mörsch, E., "Der Eisenbetonbau - Seine Theorie und Anwendung (Reinforced Concrete Construction - Theory and Application)", 5th Edition, Vol. 1, Part 2, K. Wittwer, Stuttgart, 1922.

[20] Müller, P., "Failure Mechanisms for Reinforced Concrete Beams in Torsion and Bending", International Association for Bridge and Structural Engineering, Publications, Vol. 36-II, 1976, pp. 147-163.

[21] Müller, P., "Plastische Berechnung von Stahlbetonscheiben und -balken (Plastic Analysis of Reinforced Concrete Walls and Beams)", Institute of Structural Engineering, ETH Zürich, Report No. 83, 1978, 160 pp.

[22] Nielsen, M.P., "On the Strength of Reinforced Concrete Discs", Civil Engineering and Building Construction Series, No. 70, Acta Polytechnica Scandinavica, Copenhagen, 1971, 261 pp.

[23] Nielsen, M.P., "Limit Analysis and Concrete Plasticity", Prentice-Hall, 1984, 420 pp.

[24] Potucek, W., "Die Beanspruchung der Stege von Stahlbetonplattenbalken durch Querkraft und Biegung (Stresses in Webs of Reinforced Concrete T-beams Subjected to Flexure and Shear)", Zement und Beton, Vol. 22, No. 3, 1977, pp. 88-98.

[25] Pré, M., "Etude de la torsion dans le béton précontraint par la méthode du treillis spatial évolutif (Investigating Torsion in Prestressed Concrete Using the Space Truss Method)", Annales de l'Institut Technique du Bâtiment et des Travaux Publics, No. 385, 1980, pp. 94-112.

[26] Reineck, K.H., and Hardjasaputra, H., "Zum Dehnungszustand bei der Querkraftbemessung profilierter Stahlbeton- und Spannbetonträger (Strain State Considerations in Designing Reinforced and Prestressed Concrete Girders for Shear)", Bauingenieur, Vol. 65, 1990, pp. 73-82.

[27] Ritter, W., "Die Bauweise Hennebique (Hennebique's Construction Method)", Schweizerische Bauzeitung, Vol. 17, 1899, pp. 41-43, 49-52 and 59-61.

[28] Schlaich, J., and Weischede, D., "Detailing Reinforced Concrete Structures", Canadian Structural Concrete Conference 1981, Proceedings, Department of Civil Engineering, University of Toronto, Toronto, 1981, pp. 171-198.

[29] Schlaich, J., and Schäfer, K., "Konstruieren im Stahlbetonbau (Detailing in Reinforced Concrete Design)", Betonkalender 1984, Part 2, W. Ernst, Berlin, 1984, pp. 787-1005.

[30] Schlaich, J., Schäfer, K., and Jennewein, M., "Toward a Consistent Design of Structural Concrete", Journal of the Prestressed Concrete Institute, Vol. 32, No. 3, May-June 1987, pp. 74-150.

[31] Schlaich, J., and Schäfer, K., "Zur Druck-Querzug-Festigkeit des Stahlbetons (On the Compression-Transverse-Torsion-Strength of Reinforced Concrete)", Beton- und Stahlbetonbau, Vol. 78, 1983, pp. 73-78.

[32] Schlaich, M., and Anagnostou, G., "Stress Fields for Nodes of Strut and Tie Models", Journal of Structural Engineering, ASCE, Vol. 116, No. 1, Jan. 1990, pp. 13-23.

[33] Swiss Society of Engineers and Architects, "Ultimate Strength and Plastic Design of Reinforced and Prestressed Concrete Structures", Directive 162/34 concerning the Structural Design Standard SIA 162, Zürich, 1976, 14 pp.

[34] Teutsch, M., and Kordina, K., "Versuche an Spannbetonbalken unter Kombinierter Beanspruchung aus Biegung, Querkraft und Torsion (Tests on Prestressed Concrete Girders Subjected to Flexure, Shear and Torsion)", Deutscher Ausschuss für Stahlbeton, Heft 334, 1982, 81 pp.

[35] Thürlimann, B., and Lüchinger, P., "Steifigkeit von gerissenen Stahlbetonbalken unter Torsion und Biegung (Stiffness of Cracked Reinforced Concrete Beams Subjected to Torsion and Flexure)", Beton- und Stahlbetonbau, Vol. 68, No. 6, June 1973, pp. 146-152.

[36] Thürlimann, B., Grob, J., and Lüchinger, P., "Torsion, Biegung und Schub in Stahlbetonträgern (Torsion, Flexure and Shear in Reinforced Concrete Girders)", Institute of Structural Engineering, ETH Zürich, 1975, 170 pp.

[37] Thürlimann, B., "Shear Strength of Reinforced and Prestressed Concrete Beams - CEB Approach", Concrete Design: U.S. and European Practices, SP-59, American Concrete Institute, Detroit, 1979, pp. 93-115.

[38] Thürlimann, B., "Torsional Strength of Reinforced and Prestressed Concrete Beams - CEB Approach", Concrete Design: U.S. and European Practices, SP-59, American Concrete Institute, Detroit, 1979, pp. 117-143.

[39] Thürlimann, B., Marti, P., Pralong, J., Ritz, P., and Zimmerli, B., "Anwendung der Plastizitätstheorie auf Stahlbeton (Application of the Theory of Plasticity to Reinforced Concrete)", Institute of Structural Engineering, ETH Zürich, 1983, 252 pp.

[40] Walraven, J.C., "Fundamental Analysis of Aggregate Interlock", Proceedings, ASCE, Vol. 107, No. STII, Nov. 1981, pp. 2245-2270.

STATE-OF-THE-ART OF THEORETICAL STUDIES ON MEMBRANE SHEAR BEHAVIOR IN JAPAN

HIROSHI NOGUCHI
Professor, Department of Architecture
Faculty of Engineering, Chiba University
1-33, Yayoi-cho, Chiba 260, Japan

ABSTRACT

The recent macroscopic and microscopic models for cracked concrete, reinforced concrete (RC) panels and framed shear walls which have been proposed
in Japan are introduced as the theoretical studies on membrane shear behavior. As the modeling of cracked concrete and the strength and deformation of shear panels are important for the evaluation of shear strength and
deformation of RC members, many researchers have been grappling with these
problems. There have been many papers in which the shear strength of
framed shear walls is estimated by the macroscopic model from the discussion on the shear resisting mechanisms. The deformation of the shear walls
has been recently analyzed not only by FEM microscopic models but also by
the macroscopic models.

INTRODUCTION

In Japan, the shear design equation in RC building structures has been
mainly based on the experimental studies, because the shear behavior of RC
structures is influenced by many factors and very complicated. But in JCI
colloquium on shear analysis of RC structures in June, 1982, many papers on
both macroscopic and microscopic models were presented. This colloquium
was organized by JCI Committee on Shear Strength of RC Structures
(chairman: Prof. Okamura). Especially state-of-the-art reports on macroscopic and microscopic models [1][2] were enlightening on the shear
problems, and this came to be a start of the recent theoretical studies on
the shear behavior of RC structures in Japan.

The second JCI colloquium on shear analysis of RC structures was
held in October, 1983. The main themes were modeling for shear analysis,
development of shear analytical method, comparisons of analytical results
with experimental ones and application to design. JCI colloquium on FEM
analysis of RC structures in 1984 became a preliminary seminar for preparation for US-Japan seminar on finite element analysis of RC structures which
was held in Tokyo in May, 1985.

In JCI panel discussion on macroscopic and FEM microscopic models for

RC shear walls in January, 1988, the committee reports were presented. This panel discussion was organized by JCI Committee on Finite Element Analysis of RC Structures (chairman: Prof. Noguchi). The representative macroscopic models have been systematically used in the committee, and the FEM analytical results of stress flows and the shear forces contributed by reinforcement and concrete were compared with the macroscopic models. The research results were also presented in ASCE Structures Congress in August, 1988.

In AIJ Design Guideline for Earthquake Resistant RC Buildings Based on Ultimate Strength Concept published in 1988 [3], the design equations based on the plasticity theory as a macroscopic model were introduced in the shear design provisions for beams, columns and framed shear walls. In JCI colloquium on analytical studies on shear design of RC structures in October, 1989, many papers were presented. The main themes were the evaluation of cracked concrete, the strength and deformation of RC panels, the shear strength based on the shear resisting mechanisms of RC members, the deformations of RC members and the proposals of shear design equations. The JCI Guide Line for Application of FEM Analysis to the Design of Concrete Structures was also published, and the course for the structural designers were held in March, 1989. The 2nd US-Japan seminar on finite element analysis of RC structures is held in New York in June, 1991.

In this paper, the recent macroscopic and microscopic models for cracked concrete, RC panels and framed shear walls which have been proposed mainly in JCI colloquiums are introduced.

CRACKED CONCRETE

Since the compressive deterioration of cracked concrete was grasped quantitatively from the shear test of RC panels by Vecchio and Collins in 1982 [4], many experimental and analytical research results have been reported also in Japan. As for the compressive stress-strain of cracked concrete along cracks, Maekawa indicated that the stiffness degradation, uniaxial peak compressive stress decay and the ductility degradation under strain softening were recognized as compared with the noncracked concrete [5]. He introduced the stress-strain relationship of cracked concrete element as well as that of noncracked concrete as the formulation of fracture progress caused by a crack.

Tanabe and Yoshikawa developed the constitutive equation of a cracked RC panel using damage and reinforcement tensors [6]. The derived damage tensors and reinforcement tensors expressed rationally the tension stiffness affected by the reinforcement ratio, bond slip, deterioration of concrete in compression and crack stiffness in which shear dilatancy and shear friction were incorporated.

Sumi and Kawamata tested thirty nine RC panels subjected to in-plane pure shear in order to investigate the following factors: 1) the tension stiffening effect; 2) the compressive stress-strain relation of the diagonal concrete struts; 3) the aggregate interlock action along the cracks [7]. They obtained the failure criterion of concrete struts. The criterion indicates that the maximum compressive stress increases in the stress condition of biaxial compression and decreases in the stress condition of compression-tension. They also proposed the compressive stress-strain relationship of the concrete struts which allows for the effect of the biaxial stress condition. The proposed model 1 considering the aggregate interlock was applied to the panel specimens by Vecchio et al. and could represent their test results much better than the model 2 negrecting

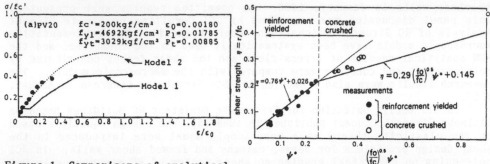

Figure 1. Comparisons of analytical
results by Sumi et al. with test
results by Vecchio et al.

Figure 2. Comparisons of analytical
results by Yoshikawa et al.
with their test results

the aggregate interlock as shown in Fig.1 [8].

Tanabe and Yoshikawa proposed a model for shear transfer across a
crack considering aggregate interlock, coupling effects (crack dilatancy
and frictional contact slip) and path-dependence [9].

Li and Maekawa proposed a contact density model in which a crack surface is idealized as being composed of stochastically distributing planes
with different inclinations and each of the constituent plane contributes
fraction of the stresses transferred across the crack surface [10]. Bujadham and Maekawa investigated the extended applicability of this model to
high strength/light weight concrete and to concrete subject to repeated
cyclic loading [11]. Watanabe proposed an aggregate interlock model considering virtual disks projected from crack surfaces [12].

STRENGTH AND DEFORMATION OF SHEAR PANELS

As for macroscopic models, Aoyagi proposed a modified Baumann's method for
the ultimate strength of RC shear panels [13]. Yoshikawa et al. evaluated
previous theories on the ultimate strength of RC panels subjected to in-
plane shear forces [14]. The calculated values by the diagonal compression
field theory proposed by Collins et al. gave good agreements with the test
results from the reinforcement yielding type to the concrete compressive
failure type. They also proposed a semi-analytical equation using the
Mohr-Coulomb criterion by Marti and the slip-free criterion by Bazant [15].
The comparisons of the proposed equation with their test results are shown
in Fig.2.

Watanabe and Muguruma tested twenty RC panels under pure shear and
indicated that the bond characteristics of reinforcing bars gave a great
influence on the compressive strength of concrete struts. He proposed an
evaluation method of the ultimate deformation of panels subjected to combined stresses on the basis of the relations between the effective compressive strength of the concrete strut and the lateral tensile strain across a
crack [16].

As for microscopic models for the strength and deformation of panels,
Noguchi and Naganuma analyzed panels parametrically for the dominant factors by FEM using a discrete crack model [17]. The shear stress-shear
strain relations of panels were compared in correspondence with each
parameter, and the formulae to evaluate the maximum shear strength were

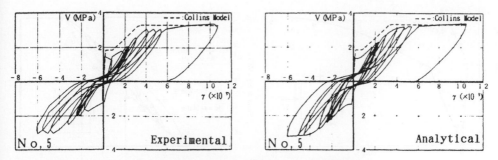

$\rho =0.85\%,\ fc=31.7MPa,\ yfsx=392MPa$

Figure 3. Comparisons of the analytical results by Kurihara et al.
with the panel test results by Saito et al.

proposed.

Aoyagi and Endoh analyzed shear transfer by FEM [18]. The analyzed specimens were push-off typed tested by Yamada. The validity and applicability of the modeling of shear transfer mechanisms using a smeared crack model were investigated.

Tsubaki introduced an approach to estimate the equivalent shear modulus using the method of equivalent elastic inclusion [19]. The shape of a through-crack in an infinite concrete slab is approximated by an elliptic cylinder of various diameter ratios. He also applied a shear transfer model based on Coulomb friction criterion to FEM and BEM and investigated the problems occurred in the application [20].

Kurihara et al. proposed a constitutive model of hysteretic behavior of cracked concrete by modifying the Collins' theory on the basis of their tests. The model is based on the hysteretic characteristics on residual stress and strain, slip stiffness, stress-strain relationships of tensile and compressive concrete and the variation of stress and strain caused by reversed cyclic shear. The cyclic behavior including the reversing loops could be well represented for previous panel tests as shown in Fig.3 [21].

Okamura et al. developed a comprehensive model for a RC element with comprising tension, compression and shear models of cracked concrete and a model for reinforcement in the concrete which are made from their basic experimental works. The proposed model predicted the panel behavior subjected to in-plane forces fairly well as shown in Fig.4 [22]. Recently Izumo and Okamura refined this model especially for unloading. The analytical results of the panels subjected to reversed cyclic loading gave a good agreement with the test results as shown in Fig.5 [23].

SHEAR STRENGTH OF FRAMED SHEAR WALLS

In JCI Committee on Finite Element Analysis of RC structures, the applicability of the four macroscopic models which were recently proposed by Shohara-Kato (1984), Minami-Murakami (1983), Shiraishi-Kanoh (1987) and Shiohara (1987) was systematically investigated through the comparisons between predicted results and analytical results by FEM or the test results [24][25][26]. All these models are based on the lower bound theory of limit analysis and consider both truss and arch mechanisms as shown in Fig.6 for Shohara-Kato model. "JCI selected specimens [27]" were used for

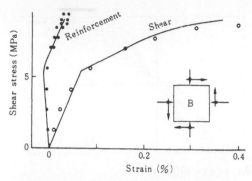

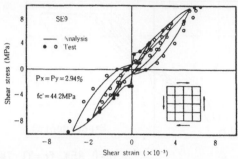

Figure 4. Comparison of analytical
results by Okamura et al. with
test results by Vecchio-Collins

Figure 5. Comparison of analytical
results by Izumo et al. with
tets results by Stevens et al.

the verification of the macroscopic models. The macroscopic models except
the Minami-Murakami model coincided with the test results better than the
conventional empirical shear strength equations. The parametric studies
were carried out using the macroscopic models for the influences of the
aspect ratio, the sectional area of columns, mean axial stress, the compressive strength of concrete and the reinforcement ratio of a wall on the
ultimate shear strength of the wall. As an example, the comparisons are
shown in Fig.7 for the influence of the mean axial stress. The shear
strength increases as the axial stress increases in each model. The axial
stress gives a larger effect in all macroscopic models than in the empirical equation. The test results coincide with the calculated results by the
macroscopic models.

    Suzuki and Inoue analyzed three-layered framed shear walls by FEM to
grasp the internal stress flow, and proposed the macroscopic model for the
ultimate strength based on the analytical results by FEM [28]. The concrete struts obtained from the FEM analytical results correspond well with
that in the macroscopic model as shown in Fig.8.

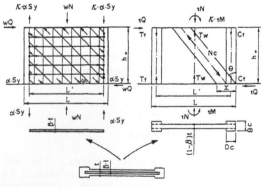

(a) Truss mechanism (b) Arch mechanism

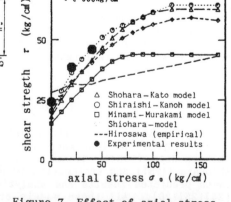

Figure 6. Shohara-Kato model

Figure 7. Effect of axial stress
on shear strength of shear walls

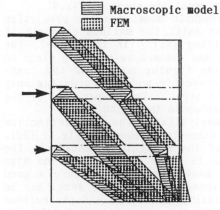

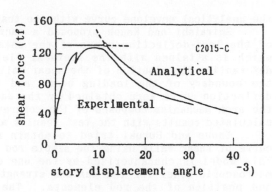

Figure 8. Comparison of concrete strut by FEM with strut by macroscopic model

Figure 9. Comparison of analytical envelope curve by macroscopic model with test result

Kabeyasawa and Kimura proposed a macroscopic model for the ultimate strength of the framed shear wall with an opening by applying the plastic theory to the side panels [29]. The analytical strength reduction factor gave a fair estimation of the observed one in their test.

STRENGTH AND DEFORMATION OF SHEAR WALLS

It has been rather difficult to predict the deformation characteristics of shear walls by the macroscopic model. But recently several researchers have tried to develop the macroscopic model to evaluate the deformation. Shiohara proposed a simple macroscopic model, in which the strain distribution on the two-dimensional plane is assumed [30]. He tried to find the rational base of the reduction factor of the effective compressive strength of concrete in the panel test and investigated the strength decay and the deformation characteristics after the maximum strength. The comparison of

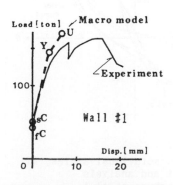

Figure 10. Comparison of calculated results with test results by Shiraishi et al.

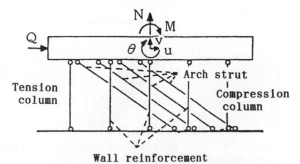

Figure 11. Rod element model by Inoue et al. when strain of arch strut is less than strain at peak compressive stress

the analytical envelope curve with the test result is shown in Fig.9.

Shiraishi and Kanoh proposed a macroscopic model for the prediction of the load-deflection curve of the shear wall, the maximum strength of which is attained with the flexural yielding [31]. In this model, the deformation distribution of the shear wall was assumed to maintain plane on the boundary of the loading beam, and the turning points of the load-deflection curve were obtained by the macroscopic model considering only the arch mechanism composed of the concrete strut. The comparisons of the calculated results with the test results are shown in Fig.10.

Inoue and Suzuki tried to obtain analytically the load-deflection curve of shear walls using the simple rod elements as shown in Fig.11 [32]. This model is characterized by the use of the analytical results by the macroscopic model for the maximum strength in order to determine the size and position of the rod elements. The analytical results gave a good agreement with the test results. They pointed out that this model has an advantage in the easier grasp of the resisting mechanisms and the balance between concrete and reinforcement than the microscopic model like FEM analysis.

There have been many contributions by FEM to the strength and defor- mation. Inoue and Noguchi introduced the applications of FEM to RC panels and shear walls in Japan in 1985 [33]. Zhang et al. analyzed multi-stories shear walls with openings by FEM with the discrete crack model for the boundary between a wall and a basement and the smeared crack model for a wall [34]. The analytical results on the load-displacement curve gave a good agreement with the test results.

Inoue et al. tried to verify macroscopic models by FEM [35]. The truss mechanism of transverse reinforcement and the arch mechanism of con- crete were separated from the analytical results of stress by using FEM. The calculated stress diagrams by FEM are compared with those calculated by macroscopic model in Fig.12. The effectiveness of lateral reinforcement

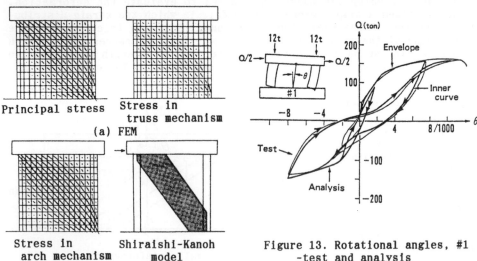

Principal stress | Stress in truss mechanism
(a) FEM

Stress in arch mechanism
(a) FEM

Shiraishi-Kanoh model
(b)Macroscopic model

Figure 12. Comparison of calculated stress diagrams by FEM with those by macroscopic model

Figure 13. Rotational angles, #1 -test and analysis (Okamura et al.)

and boundary columns was investigated.

Mikame et al. indicated the potential capacities of FEM in the parametric analyses of shear walls [36] using seven different computer programs. The effects of different reinforcement ratios in longitudinal to transverse directions, axial stress, the compressive strength of concrete, confinement by columns and openings were investigated by covering the wide range beyond the possible zone by experimental works.

Okamura and Maekawa developed the nonlinear FEM program for reinforced concrete subjected to reversed cyclic in-plane static loads [37] using the smeared crack model for RC solid elements and joint elements for discrete cracks in junction planes between members. These element models were derived form one-dimensional constitutive models concerned with tension stiffness, compression and shear transfer for cracked concrete, and from steel mechanics embedded in concrete. The ultimate capacity, deformability and hysteresis characteristics of shear walls selected by JCI Committee were fairly predicted as shown in Fig.13.

CONCLUSIONS

The recent macroscopic and microscopic models proposed in Japan for cracked concrete, RC panels and framed shear walls referring to the membrane shear behavior were reviewed.

The following items are pointed out as the future research subjects for the macroscopic models: 1) reflects of compressive deterioration of cracked concrete and bond behavior; 2) verification by investigating the internal stress flows by FEM; 3) approaches from the upper bound theory of limit analysis; 4) analysis of deformation; 5) verification and improvement of accuracy in wide range for influencing factors including high-strength concrete.

The future research subjects for the microscopic models are as follows: 1) generalization of compressive deterioration and compressive failure of cracked concrete including high-strength concrete; 2) evaluation of shear deterioration of cracked concrete subjected to reversed cyclic loading including high-strength concrete; 3) evaluation of bond-slip behavior based on the bond mechanism of deformed bars; 4) generalization of the judgment of failure modes; 5) inventions of investigating methods of the analytical results by FEM to grasp the shear resisting mechanisms and verify the macroscopic models for developing shear design equations; 6) applications of nonlinear FEM to the design of RC structures including shear walls.

REFERENCES

1. Minami, K., Limit Analysis of Shear in Reinforced Concrete Members, Proc. of JCI Colloquium on Shear Analysis of RC Structures, JCI-C4E, Sept. 1983, pp.5-24.
2. Noguchi, H. and Inoue, N, Analytical Techniques of Shear in Reinforced Concrete Structures By Finite Element Method, Proc. of JCI Colloquium on Shear Analysis of RC Structures, JCI-C4E, Sept. 1983, pp.57-96.
3. Architectural Institute of Japan, Design Guideline for Earthquake Resistant RC Buildings Based on Ultimate Strength Concept, 1988, pp.112-154.
4. Vecchio, F. and Collins, M.P., The Response of Reinforced Concrete to Inplane Shear and Normal Stress, ISBN Pub. No.82-03, Dept. of Civil Eng.,

Univ. of Toronto, March 1982.
5. Maekawa, K., Constitutive Law of Concrete based on the Elasto-Plastic and Fracture Theory, Proc. of JCI Colloquium of Shear Analysis of RC Structures, JCI-C5, Oct. 1983, pp.1-8.
6. Tanabe, T. and Yoshikawa, H., Constitutive Equations of a Cracked RC Panel, Report of IABSE Colloquium, Delft, Vol.54, 1987, pp.17-34.
7. Sumi, K. and Kawamata, S., Mechanical Characteristics of Concrete in RC Panels Subjected to In-Plane Pure Shear, Concrete Journal, JCI, Vol.26, No.10, Oct. 1988, pp.97-107.
8. Sumi, K. and Kawamata, S,. Properties of Cracked Concrete in the RC Walls Subjected to In-Plane Shear, Concrete Journal, JCI, Vol.27, No.10, Oct. 1989, pp.97-107.
9. Yoshikawa, K. and Tanabe, T., An Analytical Model for Frictional Shear Slip of Cracked Concrete, Report of IABSE Colloquium, Delft, Vol.54, 1987, pp.75-86.
10. Li, B. and Maekawa, K., Contact Density Model for Cracks in Concrete, Report of IABSE Colloquium, Delft, Vol.54, 1987, pp.51-62.
11. Bujadham, B, Li, B. and Maekawa, K., Path-Dependent Stress Transfer along Crack in Concrete, Proc. of JCI Colloquium on Shear Design Method of RC Structures, JCI-C18, Oct. 1989, pp.65-72.
12. Watanabe, F., Kono, S. and Muguruma, H., Aggregate Interlock along a Cracked Surface and Its Modeling, Proc. of JCI Annual Convention, Vol.11-2, 1989, pp.311-316.
13. Aoyagi, Y., Analytical Approach to the Design of Orthogonally Reinforced Concrete Shell Elements Subjected to In-Plane Shear, Proc. of JCI Colloquium on Shear Analysis of RC Structures, JCI-C4E, Sept. 1983, pp.111-130.
14. Yoshikawa, H., Umehara, H., and Tanabe, T., Comprehensive Evaluation of Major Theories on the Ultimate Strength of RC Panels Subjected to In-Plane Shear Forces and the Proposed Semi-Analytical Method for Estimation of Ultimate Strength, Proc. of 2nd JCI Colloquium on Shear Analysis of RC Structures, JCI-C5, Oct. 1983, pp.69-78.
15. Yoshikawa, H., Kodama, K,. and Tanabe, T., Ultimate Strength of RC Members Subjected to In-Plane Shear Stresses Estimated by Limit Analysis, Proc. of JCI Colloquium on Analytical Studies on Shear Design Method of RC Structures, JCI-C18, Oct. 1989., pp.87-94.
16. Watanabe, F. and Muguruma, H., Ultimate Strength and Deformation of RC Panels, Proc. of ASCE Structures Congress, Structural Design, Analysis and Testing, 1989, pp.31-38.
17. Noguchi, H. and Naganuma, K., Nonlinear Parametric Analysis of RC Panels, Proc. of 2nd JCI Colloquium on Shear Analysis of RC Structures, JCI-C5, Oct.1983, pp.147-154.
18. Aoyagi, Y. and Endo, T., Properties of Shear Transfer along a Crack in Reinforced Concrete Evaluated by FEM, Proc. of JCI Colloquium on Finite Element Analysis of RC Structures, JCI-C8, Dec. 1984, pp.53-60.
19. Tsubaki, T., Modeling of Shear Transfer across Cracks, Proc. of JCI Colloquium on Shear Analysis of RC Structures, JCI-C4E, Sept. 1983, pp.131-140.
20. Tsubaki, T., Analysis of Concrete Plates Subjected to In-Plane Forces - Finite Element Method and Boundary Element Method, Proc. of 2nd JCI Colloquium on Shear Analysis of RC Structures, JCI-C5, Oct. 1983, pp.29-36.
21. Kurihara, K., Ohmori, N., Takahashi, T., Tsubota, H., Inoue, N. and Watanabe, S., Analytical Hysteresis Model for Reinforced Concrete Panels Subjected to Cyclic In-Plane Shear, Jour. of Structural and Construction Engrg., AIJ, No.410, April 1990, pp.93-105.
22. Okamura, H., Maekawa, K. and Izumo, J., Reinforced Concrete Plate Element Subjected to Cyclic Loading, Report of IABSE Colloquium, Delft,

Vol.54, 1987, pp.575-590.

23. Izumo, J. and Okamura, H., Ultimate Strength and Deformation of RC Panels Subjected to In-Plane Stresses, Proc. of SCI-C, Computer Aided Analysis and Design of Concrete Structures, April 1990, pp.177-188.

24. Shiraishi, I, Shirai, N., Murakami, T. and Minami, K., Macroscopic Models for RC Shear Walls, Proc. of ASCE Structures Congress, Structural Design, Analysis and Testing, May 1989, pp.271-280.

25. Shohara, R., Shirai, N. and Noguchi, H., Verification of Macroscopic Models for RC Walls, Proc. of ASCE Structures Congress, Structural Design, Analysis and Testing, May 1989, pp.281-290.

26. Inoue, N., Shiraishi, I., and Noguchi, H., Verification of Macroscopic Models for RC Walls, Proc. of ASCE Structures Congress, Structural Design, Analysis and Testing, May 1989, pp.291-300.

27. JCI Committee on Shear Strength of RC Structures, Collected Experimental Data of Specimens for Verification of Analytical Models, Proc. of JCI Colloquium on Shear Analysis of RC Structures, JCI-C6, Oct. 1983, pp.43-53.

28. Suzuki, N. and Inoue, N., Study on the Method to Estimate Maximum Strength of Multi-Storied RC Shear Walls by Macroscopic Models, Proc. of JCI Colloquium on Analytical Studies on Shear Design Method of RC Structures, JCI-C18, Oct. 1989., pp.157-164.

29. Kabeyasawa, T. and Kimura, T., Ultimate Strength Reduction of Reinforced Concrete Shear Walls with an Opening, Trans. of the JCI, Vol.11, 1989, pp.401-408.

30. Shiohara, H., Simple Estimation of Load-Deflection Relation of RC Shear Wall Considering Reduction of Effective Compressive Strength of Concrete, Proc. of JCI Colloquium on Analytical Studies on Shear Design of RC Structures, JCI-C18, Oct. 1989, pp.1-8.

31. Shiraishi, I. and Kanoh, Y., Analytical Study on Load-Deflection Relationship of Reinforced Concrete Shear Walls, Proc. of JCI Colloquium on Analytical Studies on Shear Design of RC Structures, JCI-C18, Oct. 1989, pp.171-178.

32. Inoue, N. and Suzuki, N., Rod Element Model for Analyzing Displacement of Reinforced Concrete Shear Walls, Proc. of JCI Colloquium on Analytical Studies on Shear Design of RC Structures, JCI-C18, Oct. 1989, pp.179-186.

33. Inoue, N. and Noguchi, H., Finite Element Analysis of Reinforced Concrete in Japan, Proc. of US-Japan Seminar, Finite Element Analysis of RC Structures, ASCE, 1986, pp.25-46.

34. Zhang, A., Okabe, T., Tsuda, K. and Noguchi, H., Nonlinear Analysis of Main Structural Frames in a RC Multi-Stories and Multiple-Unit Dwelling, Trans. of JCI, Vol.10, 1988, pp.305-312.

35. Inoue, N., Shiraishi, I. and Noguchi, H., Verification of Macroscopic Models for RC Walls, Proc. of ASCE Structures Congress, Structural Design, Analysis and Testing, May 1989, pp.291-310.

36. Mikame, A., Yoshikawa, H., Kamiyama, Y., Iizuka, S., Sato, K., Kawasaki, K. and Noguchi, H., Parametric Analyses of RC Shear Walls by FEM, Proc. of ASCE Structures Congress, Structural Design, Analysis and Testing, May 1989, pp.301-310.

37. Okamura, H. and Maekawa, K., Non-Linear Analysis and Constitutive Models of Reinforced Concrete, Proc. of SCI-C, Computer Aided Analysis and Design of Concrete Structures, Pineridge Press, April 1990, pp.831-850.

AN ANALYTICAL MODEL FOR RC PANELS SUBJECTED TO IN-PLANE STRESSES

JUNICHI IZUMO
Department of Civil Engineering, Kanto Gakuin University, 4836 Mutsuura-chyo,
Kanazawa-ku, Yokohama, 236, Japan

HYUNMOCK SHIN
Department of Civil Engineering, Sung Kyun Kwan University, 53, 3-ka, Myungryun-dong,
Chongro-ku, Seoul, 110-745, Korea

KOICHI MAEKAWA
Asian Institute Technology, G.P.O., Box 2754, Bangkok, 10501, Thailand

HAJIME OKAMURA
Department of Civil Engineering, The University of Tokyo, 7-3-1, Hongo, Bunkyo-ku,
Tokyo, 113, Japan

ABSTRACT

This study aims at the development of the analytical model for reinforced concrete (RC) panels subjected to in-plane stresses including not only monotonic loading but also reversed cyclic loading. The analytical model for RC panels has been established by the constitutive models for cracked concrete and reinforcement. Further, through verification with the tested specimens, it is confirmed that the proposed model can predict the behavior of the RC panel under reversed cyclic loading.

NOTATION

ε_x	: average strain in x-direction
ε_y	: average strain in y-direction
γ_{xy}	: average shear strain with respect to x, y axes
ε_{tu}	: strain; two times of cracking strain.
ε_{co}	: average strain at crack contacting
ε_{cpc}	: average strain at crack perfectly contacting
ε_{xmax}	: maximum average tensile strain experienced
ε'_{ymax}	: maximum compressive strain experienced
ε'_{yp}	: compressive plastic strain
ε_s	: average strain of reinforcement
ε_{smax}	: maximum average strain of reinforcement experienced
ε_b	: strain given by subtracting the plastic strain of reinforcement from ε_{smax}
σ_{cx}	: average tensile stress of concrete in x-direction
σ'_{cy}	: average compressive stress of concrete in y-direction
τ_{cxy}	: average shear stress with respect to x, y axes

σ'_c : compressive stress generated by shear transfer along the cracked plane
σ_{cc} : average stress generated by the crack surface contact
σ_{cb} : average stress generated by the bond action.
σ_{cbmax} : average stress generated by the bond action corresponding to ε_{xmax}
σ'_{cymax} : compressive stress corresponding to ε'_{ymax}
σ_s : average stress of reinforcement
σ_{scr} : stress at the cracked plane
σ_{smax} : average stress corresponding to ε_{smax}
ω : reduction coefficient of fracture parameter or crack width
ω_θ : displacement in x-direction
$\omega_{\theta p}$: plastic displacement of ω_θ
δ : shear displacement along the cracked plane
σ_1 : principal stress in concrete
σ_2 : principal stress in concrete ($\sigma_2 \le \sigma_1$)
f_t : tensile strength of concrete (MPa)
f_{tb} : tensile strength under biaxial stress condition (MPa)
f'_c : cylinder compressive strength of concrete (MPa)
f_y : yield point of reinforcement
E_c : elastic modulus of concrete
E_{cc} : stiffness of concrete at the crack contacting state
E_s : elastic modulus of reinforcement
K : fracture parameter
c : parameter that describes the bond characteristics
θ : angle between X-direction and x-direction, or
 : direction that contact stress acts on the cracked plane
$\{\sigma_{RC}\}$: reinforced concrete stress matrix
$\{\sigma_c\}$: concrete stress matrix
$\{\sigma_s\}$: reinforcement stress matrix
$[T]$: coordinate transformation matrix
The sign (') means that the compression is plus.

INTRODUCTION

Finite Element Method (FEM) is considered very effective as an analytical method for reinforced concrete. However, in order to get sufficient accuracy in tracing the response of RC members and structures, FEM requires an analytical model that can accurately describe the behavior of reinforced concrete elements that compose the structures. Moreover, in the earthquake design, it is important to evaluate the earthquake resistance of reinforced concrete under cyclic loading. Therefore, it is necessary to develop the analytical model that can predict the response of reinforced concrete under cyclic loading at the range from its cracking to failure.

This study aims at the development of the analytical model for RC panels subjected to reversed cyclic in-plane stresses. Through comparison of the calculated results with the experimental data, the proposed model for RC panels has been evaluated and discussed.

ANALYTICAL MODEL FOR RC PANELS

This study treats RC panels subjected to in-plane stresses, which have reinforcing bars distributively arranged and where smeared cracks will be generated. Therefore, the constitutive equations are always expressed by the average stress and average strain. As the stresses of concrete and those of reinforcements should resist the stresses that act on the RC panel, the stresses of the RC panel is expressed as follows:

$$\{\sigma_{RC}\} = [T(\theta)]\{\sigma_c\} + \{\sigma_s\} \tag{1}$$

where, θ is the inclination of the direction orthogonally crossing the cracked plane against the X-axis (Fig. 1).

It is assumed that the shear resistance of the reinforcement is so small to be ignored when it is compared with that of concrete. In order to describe the stresses of cracked concrete, the stresses are given with respect to the coordinate formulated by the x-axis parallel to cracked plane and the y-axis orthogonally crossing it (Fig. 1).

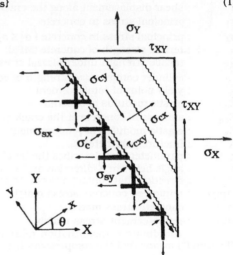

Figure 1. Stresses in RC panel

Criteria for Cracking

After cracking, concrete becomes anisotropic. The cracking criteria of the RC panel under biaxial stress condition follow the Niwa's equation [1] for compression-tension and the Yamada's equation [2] for tension-tension.

$$f_{tb} = f_t \sqrt[3]{1 - (\sigma_2/f_c)} \qquad (\sigma_2 \leq 0) \tag{2.1}$$

$$f_{tb} = f_t \left\{ 1 - 0.3 \, (\sigma_2/\sigma_1)^2 \right\} \qquad (\sigma_2 > 0) \tag{2.2}$$

Tension Stiffening Model

Concrete in the direction orthogonally crossing the cracked plane has the tensile stress after cracking. To express the tension stiffening effect, the Okamura and Maekawa model [3] for envelope curve and the Shima model [4] for unloading and reloading curve have been adopted. In the Shima model obtained from the his experimental results, the stress acting on concrete has been assumed to consist in the stress transferred from reinforcement by the bond action and transferred by the contact of the cracked plane and these have been individually modeled. The tension stiffening is expressed as the following equations.

$$\sigma_{cx} = f_{tb} \left(\varepsilon_{tu}/\varepsilon_x \right)^c \qquad \text{for envelope curve} \qquad (3)$$

$$\sigma_{cx} = \sigma_{cc} + \sigma_{cb} \qquad \text{for unloading and reloading curve} \qquad (4)$$

where,

$$\sigma_{cc} = E_{cc} \left(\varepsilon_x - \varepsilon_{cc} \right) \le 0 \qquad (\varepsilon_{cpc} \le \varepsilon_x \le \varepsilon_{cc}) \qquad \text{for unloading} \quad (5.1)$$

$$\sigma_{cc} = E_c \varepsilon_x \qquad (\varepsilon_x \le \varepsilon_{cpc}) \qquad \text{for unloading} \quad (5.2)$$

$$\sigma_{cc} = E_c \varepsilon_x \le 0 \qquad (\varepsilon_x \le 0) \qquad \text{for reloading} \quad (5.3)$$

$$\sigma_{cb} = \frac{\sigma_{cbmax} + 0.0016 E_c \varepsilon_{xmax}}{\varepsilon_{xmax}^2} \varepsilon_x^2 - 0.0016 E_c \varepsilon_{xmax} \qquad \text{for unloading} \quad (6.1)$$

$$\sigma_{cb} = \frac{\sigma_{cbmax} + 0.0016 E_c \varepsilon_{xmax}}{\varepsilon_{xmax}} \varepsilon_x - 0.0016 E_c \varepsilon_{xmax} \qquad \text{for reloading} \quad (6.2)$$

The parameter, c in Eq. (3) represents the bond characteristics. The value of the parameter is appropriate as 0.4 for ordinary deformed bars and 0.2 for welded meshes that Vecchio and Collins employed in their test specimens. The tension stiffening model that has been adopted in the model for RC panels is illustrated in Fig. 2.

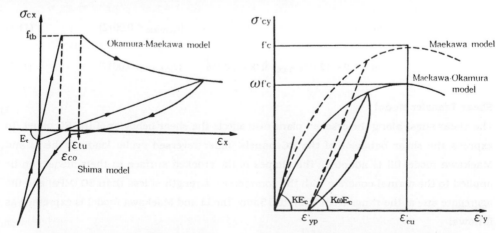

Figure 2. Tension stiffening model

Figure 3. Cracked concrete model for
compression

Cracked Concrete Model for Compression

The compressive stiffness of cracked concrete in the direction parallel to the cracked plane has been experimentally proved to be smaller than that of uncracked concrete [5][6]. This phenomenon may be attributed mainly to the stress relaxation that takes place in the vicinity of cracks and it could be explained by reducing the fracture parameter proposed by Maekawa [7]. To describe the deteriorated stiffness of cracked concrete, the reduction coefficient, ω has been introduced in the Maekawa model. Namely, from the observation of the experimental data [5][6], ω has been determined to be set at 1.0 for the tensile strain less than 0.12%, constant at 0.6% for the tensile strain of over 0.44% and change linearly between the strains of 0.12% and 0.44%. On the other hand, the compressive stiffness becomes constant, as the fracture parameter and the plastic strain are assumed not to change upon unloading and reloading. However, the unloading curve generally has a tendency to get convex toward the bottom. In order to give enough convex to the analysis, the circular arc has been used as an analytical model for unloading. Therefore, the compressive stress of cracked concrete is shown in Fig. 3 and is expressed as follows:

$$\sigma'_{cy} = f'_c K \omega E_c \left\{ (\varepsilon'_y/\varepsilon'_{cu}) - (\varepsilon'_{yp}/\varepsilon'_{cu}) \right\} \quad \text{for envelope and reloading} \quad (7)$$

$$\sigma'_{cy} = \sigma'_{cymax} \left\{ 1 - \sqrt{\frac{\varepsilon'_{ymax} - \varepsilon'_y}{\varepsilon'_{ymax} - \varepsilon'_{yp}}} \right\} \quad \text{for unloading} \quad (8)$$

where,

$$K = \exp[-0.73(\varepsilon'_{ymax}/\varepsilon'_{cu})\{1-\exp(-1.25\varepsilon'_{ymax}/\varepsilon'_{cu})\}] \quad (9)$$

$$\varepsilon'_{yp} = \varepsilon'_{ymax} - 2.86\varepsilon'_{cu}\{1-\exp(-0.35\varepsilon'_{ymax}/\varepsilon'_{cu})\} \quad (10)$$

$$\omega = 1.0 \quad (\varepsilon_{xmax} \leq 0.0012) \quad (11.1)$$

$$\omega = 1.0 - 125(\varepsilon_{xmax} - 0.0012) \geq 0.6 \quad (\varepsilon_{xmax} > 0.0012) \quad (11.2)$$

Shear Transfer Model

The shear stress along the cracked plane also affects the shear behavior of the RC panel. To express the shear behavior of the RC panels under reversed cyclic loading, the Li and Maekawa model [8] is adopted. The shapes of the cracked surface in their model can be applied to the normal concrete with the compressive strength of less than 50 (MPa) and the aggregate size at the range from 5mm to 15mm. The Li and Maekawa model is expressed as follows:

$$\tau_{cxy} = \int_{-\pi/2}^{\pi/2} 4400f'_c{}^{1/3}(\omega_\theta - \omega_{\theta p})\sin\theta\cos\theta d\theta \quad (12)$$

$$\sigma'_c = \int_{-\pi/2}^{\pi/2} 4400f_c^{1/3}(\omega_\theta - \omega_{\theta p})\sin^2\theta d\theta \tag{13}$$

where,
$$\omega_\theta = -\delta\sin\theta + \omega\cos\theta \tag{14}$$

As the Li and Maekawa model has been formulated by focusing on the single crack, the average crack spacing is needed to express δ and ω with the average strain. Fortunately the effect of the average crack spacing on the analysis is too small to be considered in this model. The compressive stress along the cracked plane should be generated due to the shear displacement and this is also formulated in the Li and Maekawa model. As a result, the stress in the reinforcing bars increases and the yielding is accelerated. This is taken into consideration for the constitutive equation of reinforcement.

Constitutive Model for Reinforcement
As the stress of the reinforcement between cracks is not uniform due to the bond action, the yielding of the reinforcement will occur when the stress of the reinforcing bars at the cracked plane reaches the yield point. Further, the stress of the yielded reinforcement rapidly reaches the strain hardening because the bond action is expected to exist even after yielding. The bond effect should be taken into consideration when the constitutive equation for reinforcement is formulated.

The constitutive model for the reinforcement is derived from the assumption that stress distribution of the reinforcement between cracks is expressed by a cosine function [9]. Shima [4] reported that the constitutive model for the reinforcement upon reversed cyclic loading proposed by Kato [10] can be used for the average stress and strain relationship of the reinforcing bars in concrete. The constitutive equation for reinforcement is expressed as follows:

$$\sigma_s = E_s\varepsilon_s \qquad (\sigma_{scr} < f_y) \qquad \text{for envelope before yielding} \tag{15.1}$$

$$\sigma_s = f(\varepsilon_s) \qquad (\sigma_{scr} \geq f_y) \qquad \text{for envelope after yielding} \tag{15.2}$$

$$\sigma_s = E_s(\varepsilon_s - \varepsilon_{smax}) + \sigma_{smax} \qquad \text{for cyclic loading on the straight line} \tag{15.3}$$

$$\sigma_s = \frac{-a(a+1)E_s\sigma_{smax}\varepsilon_{smax}}{\varepsilon_s - \varepsilon_{smax}(a+1)} - a\sigma_{smax} \qquad \text{for cyclic loading on the curved line} \tag{15.4}$$

where,
$$a = E_s/\{E_s - (\sigma_{smax}/\varepsilon_b)\} \tag{16}$$

The stress and strain relationship of the reinforcement used in the model is shown in Fig. 4.

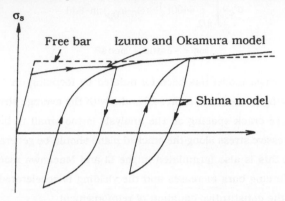

Figure 4. Model for reinforcement under reversed cyclic loading

ANALYTICAL RESULTS AND REMARKS

As the obtained constitutive equations for RC panels results in the three-element non-linear simultaneous equations, the strains corresponding to the stresses of the RC panel are obtained by using the Newton-Raphson's method. In the case of the reversed cyclic loading, the dominant cracks will occur in two directions of the RC panel. The concrete stresses in the RC panel are always given in regard to the coordinate that contains the larger tensile strain in the x-axis. Under ordinary loading condition, when cracks in one direction open, cracks in the other direction close. Under lower stress condition upon unloading and reloading, cracks in the both directions might open. However, in this case, as the loading level is comparatively low, the neglect of one crack does little affect the analytical results.

Envelopes

The calculated envelopes of the RC panel have been evaluated by the 17 panels from the Vecchio and Collins' experiments [5], those of which investigators judged improper were excluded and the 5 panels from the Aoyagi and Yamada's experiments [2]. In the analysis, the tensile strengths that were good accordance with the experimental values were used. The analytical result is quite satisfactory as one of the example is shown in Fig. 5. Similarly, the good agreements have been obtained from the other experimental results. From the Vecchio and Collins' experiments, the average ratios of the observed shear to the calculated ones corresponding to the shear strains, 0.1%, 0.2%, 0.4% are 0.99, 1.02 and 1.03, where the coefficients of variation are 4.9%, 6.7% and 4.2%, respectively. From the Aoyagi and

Yamada's experiments, the average ratios of the observed tensile stress to the calculated ones corresponding to the tensile strains, 0.1% and 0.2% are 1.03 and 1.06, where the coefficients of variation are 5.0% and 5.0%, respectively.

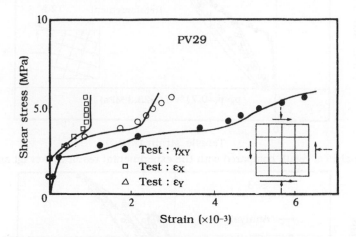

Figure 5. Analytical result compared with the experimental result by Vecchio and Collins

Reversed Cyclic Loading

The analytical results under reversed cyclic loading have been compared with the experimental data done by Aoyagi and Yamada [4], by Yoshikawa et al. [11] and by Stevens and Collins [12].

The analytical result of the Aoyagi and Yamada's specimen under cyclic loading faithfully describes the test result as one of the examples is shown in Fig. 6. The analytical result describes well the test result of the Yoshikawa's specimen that was the hollowed reinforced concrete cylinder subjected to pure torsion as shown in Fig. 7. Moreover, the analytical result of the Stevens and Collins' specimen with the reinforcing bars anisotropically arranged, where the reversed cyclic shear and compression acted, is judge to be good agreement with the test result as one of the examples is shown in Fig. 8.

As far as the verification has been done, the proposed model for RC panels has been confirmed to be effective for the analysis of the RC panels subjected to in-plane stresses including not only monotonic loading but also reversed cyclic loading. Further, in this study, it is considered that the constitutive model for the materials of concrete and reinforcement considerably affect the characteristics of the restoring force of the RC panels. However, the effects of the loading speed or the loading time on the materials should be further studied.

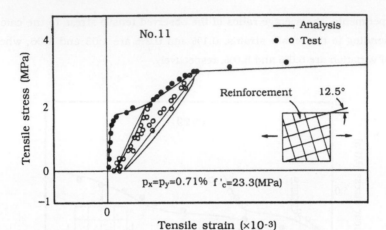

Figure 6. Analytical result compared with the experimental result by Aoyagi and Yamada

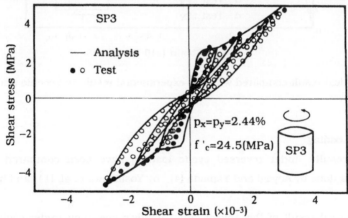

Figure 7. An analytical result compared with the experimental result by Yoshikawa et al.

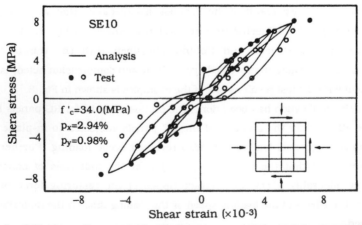

Figure 8. Analytical result compared with the experimental result by Stevens and Collins

CONCLUSIONS

The analytical model for RC panels has been developed with the existing constitutive equations for cracked concrete and reinforcement. The proposed model has been verified through comparison with the experimental results of the RC panels. As far as verification has been done, the proposed model can predict the behavior of the RC panel subjected to in-plane stresses including not only monotonic loading but reversed cyclic loading fairly well.

REFFERENCES

1. Niwa, J, Maekawa, K. and Okamura, H., Nonlinear Finite Element Analysis of Deep Beams, Final Report IABSE Colloquium, Delft, 1981, pp. 625-638

2. Aoyagi, Y. and Yamada, K., Strength and Deformation Characteristics of Reinforced Concrete Shell Elements Subjected to In-plane Forces, Proc., of JSCE, No.331, Mar., 1983, pp. 167-190, Concrete Library International, JSCE, No.4, Dec., pp. 129-160

3. Okamura, H., Maekawa, K. and Sivasubramaniyam, S., Verification of Modelling for Reinforced Concrete Finite Element, Finite Element Analysis of Reinforced Structures, Proc. of Seminar ASCE, 1985, pp.528-543

4. Shima, H., Chou, L. and Okamura, H., Micro and Macro Models for Bond Behavior in Reinforced Concrete, Journal of the Faculty of Engineering, The University of Tokyo (B), Vol.39, No.2, 1987, pp. 133-194

5. Vecchio, F., J. and Collins, M., P., Response of Reinforced Concrete to In-plane Shear and Normal Stresses, Publications, No.82-03, Department of Civil Engineering, University of Toronto, Mar., 1982

6. Miyahara, T., Kawakami, T. and Maekawa, K., Nonlinear Behavior of Cracked Concrete Plate Element under Uniaxial Compression, Proc. of JSCE, No.378, Aug., 1987, pp. 249-258, Concrete Library International, JSCE, No.11, June, 1988, pp. 131-144

7. Maekawa, K. and Okamura, H., The Deformational Behavior and Constitutive Equation of Concrete using the Elasto-plastic and Fracture Model, Journal of the Faculty of Engineering, The University of Tokyo (B), Vol.37, No.2, 1983, pp.253-328

8. Li , B. and Maekawa, K. and Okamura, H., Contact Density Models for Stress Transfer across Cracks in Concrete, Journal of the Faculty of Engineering, The University of Tokyo (B), Vol.40, No.1, 1989, pp. 9-52

9. Izumo, J., A Parametric Study of the RC panel ductility by using the Analytical Model, Concrete Shear in Earthquake, 1991

10. Kato, B., Mechanical Properties of Steel under Load Cycles Idealizing Seismic Action, Bulletin D'information, No.131, CEB, IABSE-CEB Symposium, May, 1979, pp.7-27

11. Yoshikawa, H. et al., Study on Shear Behavior of Concrete Container Vessel, Hazama Corporation Annual Report, 1983 (in Japanese)

12. Stevens, N., J., Uzumeri, S., M. and Collins, M., P., Analytical Modelling of Reinforced Concrete Subjected to Monotonic and Reversed Loading, University of Toronto, Dec., 1986

COMPARISON OF FIXED AND ROTATING CRACK MODELS IN THE ANALYSIS OF PANELS, PLATES, AND SHELLS SUBJECTED TO SHEAR

JOHANN KOLLEGGER
Ingenieurbüro Mehlhorn, Kassel, Germany

ABSTRACT

A simple material model for concrete is used in order to facilitate the comparison of fixed and rotating crack models. Based on comprehensive experimental results obtained from tests carried out at the University of Kassel and at the University of Toronto an effective concrete strength of 80% of the cylinder crushing strength is assumed in the analysis of panels. For the analysis of plates and shells the effective concrete strength is further reduced to 70% of the cylinder crushing strength.

Results from comparison analyses on panels subjected to shear, plates subjected to twisting moment, and shell elements subjected to shear and bending moment are given in the paper. Considerable differences between results from fixed and rotating crack models may occur in the analysis of panels. The effect of the crack re-orientation is less pronounced in the analysis of plate and shell elements.

INTRODUCTION

The application of the finite element method to reinforced concrete structures started out with models which tried to represent each individual crack [1]. However, in the analysis of reinforced concrete surface structures - panels, plates, and shells - it is advantageous not to model the propagation of each single crack but to use stress-strain characteristics for a representative volume element of reinforced concrete. This treatment of cracked reinforced concrete as a continuum is called "smeared crack technique" in the literature [2,3].

In the analysis of plane reinforced concrete elements with smeared cracks it usually was assumed that the direction of a crack was fixed once it had occurred. A second crack at an integration point was permitted to occur in the orthogonal direction only. Since analyses employing fixed-orthogonal crack models may overestimate experimental failure loads, non-orthogonal crack models were introduced. In the non-orthogonal crack models, for example those described in [4,5], the second crack at an integration point may attain any direction between a certain threshhold angle and 90 degrees. While better correlation with experimental data was achieved with these models, the numerical stability of these models is sometimes unreliable.

Another approach of modelling cracked reinforced concrete was the introduction of the swinging crack model by Cope et al. [6] and the rotating crack model by Akbar and Gupta [7]. The basic assumption of the swinging or rotating crack model is that the direction of the principal tensile strain is orthogonal to the principal compressive stress of concrete. The direction of the principal tensile strain calculated by the swinging or rotating crack model turns out to be the same as the direction obtained from an equation derived by Baumann [8]. Baumann derived his expression for the principal tensile strain direction using the principle of the mimimum of the internal work following ideas published by Kupfer [9].

Results obtained from the fixed and the rotating crack model will be compared to experimental data of panel, plate, and shell tests in the paper. In order not to diffuse the results from the different models, only experiments with uniform loading, e.g. pure shear or pure twisting moment, were selected for the comparison analyses.

LAYERED SHELL ELEMENT

Since only test specimens with uniform loading will be analyzed, it is not necessary to use finite elements for the analyses. It is sufficient to work with a layered element (Fig. 1) with a linear strain variation through the thickness and a state of plane stress in each layer. The element is divided into concrete layers and reinforcement layers. The strain and stress state within each layer is uniform. Knowing the strains and curvatures of the middle surface the strain in each layer may be calculated using the hypothesis of Bernoulli.

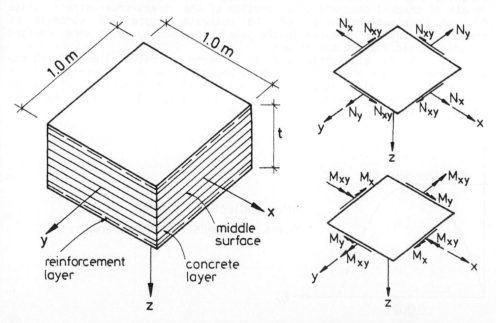

Figure 1. Layered shell element (left), applied forces and moments (right)

Concrete Layer

Under loading three states - uncracked, cracks in one direction, cracks in two directions - are possible in each concrete layer as is shown in Fig. 2, where axes 1 and 2 refer to the crack-oriented coordinate system.

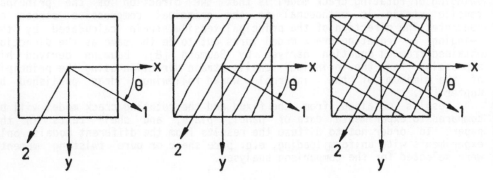

Figure 2. Possible states of a concrete layer: Uncracked, cracks in one direction, cracks in two directions

A simple material model is used for the concrete. The uniaxial stress-strain diagram is shown in Fig. 3. The compressive stress varies according to a quadratic parabola as a function of the compressive strain. The peak strength of concrete is assumed to occur at a strain of 0.0025.

Based on comprehensive experimental results obtained from tests on reinforced concrete panels carried out at the University of Kassel and at the University of Toronto it has been suggested to evaluate the compressive strength of cracked concrete as a function of the transverse stress state [10]. A maximum reduction of the concrete compressive strength of approximately 20% was measured in the panel experiments, which were carried out under static loading conditions.

While it is straightforward to program a material model with the compressive strength of cracked concrete depending on the transverse stress state [11], it will be assumed here that the effective strength of cracked reinforced concrete is reduced to 80% of the uniaxial strength for the analysis of panels.

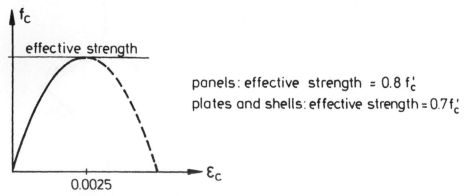

Figure 3. Uniaxial stress-strain diagram of concrete

The effective concrete strength for the analysis of plate and shell elements is further reduced to 70% of the cylinder crushing strength because the top concrete is usually weaker than the bottom concrete in plate specimens cast in a horizontal position. While the effect of the varying concrete strength through the thickness is somehow averaged in panel tests, the weaker top concrete may become the compression zone in a plate test.

Tensile strength of concrete is set to zero and tension stiffening is also neglected in order to simplify the analysis and to facilitate the comparison of fixed and rotating crack models. The increase of the concrete strength under biaxial compression will not be relevant in any of the analyzed examples and is therefore not considered. Uncracked concrete under biaxial compression is analyzed with an orthotropic material model using the stress strain diagram of Fig.3 in each principal strain direction and a shear modulus based on the tangent moduli in the principal strain directions.

Fixed crack model: In the fixed crack model the cracking direction will be fixed to the direction of the first crack, which will form in the direction normal to the major principal stress in concrete as soon as this stress becomes positive. A second crack in a layer is restricted to open in the orthogonal direction. The shear mdulus of cracked concrete is set to 10% of the initial shear modulus of cracked concrete.

Rotating crack model: Progressive cracking with cracks forming in a new direction will cause a shift in the principal tensile strain direction of reinforced concrete. In the rotating crack model this shift of the principal tensile strain direction is reproduced and not an actual rotation of the cracks, as the name of the model might indicate. Initially the crack forms in a direction normal to the principal tensile stress in concrete. Upon increased loading the principal tensile strain direction is adjusted according to the principle of the minimum of the internal energy. The differences in the programming of the fixed and rotating crack models have been discussed in [11].

Reinforcement Layer
A reinforcement layer is described by the area of reinforcement, the angle between x-axis and the reinforcing direction, and the distance of the reinforcement layer to the middle surface of the shell element (Fig. 4). An elastic-plastic stress-strain relationship with a hardening modulus is used for tensile and compressive strains as is shown in Fig. 4.

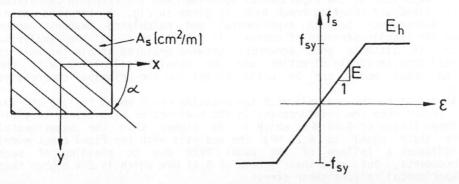

Figure 4. Reinforcement layer and material model for the reinforcement

Numerical Algorithm

The layered shell element of Fig. 1 may be subjected to axial forces in the x- and y-directions, a shear force, bending moments in the x- and y-directions, and a twisting moment. The strains in the middle surface and the curvatures of the layered shell element are calculated for any set of applied external forces and moments in an iterative procedure.

Starting with linear elastic material properties in the first iteration step, strains and curvatures are calculated. The strains in the individual layers are then determined, and using the material models for the concrete and the reinforcement described above the internal stresses of each layer are calculated. Integrating the internal stresses over the thickness of the element yields the internal forces and moments. Unbalanced forces and moments are calculated as the difference of the externally applied and the internal forces and moments of the element. The unbalanced forces and moments are applied to the layered shell element in the next iteration step. This procedure is repeated until the unbalanced member forces and the differences of the current strains and curvatures with respect to the ones of the last iteration step are smaller than required convergence limits. Then the external member forces of the next load step are applied to the layered shell element and the procedure described above is repeated starting with the element stiffness of the previous load step.

ANALYSIS OF PANELS SUBJECTED TO SHEAR

In 1982 Vecchio and Collins published experimental results on the reduction of the compressive strength of cracked reinforced concrete [12]. The experiments were carried out with a sophisticated test set-up which permitted any combination of shear and axial stresses to be applied to the specimen. The loads were introduced into the panel with the aid of "shear keys", pins, and anchor plates. The panel dimensions were 890 mm times 890 mm and the thickness was 70 mm. The reinforcement consisted of two orthogonal layers of welded wire mesh.

While a comparison analysis of the complete test series by Vecchio and Collins [12] is contained in [10], only the panels subjected to pure shear with a reinforcement ratio of 1.79% in the x-direction and different reinforcement ratios in the y-direction will be considered here. The material properties of this sub-series consisting of six panels are given in Table 1. Also included in Table 1 are the material properties of a panel tested by Andre and Collins [13] on the same test set-up.

A comparison of the experimental behaviour and the response calculated with the fixed and rotating crack model is given in Fig. 5 for the panel PV20. Differences between experimental and calculated response occur because the tensile strength of concrete is set to zero in the analysis. However, at ultimate good agreement between measured strains as well as principal tensile strain direction and the results calculated with the rotating crack model can be noted, except for the principal compressive strain.

According to the analysis with the rotating crack model failure of the panel occurs when the reinforcement in the x-direction also starts to yield at a shear stress of 4.64 MPa, which is 9% higher than the experimental failure shear equal to 4.26 MPa. The analysis with the fixed crack model also indicates a failure of the panel PV20 due to yielding of both reinforcements, but at a shear stress of 5.41 MPa which is 27% higher than the experimental failure shear stress.

TABLE 1
Material properties of panel tests [12,13], comparison of experimental
and calculated ultimate shear stresses and principal strain directions

	x-reinf.		y-reinf.		ultimate shear [MPa]			ultimate theta [°]	
test	fc [MPa]	μ [%]	fy [MPa]	μ [%]	fy [MPa]	experi- ment	rotating cracks	fixed cracks	experi- ment	rotating cracks
PV12	16.0	1.79	469	0.45	269	3.13	3.16	4.78	65.1	69.2
PV19	19.0	1.79	458	0.71	299	3.95	4.17	5.15	62.7	62.8
PV20	19.6	1.79	460	0.89	297	4.26	4.64	5.41	59.7	60.5
PV21	19.5	1.79	458	1.30	302	5.03	5.63	6.04	57.6	55.3
PV22	19.6	1.79	458	1.52	420	6.07	7.23	7.28	47.9	48.5
PV27	20.5	1.79	442	1.79	442	6.35	7.89	7.89	45.1	45.0
TP4A	24.9	2.04	453	2.04	453	8.79	9.23	9.23	45.0	45.0

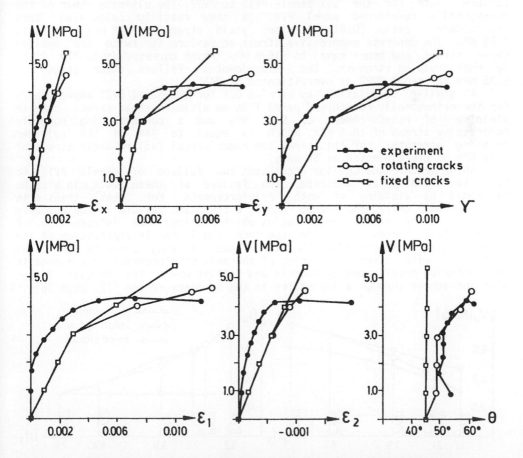

Figure 5. Comparison of experimental results of panel PV20 [12] with
analytical results from rotating and fixed crack model

The ultimate shear stresses and the inclinations of the principal tensile strain direction are given in Table 1 for the six panels tested by Vecchio and Collins [12] and for the panel tested by Andre and Collins [13]. The comparison of ultimate shears and the inclinations of the principal tensile strain is also drawn Fig. 6 for the Vecchio and Collins tests.

The difference in the ultimate shear calculated by the rotating and the fixed crack model is large for weak reinforcement, e.g. 3.16 MPa versus 4.78 MPa for panel PV12. The difference in the ultimate shear calculated by the two models becomes smaller for increasing reinforcement ratio of the y-reinforcement. Identical results are obtained with the fixed and rotating crack model for equal reinforcement in both directions, e.g. panel PV27.

The angle between the normal to the cracks and the x-axis is 45° for all panels analyzed with the fixed crack model. Using the rotating crack model this angle changes from 69.2° for PV12 to 45° for PV27 (Fig. 6).

While the correlation of the principal strain direction of the experiments and the analysis with the rotating crack model is quite good for the six analyzed panels, the difference in the ultimate shear stress becomes larger for increasing transverse reinforcement ratios.

The analysis indicates that yielding of both reinforcements is the failure cause for the six panels PV12 to PV27. The ultimate shear of the orthogonally reinforced panel PV27 is then readilly calculated from reinforcement ratio (0.01785) times yield strength (442 MPa) equal to 7.89 MPa. The concrete compressive stress at failure is twice the applied shear stress and hence equal to 15.8 MPa, which corresponds to 77% of the cylinder crushing strength. The experimental failure shear stress of 6.35 MPa for panel PV27 is overestimated by 24%.

A similar calculation as carried out for the panel PV27 above yields for the orthogonally reinforced panel TP4A an ultimate shear stress due to yielding of reinforcement of 9.23 MPa and a corresponding concrete compressive stress of 18.5 MPa, which is equal to 74% of the cylinder crushing strength. For this panel the experimental failure shear stress of 8.79 MPa is overestimated by 5% only.

While Vecchio and Collins attributed the failure of panels PV12 to PV27 to crushing of concrete, the failure of these tests can also be explained by yielding of both reinforcements for weak transverse reinforcement (PV12 and PV19) and by a premature failure in the boundary region for larger reinforcement ratios which required the introduction of larger shear forces into the specimen. The latter interpretation of the failure cause is supported by the higher load carrying capacity of panel TP4A [13] with direct anchoring of the main reinforcement as compared to the anchoring detail used by Vecchio and Collins where the tensile forces were introduced through a lap splice in the boundary region [12, page 36].

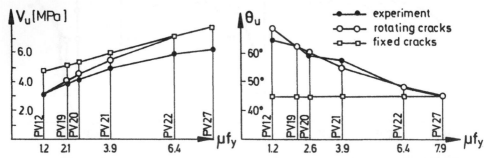

Figure 6. Ultimate shear (left) and inclination of crack direction (right) as a function of ratio times yield strength of y-reinforcement

ANALYSIS OF PLATES SUBJECTED TO TWISTING MOMENT

The plates ML1 to ML9 were tested by Marti et al. as part of a research program on reinforced concrete plates subjected to torsion [14]. The orthogonally reinforced slab elements (1700 mm times 1700 mm, thickness of 200 mm) were subjected to two upward and two downward acting corner forces thus producing a state of pure twist in the interior of the test specimens. The material properties as well as the experimental and calculated ultimate twisting moments are shown in Table 2.

For the five specimens with equal reinforcement ratios in both directions there was no re-orientation of the cracking direction and therefore the same results are calculated using the rotating and the fixed crack model. Yielding of reinforcement was the failure cause of plates ML1 and ML7 while the other three plates failed due to crushing of concrete. Some of the plates of the test series experienced a premature failure according to [14]. The analytical results indicate that the failure of plates due to yielding of reinforcement or crushing of concrete must have been imminent when the edge failure occured. In the case of plate ML3 the analysis predicts a failure at 89 kNm/m whereas an edge failure is reported at 93 kNm/m.

Of the four plates with weaker reinforcement in the y-direction than in the x-direction only for the plate ML8 a noticeable difference in the ultimate moments calculated with the rotating and the fixed crack occurs. In this experiment excessive yielding of the transverse reinforcement took place prior to failure. The rotating crack model is able to better adjust to this re-orientation of the principal tensile strain direction and therefore predicts a lower failure load, which is still 12% higher than the experimental ultimate moment.

While the shape of the stress-strain diagram of concrete after passing the effective compressive strength did not matter in the analysis of panels, an influence of the post peak behaviour on the ultimate moment is of course present in the analysis of plates. A complete loss of strength for strains smaller than -0.0025 was assumed for the current analysis.

TABLE 2

Material properties of plate tests [14], comparison of experimental and calculated ultimate moments

test	fc [MPa]	μ [%]	fy [MPa]	μ [%]	fy [MPa]	experiment	rotating cracks	fixed cracks	failure cause in analysis
	x-reinf.		y-reinf.			ultimate moments [kNm/m]			
ML1	46.7	0.25	551	0.25	551	44.4*	50.	50.	yielding of r.
ML7	44.7	0.25	479	0.25	479	42.5*	43.	43.	yielding of r.
ML2	36.2	0.50	551	0.50	551	69.9*	75.	75.	crushing of c.
ML3	37.5	1.00	481	1.00	481	93.4*	89.	89.	crushing of c.
ML9	44.4	1.00	412	1.00	412	101.5	102.	102.	crushing of c.
ML4	44.7	0.50	551	0.25	551	50.8*	66.	70.	yielding of r.
ML6	23.3	1.00	481	0.50	551	63.6	56.	56.	crushing of c.
ML5	35.6	1.00	481	0.25	551	60.6*	66.	68.	crushing of c.
ML8	49.1	1.00	412	0.25	479	64.8	73.	83.	crushing of c.

* denotes an edge failure in the experiment

ANALYSIS OF SHELL ELEMENTS SUBJECTED TO SHEAR FORCE AND BENDING MOMENT

Tests on elements subjected to shear forces and bending moments are described in [15]. Exemplarily some of the experimental and calculated results are compared for the element SE7 in Fig.7. The element (1625 mm times 1625 mm, thickness of 285 mm) was subjected to a shear force and a bending moment. Shear force and bending moment were monotonically increased with a constant ratio of bending moment divided by shear force equal to 0.13 m. The comparison of the strains of the middle surface of the shell element in Fig. 7 shows that the results calculated with the rotating crack model are in good agreement with the experiment and that the fixed crack model overestimates the experimental failure load. The failure cause indicated by the analysis was yielding of both reinforcements on the side of the specimen which was additionally subjected to tension by the bending moment.

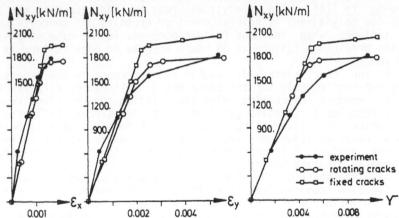

Figure 7. Comparison of experimental results of shell element SE7 [15] with analytical results from rotating and fixed crack model

CONCLUSIONS

Ultimate loads and corresponding strains obtained from the rotating crack model are in better agreement with experimental behaviour than results calculated with the fixed crack model. In the analysis of elements subjected to shear large differences between the results from rotating and fixed crack model can only be expected if the transverse reinforcement ratio is less than 50% of the main reinforcement ratio and if yielding of the transverse reinforcement leads to large transverse strains.

Using an effective concrete strength of 80% of the cylinder crushing strength for the analysis of panels and an effective concrete strength of 70% of the cylinder strength for the analysis of plates and shells subjected to static shear and twist loading good agreement between experimental and analytical failure loads was obtained.

ACKNOWLEDGEMENT

Support from Deutsche Forschungsgemeinschaft is gratefully acknowledged.

225

REFERENCES

1. Ngo, D. and Scordelis, A.C., Finite element analysis of reinforced concrete beams. J. ACI, 1967, 152-163.

2. Rashid, Y.R., Analysis of prestressed concrete pressure vessels. Nucl. Eng. Des., 1968, 334-344.

3. Cervenka, V. and Gerstle, K.H., Inelastic analysis of reinforced concrete panels: theory (1), and experimental verification and application (2). Publ. of IABSE, Zürich, Vol. 31-00, 1971 and Vol. 32-11, 1972.

4. Klein, D., Kristjansson, R., Link, J., Mehlhorn, G. and Schäfer, H., Zur Berechnung von dünnen Stahlbetonplatten bei Berücksichtigung eines wirklichkeitsnahen Werkstoffverhaltens. Forschungsber. 25, Inst. Massivbau, TH Darmstadt, 1975.

5. De Borst, R. and Nauta, P., Non-orthogonal cracks in a smeared finite element model. Eng. Comput., 1985.

6. Cope, R.J., Rao, P.V., Clark, L.A. and Norris, P., Modelling of reinforced concrete behaviour for finite element analysis of bridge slabs. In Numerical methods for nonlinear problems, ed. Taylor, D., Pineridge Press, Swansea, 1980, 457-470.

7. Akbar, H. and Gupta, A., Membrane reinforcement in concrete shells - design versus nonlinear behavior. Dept. Civil Eng., North Carolina State University, 1985.

8. Baumann, T.,Tragwirkung orthogonaler Bewehrungsnetze beliebiger Richtung in Flächentragwerken aus Stahlbeton. Deutscher Ausschuß für Stahlbeton, Heft 217, 1982

9. Kupfer, H., Erweiterung der Mörsch'schen Fachwerkanalogie mit Hilfe des Prinzips vom Minimum der Formänderungsarbeit. Bul. 40, CEB, Paris, 1964

10. Kollegger, J. and Mehlhorn, G., Experimentelle Untersuchungen zur Bestimmung der Druckfestigkeit des gerissenen Stahlbetons bei einer Querzugbeanspruchung. Deutscher Ausschuß für Stahlbeton, H. 413, 1990.

11. Kollegger, J. and Mehlhorn, G., Material model for the analysis of reinforced concrete surface structures. Comp. Mech., 1990

12. Vecchio, F. and Collins, M.P., The response of reinforced concrete to in-plane shear and normal stresses. Publ. 82-03, Dept. Civil Eng., University of Toronto, 1982

13. Andre, H., Toronto/Kajima study on scale effects in reinforced concrete. Masters thesis, Dept. Civil Eng., Univ. of Toronto, 1986

14. Marti, P., Leesti, P. and Khalifa, W.U., Torsion tests on reinforced concrete slab elements. J. Struct. Div. ASCE, 1987, 994-1010.

15. Kirschner, U. and Collins, M.P., Investigating the behaviour of reinforced concrete shell elements. Publ. 86-09, Dept. Civil Eng., University of Toronto, 1986.

MODELLING OF SHEAR TRANSFER IN CONCRETE USING CONTACT DENSITY FUNCTION

BAOLU LI   KOHICHI MAEKAWA   HAJIME OKAMURA
Tsinghua Univ.   Tokyo Univ.   Tokyo Univ.
Beijing, CHINA   Tokyo, JAPAN   Tokyo, JAPAN

ABSTRACT

Deformational behavior and ultimate capacity of reinforced concrete structures subject to earthquake loading may considerably be affected by occurrence of cracks in concrete. The subject of this study is to propose a stress transfer model to predict the mechanical behavior across cracks in concrete for various stress paths including reversed cyclic loading.

The stress transfer model for a single crack is derived by idealizing the rough crack surface as a set of numerous contact units with various inclinations. The contact units in all directions transfer normal and shear stresses, which are formulated by the elasto–plasticity model, and the distribution of their directions is expressed by the "contact density function". Discussion on the application of the model to reinforced concrete with discrete crack is given finally.

INTRODUCTION

Stress transfer across existing cracks in concrete is an important problem to be formulated in the seismic analysis of reinforced concrete structures since the structural energy absorption is highly related to the energy consumption along cracks. The deformational behavior and ultimate strength of structures, such as reinforced concrete corbel or brackets, shearwall and containments, cannot be predicted unless the behavior of their component elements before and after cracking is well known. When Finite Element Method(FEM) is used for analyzing the behavior of reinforced concrete members, either in discrete or in smeared crack approach, the modelling of stress transfer along cracks is also one of the essential factors.

Studies on shear transfer problem have frequently been performed based on different kinds of experiments, and a number of models have been proposed. Paulay[2] studied the effect of aggregate interlock on shear transfer experimentally in 1972, Jimenez and White[7] etc. have proposed an analytical model taking into account of initial crack width and normal restraint stiffness. Some microscopic physical models from a view point of crack surface contact have also been proposed. Walraven and Reinhardt's model[4] was derived by idealizing the crack surface as a set of circular aggregates, Gambarova and Bazant's model[5] was introduced by assuming the multidirectional stress distributions on the isotropically arranged microplane in the crack band. However, the elastic and plastic behavior, the shear dilatancy, the ratio of transferred stresses under cyclic loading and the influence of crack surface asperity have never been unifiedly simulated with reasonable accuracy. Constitutive equations having capability of dealing with any kind of stress path and wide applicability are still far from satisfactory.

EXPERIMENTAL OUTLINE

The specimens used in this study had $150 \times 300mm$ cross section and were $600mm$ long. A single crack was introduced before testing. The specimen set−up is shown in Figure 1.

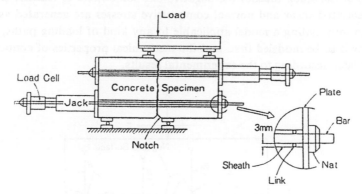

Figure 1. Specimen and loading system

Two steel bars passing through the inner sheaths set in the longitudinal direction were provided to apply axial compressive force upon cracked plane with two center−hole jacks. Dowel action and unnecessay shear resistance due to the steel bars were avoided by making the sheaths $6mm$ larger in diameter and placing the bars in their centers, and by setting two rings between the sheaths and the PC−bars at both ends of each bar. Both shear stress τ and compressive stress σ' were measured with load cells. Crack width ω was measured with four π−gages stuck across the crack, shear displacement δ was measured with two transducers on both sides of the specimen.

Systematical experiments were carried out under various types of loading paths as shown in TABLE 1.

TABLE 1   Mix proportion and properties of specimens

Series	W / C(%)	S / A(%)	W(kg / m³)	fc′(MPa)	G(mm)	Loading path
A	67	52	190	23.3	5–15	ω=const
B	40	46	190	49.0	5–15	ω=const
C	67	52	190	32.0	5–25	ω=const
D	67	52	190	29.4	15–25	ω=const
E	52	52	190	38.0	5–15	ω=const
F	75	49	190	22.1	5–15	τ / ω=const
G	52	52	190	39.0	5–15	cyclic

MODEL FOR CONTACT MECHANICS AND GEOMETRY

When a crack width and a shear displacement occur in the rough crack surface, positive surface of the crack touches the negative one somewhere to transfer the stresses, and the associated shear and normal compressive stresses are generated as shown in Figure 2. In formulating a model applicable to any kind of loading paths, crack surface asperity is to be modeled first, and the mechanical properties of contact area are formulated later according to the experimental results.

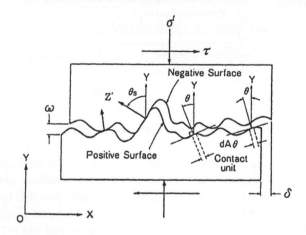

Figure 2. Definition and notations in contact plane

Geometry of Crack Surface

To investigate the geometrical characteristics of concrete cracks, precracked blocks with cracks were cut through aggregates and mortar using concrete cutter, the crack edge was traced and measured with a digitizer connected to a computer. Typical two–dimensional projection of crack surface of concrete with different maximum sizes of coarse aggregates G_{max} are shown in Figure 3(a). The complicated asperity of crack surface can be divided into infinite units having inclinations θ between $-\pi/2$ and $\pi/2$ with respect to the horizontal direction, and the histogram of crack surface corresponding to each inclination can be obtained and described as Figure 3(b).

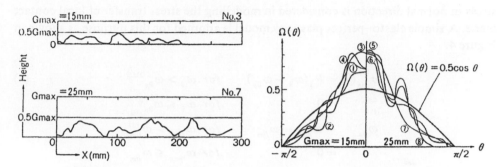

Figure 3. (a) Geometry of crack surface   (b) Contact density

It is obvious that there exist more horizontal units compared with vertical ones, and little difference between the concretes with different maximum aggregate sizes can be recognized. No simple density function seems to be reasonable to represent the distribution of crack surface directions. For the sake of brevity, a trigonometric function was decided as the "contact density function" which is also obtained by fitting them to the experiment results.

$$\Omega(\theta) = 0.5cos\theta \tag{1}$$

The surface area of crack plane between direction θ and θ+dθ is given as,

$$dA(\theta) = A_i\Omega(\theta)d\theta \tag{2}$$

A_t refers to the surface area of crack plane having a unit projection area.

Contact Stress
When crack opening ω and shear displacement δ occur in a single crack, local compressive displacement ω_θ' and shear displacement δ_θ upon those units in the direction between θ and θ+dθ accompany,

$$\omega_\theta' = \delta sin\theta - \omega cos\theta, \qquad \delta_\theta = \delta cos\theta + \omega sin\theta \tag{3}$$

where ω_θ' is defined positive for compression. If ω_θ' is minus, or θ is less than arctan (ω/δ), no stress is transferred as there is no contact in these directions. Units between $\arctan(\omega/\delta)$ and $\pi/2$ are the potential ones to transfer stress. To calculate their contribution to the shear stress τ and corresponding compressive stress σ', local compressive stress σ'_{con} due to ω'_θ and frictional stress f due to δ_θ are conceivably related. The local frictional shear stress f takes the direction opposite with the local shear displacement δ_θ and the change of its direction due to unloading may result in a noncontinuous change of the ratio of shear stress to compressive stress. However, this phenomenon was not observed significantly in the tests, and thus only the contact stress in normal direction is considered in modelling the stress transfer of local contact plane. A simple elasto–perfect plasticity model is adopted for this purpose as shown in Figure 4.

$$\sigma'_{con}(\theta) = R_s(\omega'_\theta - \omega'_{\theta p}) \qquad for \ \omega'_\theta > \omega'_{\theta p} \qquad (4)$$
$$= 0 \qquad for \ \omega'_\theta \leqslant \omega'_{\theta p}$$
$$\omega'_{\theta p} = \omega'_{max} - \omega'_{lim} \qquad for \ \omega'_{max} > \omega'_{lim}$$
$$= 0 \qquad for \ \omega'_{max} \leqslant \omega'_{lim}$$

where ω'_{max} is the maximum local compressive displacement experienced in direction θ and defined as positive for compression, ω'_{lim} is the elastic limit.

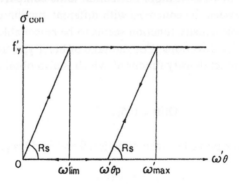

Figure 4. Contact stress

Effective Ratio of Contact

The contact density function represents the crack surface shape but no information concerning the height of crack plane, which obviously influences the effective contact area when crack opens. The number of effective contact units decreases as crack width increases. Considering that the crack width is usually much smaller than maximum size of aggregate G_{max}, which was 15mm or 25mm in this experimental study, it seems to be reasonable to idealize this characteristics approximately by a simple line,

$$K(\omega) = 1 - 2\omega / G_{max} \geqslant 0 \qquad (5)$$

Contact Force

By summarizing Equations(2,4,5), the contact compressive force between θ and $\theta+d\theta$

can be obtained,

$$Z'(\omega,\delta,\theta) = \sigma'_{con}(\theta) \cdot dA(\theta) \cdot K(\omega) \tag{6}$$

and the shear stress τ and compressive stress σ' are given as,

$$\tau = \int_{-\pi/2}^{\pi/2} Z'(\omega,\delta,\theta)\sin\theta \tag{7}$$

$$\sigma' = \int_{-\pi/2}^{\pi/2} Z'(\omega,\delta,\theta)\cos\theta$$

By fitting the experimental results of all specimens, the material–dependent constants needed in Equation(7) were decided as

$$A_t R_s = 436 f_c'^{1/3} (MPa / mm), \qquad \omega'_{lim} = 0.04(mm) \tag{8}$$

f_c' is the cylindrical compressive strength of concrete in MPa.

EXPERIMENTAL VERIFICATIONS

Cyclic Loading
As shown in Figure 5, the analytical results agree well with the experimental results of cyclic loading under constant crack width loading paths. The non–linearity appearing in the relations between the shear stress and shear displacement including the unloading process are recognized and the proposed model succeeds in predicting this property. One example of reversed cyclic loading is shown in Figure 6.

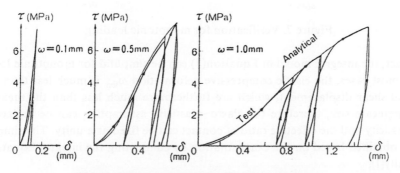

Figure 5. Experimental verification, Series A

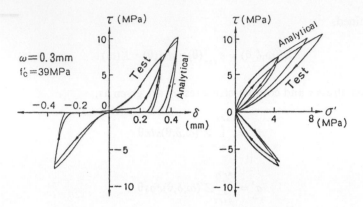

Figure 6. Experimental verification, Series G

Monotonic Loading

Stress transfer across cracks under monotonic loading should be of path independence as the contact stress in all directions and the effective ratio of contact are determined only by crack width ω and shear displacement δ. Figure 7 shows the comparison between calculated and experimental results of test series B to F in TABLE 1 including constant crack width loading ($\omega = 0.1$, 0.3, $1.0mm$), and constant shear stress crack width ratio loading ($\tau / \omega = 9.8$, $19.6 MPa / mm$).

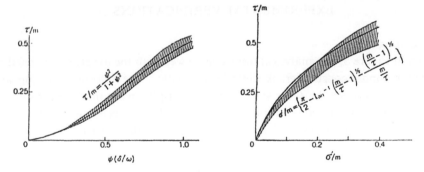

Figure 7. Verification for monotonic loading

In fact, the integral model in Equation(7) can be simplifid for monotonic loading path. In most cases, the elastic compressive deformation ω'_{lim} is much less than crack width and shear displacement, which are furthermore much less than the maximum coarse aggregate size, therefore the elasto−plasticity assumption can be replaced by rigid−plasticity and the effective ratio of contact can be fixed as a unity. The simplified model is obtained based on the exact solution by conducting the integral Equation(7) analytically[1].

$$\sigma = m[\pi / 2 - arccot\psi - \psi / (1 + \psi^2)] \tag{9}$$

$$\tau = m[\psi^2 / (1 + \psi^2)], \qquad \psi = \delta / \omega$$

where $m = 3.83 f_c'^{1/3} (MPa)$, which is in fact the maximum shear strength transferred by a single crack and independent of crack opening and shear displacement. Equation(9) gives precise prediction which almost coincides with integral model for monotonic loading (Figure 7).

Application to Reinforced Concrete

The motive of developing the constitutive model for stress transfer originates from the structural shear analysis of reinforced concrete by FEM. For monotonic loading, the shear and compressive stresses are functions of $\psi = \delta / \omega$ as shown in Equation(9), hence they have the same form with $\psi = \gamma / \varepsilon$ when used as a smeared crack model where γ is the average shear strain and ε the average tensile strain corresponding to the crack plane in the element concerned. On the other hand, the model itself appears to be a discrete crack model relating stresses to deformations. In the following, application to precracked reinforced concrete is discussed.

Let us consider that the shear force is mainly carried by the aggregate interlock effect and shear failure is reached after the yielding of reinforcement which provides compressive stress upon the crack plane and thus transfer shear stress, the shear strength can be estimated easily with the $\tau - \sigma$ relation obtained from Equation(9).

$$\sigma' / m = \pi / 2 - \tan^{-1} q^{1/2} - q^{1/2} / (m / \tau), \qquad q = m / \tau - 1 \tag{10}$$

Figure 8 shows the comparison between the predictions and experimental results by Mattock etc.[6], good agreements were reached for specimens with or without external force. Here σ' in Equation(10) was replaced by $(N / A_{sh} + \rho f y)$ in the calculation according to the equilibrium in the direction normal to the crack plane, N is the external force with compression being positive and tension being negative, A_{sh} is the cross sectional area of shear plane, ρ the reinforcement ratio and $f y$ the yielding strength of reinforcement.

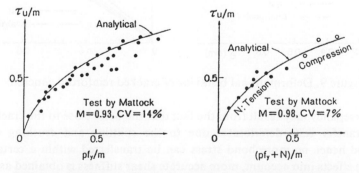

Figure 8. Shear strength along cracks in reinforced concrete

As for the deformational behavior of cracked reinforced concrete, the bond model relating the bond stress to slip between concrete and reinforcement proposed by Okamura etc.[3] is introduced.

$$S / D = (20 / fc')^{2/3} \cdot \varepsilon \cdot (2 + 3500\varepsilon) \tag{11}$$

where S is the slip in millimeter, $D(mm)$ the diameter of reinforcement and ε the strain of reinforcement at the position under consideration. Supplying this model for the crack plane in shear transfer test of reinforced concrete, relations between crack width ($\omega = 2S$) and compressive stress of concrete ($\sigma' = \rho\sigma_s$) can be obtained, where σ_s is the tensile stress of reinforcement. Then the shear stress–shear displacement and crack opening–shear displacement relations can be calculated by combining Equation(9) with the bond model Equation(11). As the result, the predicted shear stiffness is a little bit higher than the experimental ones as shown in Figure 11, the same specimens and testing method explained in chapter two were used and their properties are shown in TABLE 2.

TABLE 2   Properties of reinforced concrete specimens

No.	Section (mm)	Length (mm)	ρ (%)	fy (MPa)	Es (MPa)	fc' (MPa)
RC–1	150 × 300	600	0.71	389	1.92×10^5	41.3
RC–2	150 × 300	900	0.71	429	2.05×10^5	32.8
RC–3	200 × 300	1500	1.88	456	1.91×10^5	28.0

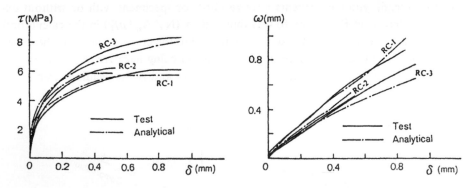

Figure 9. Deformational behavior of cracked reinforced concrete

The discrepancy may result from the fact that the bond close to the crack plane in the shear transfer tests deteriorates due to the splitting and crushing effects of concrete, and hence no more bond stress can be transferred within a certain range. Taking these effects into account, more accurate shear stiffness is obtained as shown in Figure 9, where $\omega = 2[S+2D\varepsilon]$ was used instead of $2S$. $2D$ was taken as the bond–deteriorating range. Their deformational paths were also predicted well.

CONCLUSIONS

Stress transfer across crack in concrete can simply be predicted by a physical model, which is completed by taking account of the geometrical asperity of cracked plane, using a suitable contact density function and a simple assumption of elasto–plastic hypothesis for contact stress. The model has good agreement with experimental results for ordinary loading paths. For monotonic loading path, a simplified model is derived in terms of the exact solution of the original integral equation. This model can easily be used for not only a discrete but also a smeared crack model in FEM analysis because of its independency of stiffness on the crack spacing.

Shear strength simulation of cracked plane developed from plain concrete can be extended to initially cracked reinforced concrete with reinforcement embedded in the direction normal to the crack. The ultimate shear strength is achieved after yielding of the reinforcement. The discrete shear behavior of cracked reinforced concrete plane can be simulated appropriately by combining the bond model with the shear transfer model.

REFERENCES

1. Li, B.L., Maekawa, K. and Okamura, H., Contact density model for stress transfer across cracks in concrete, Journal of the Faculty of Engineering, University of Tokyo(B), Vol.xxxx, Vo.1(1989), pp.9–52.

2. Paulay, T.and Leober, P.J., Shear transfer by aggregate interlock. Shear in Reinforced Concrete, ACI, SP–42, Vol.1, 1974, pp.1–15.

3. Shima, H., Chou.L. and Okamura, H., Micro and macro models for bond behavior in reinforced concrete, Journal of the Faculty of Engineering, University of Tokyo(B), Vol.39, No.2, 1987, pp.133–194.

4. Walraven, J.C. and Reinhardt, H.W., Theory and experiments of the mechanical behavior of crack in plain and reinforced concrete subjected to shear loading. HERON, Vol.26, No.1A, 1981.

5. Gambarova, P.and Bazant, Z.P., Crack shear in concrete: crack band microplane model. ASCE, Vol.110, No.9, Sep.1984, pp.2015–2035.

6. Mattock, A.H. and Hawkins, N.M., Shear transfer in reinforced concrete recent research. PCI Journal, Mar.–Apr. 1972, pp.55–75.

7. Jimenez, R., White, R.N. and Gergely, P., Cyclic shear and dowel action models in R / C. ASCE, Vol.108, ST5, May 1982, pp.1106–1123.

ROUGH CRACKS SUBJECTED TO EARTHQUAKE LOADING

JOOST WALRAVEN
Delft University of Technology
Faculty of Civil Engineering
Stevinweg 1, 2628 CN Delft, The Netherlands

ABSTRACT

The behaviour of cracks under cyclic shear loading is characterised by a considerable irreversible damage of the crack faces. Therefore the response of cracks to
reversed and alternative actions can only be well described if load-history effects
are taken into account.

A model is presented with which it is possible to explain and predict the
behaviour of cracks under cyclic loading. This model reflects the actual behaviour,
by a realistic implementation of the concrete properties.

INTRODUCTION

In structures subjected to dynamic loading, aggregate interlock in cracks may play
an important role. For instance under earthquake conditions shear loading may
occur across cracks existing already as a result or other types of loading. An
example is given in Fig. 1.

In this case, a reinforced concrete nuclear containment vessel, it is required
that the structure be designed to withstand the simultaneous occurrence of an
internal pressurization and the inertia forces generated by a strong motion earthquake [1]. From tests on cracked specimens subjected to earthquake loading it was
observed that there is a considerable difference in behaviour between the first and
the subsequent loading cycles. Furthermore it turned out that each new cycle of
loading leads to further damage of the crack faces, resulting in steadily increasing
values of the shear displacements and the crack width at peak loading.

Fig. 2 shows the relation between shear stress and shear displacement for a dynamic shear test [2]. It is shown that further damage occurs even after 55 loading cycles.

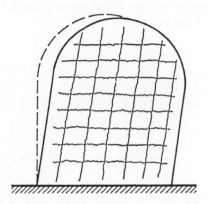

Figure 1.   Nuclear containment vessel under earthquake loading [3].

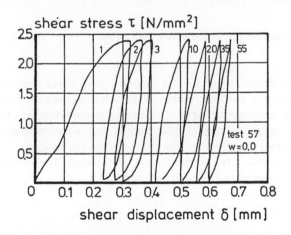

Figure 2.   Shear stress to shear displacement relation for a crack under
dynamic loading [2].

Up to now only tendencies of behaviour were known. No physical model able to explain and analyze the behaviour and to give quantitatively reliable results is available. The development of such a model is the subject of this paper.

MODELLING THE BEHAVIOUR

In order to model the behaviour, the particles are simplified to spheres which are randomly distributed over the volume considered. A further simplification is obtained by considering a representative slice of the volume with a finite width Δx, so that a 2-dimensional analysis can be carried out.

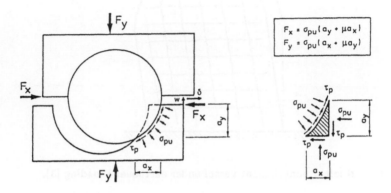

Figure 3.   Contact mechanism at shear displacement.

Fig. 3 shows the behaviour of a crack during a pure shear displacement, as illustrated for one specific particle section. The matrix is considered to be perfectly brittle, with a crushing strength σ_{pu}. During penetration of the particle into the matrix not only crushing of the matrix, but simultaneously friction between particle and matrix occurs. For the contact area the relation $\tau_p = \mu \sigma_{pu}$ is valid, where μ is the coefficient of friction. The stresses at the contact area can be composed to forces. It is found that:

$$F_x = \sigma_{pu} (a_y + \mu a_x) \tag{1}$$

$$F_y = \sigma_{pu} (a_x - \mu a_y)$$

where σ_{pu} and μ are material constants, which are not yet known, and a_x, a_y are the projected contact areas.

In order to describe the behaviour of a crack subjected to an arbitrary sequence of loadings, the model shown in Fig. 4. is adopted. The particles are represented by a single particle D (i, j) with a diameter $D_i = (0.1i-0.05)D_{max}$ and an embedment depth $t = 0.1j(\frac{1}{2}D_i)$. The parameters i and j may adopt values between 1 and 10, so that 10 particle sizes and 10 embedment depths are considered. The most probable number of particles with a diameter D_i, intersected over a unit crack length, can be shown to be

$$n(D_i) = 0.127 \ p_c(D_i) \ \frac{D_{max}}{D_i} \tag{2}$$

with

$$p_c(D_i) = p_K(0.53D_i^{-0.5}D_{max}^{-0.5} - 0.21D_i^3D_{max}^{-4} - 0.075D_i^5D_{max}^{-6})$$

[3] where p_k is the ratio aggregate volume - total volume, which is normally about 0.75. The reinforcement crossing the crack is represented by springs.

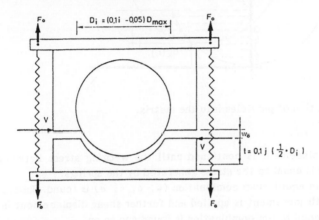

Figure 4.   Basic model.

The calculation procedure is as follows. Starting from an initial crack width w_o a crack increment Δw is applied. This leads to a certain force in the springs, which follows from the spring stiffness. Subsequently an incremental shear displacement $\Delta\delta$ is applied, during which the particle penetrates into the matrix (Fig. 5).
The contact areas $a_x(i,j)$ and $a_y(i,j)$ are calculated.
Numerical integration over i and j from 1 to 10, taking account of the number of particles $n(D_i)$ given in Eq. 1 gives the total projected contact areas

$$\Sigma a_x \quad \text{and} \quad \Sigma a_y$$

Substitution of these values in Eq. 1 gives the shear and normal stresses:

$$\sigma = \sigma_{pu}(\Sigma a_y - \mu\Sigma a_x) \tag{3a}$$

$$\tau = \sigma_{pu}(\Sigma a_x + \mu a_y) \tag{3b}$$

for the displacement combination (w_o + $\Delta w, \Delta \delta$) considered.

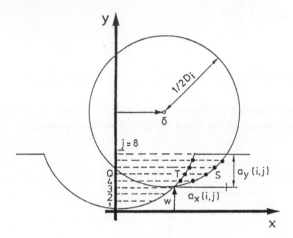

Figure 5.   Penetration of particles into the matrix.

The shear displacement is continued until the wedging stress between the crack faces (Eq. 3a) is equal to the stress generated by the springs.

If this is the case, an equilibrium combination (w, δ, τ, σ) is found. Subsequently a new crack width increment is applied and further shear displacement is conducted until a next equilibrium combination is found and so on.

By carrying out this calculation on a micro-computer, the crack damaging process can be followed step by step as a function of the load history (Fig. 6).

Up to now it has been assumed that the aggregate is so strong, that no aggregate particle will fracture during cracking of the concrete, so that all aggregate particles in the crack contribute to the aggregate interlock mechanism, Eq. 3. Whether this ideal situation will apply, depends on the ratio between particle strength and matrix strength. In high-strength concrete a significant part of the aggregate particles will fracture and does not contribute to the interlocking resistance. The same will be true for an intermediate-strength concrete with low-strength aggregate particles. In order to take this effect into account, the fracture index C_f is introduced: now the projected contact areas are expressed by $C_f \Sigma a_x$ and $C_f \Sigma a_y$ so that the constitutive relations become:

$$\tau = C_f \cdot \sigma_{pu}(\Sigma a_y + \mu \Sigma a_x) \tag{4a}$$

$$\sigma = C_f \cdot \sigma_{pu}(\Sigma a_x - \mu \Sigma a_y) \tag{4b}$$

where $0 < C_f < 1$ and Σa_x and Σa_y are functions of w and δ , and the load history.

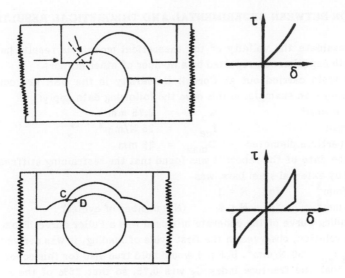

Figure 6.   Various stages of the crack damaging process.

Although C_f is an additional material parameter, the effective number of material characteristics in Eq. 4a,b is still limited to two: the values $C_f\sigma_{pu}$ and μ . These values are characteristic for a specific concrete mixture. They cannot be determined on the basis of classical control tests, like the crushing or splitting test on cylinders. The easiest way to find those material characteristics is to do an aggregate interlock test on a crack, i.e., like shown in Fig. 2 or like described in [5,6,8]. The specimen is subjected to a monotonically increasing shear load: during loading the development of the displacements w and δ of the crack faces are measured and the restraining stress σ , normal to the crack plane, is calculated from the strain measurements on the restraint rods.

On the basis of those measurements the τ–δ and σ–δ relations are known. Subsequently the values $C_f\sigma_{pu}$ and μ are chosen so, that the τ–δ and σ–δ curves are correctly described by eq. 4a,b.
Tests described in [3,5,6] yielded as a result $\mu = 0.4$ and $C_f\sigma_{pu} = 8.0\, f_{cc}^{0.5}$, where f_{cc} = concrete crushing strength of 150 mm cubes. Those values were valid for a range $15 < f_{cc} < 60$ N/mm, for concretes made with blast furnace cement and glacial river aggregate. A visual inspection of the crack faces showed that hardly no particle broke through, so that $C_f \approx 1.0$ and

$$\sigma_{pu} = 8.0\, f_{cc}^{0.5} \tag{5}$$

COMPARISON BETWEEN EXPERIMENTAL AND THEORETICAL RESULTS

In order to evaluate the validity of the theoretical model, its results have been compared with experiments, reported by a number of other authors.

At first the tests carried out at Cornell University in the USA are considered [1,7]. Fig. 7 shows an example. In this case the following data apply:
- initial crack width w_o = 0.75 mm
- cube strength f_{cc} = 26 N/mm^2
- maximum particle diameter D_{max} = 38 mm

From the data of the report it was found that the restraining stiffness which was supplied by external steel bars, was

F = 2.6 N/mm^3 for $N = 1$

F =1.4   N/mm^3 for $N \leq 2$ (N = number of cycles)

The grading curve of the concrete mix used was a Fuller curve. From the $\tau-\delta$ and $\sigma-\delta$ relation, observed in the first cycle of loading, it was derived that μ = 0.20 and $C_f \sigma_{pu}$ = 30 N/mm^2. If Eq. 5 would hold true also for this concrete, this would mean that the fracture index C_f was 0.75, so that 25% of the aggregate particles would have been broken. (In the report the strength of the aggregate was described as "medium".)

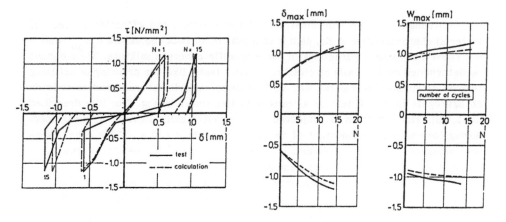

Figure 7.   Comparison between experimental curves from [1] and theoretical
curves, using Eq. 4a,b.

Fig. 7a shows the predicted $\tau-\delta$ relations in comparison with the relations found experimentally. Only the first and the 15th cycle are shown. Fig. 7b and c show the development of the maximum values of δ and w as a function of the number of loading cycles. The agreement between theory and experiment is shown to be very good. Comparisons with other tests of the same series further confirmed the validity of the model.

Fig. 8 shows an example of a test, carried out by Brisegella and Gori [2]. The test was also carried out on a specimen which was precracked: the initial crack was set to be w_o = 1.0 mm. The concrete cylinder strength was reported to be 36 N/mm^2. From the first cycle of a representative test, reported in the paper, it was found that the ascending branch of the τ-δ and σ-δ relations are well described by Eq. 4 if $C_f \sigma_{pu}$ = 24 N/mm^2 and μ = 0.4.

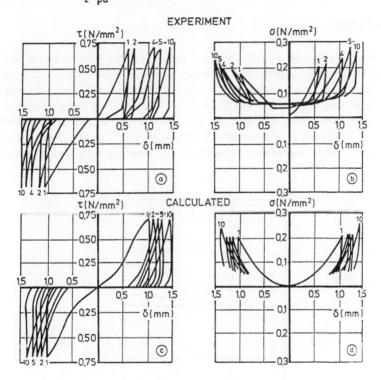

Figure 8   a,b   Cyclic loading response as reported by Briseghella and Gori [2] for w_o = 1.0 mm and f_c = 36 N/mm^2

         c,d   Calculated relations for the same conditions.

Assuming that the matrix strength follows from the cube crushing strength according to Eq. 5, and that the cube crushing strength is f_{cc} = 36/0.8 = 45 N/mm^2, it is found that the fracture index must have been 0.45. This means that 55% of the aggregate particles would have been broken at precracking of the specimen. Apparently this agrees with the observation of the authors, who report that "the fracture line obtained (by traction) passes through the cement layer and also numerous elements of gravel which break as a result" [2].

Fig. 8a, b show the cyclic response of the crack as observed in the test. Fig 8c, d show the calculated response. Also here the agreement is seen to be very good. The positive and negative cycles in the experiment were not symmetric. The negative cycles in experiment and calculation nearly coincide. With regard to the positive cycles, the first four, obtained by calculation, slightly overestimate the shear displacements, however, from the fifth cycle on, the agreement also is very good.

APPLICATION

The theoretical model can be used for parameter studies, revealing the role of certain influencing factors. One example is given below. The example deals with the type of concrete as used at Cornell University in the tests, described in [1] (D_{max} = 38 mm, f_{cc} = 26 N/mm^2, μ = 0.2, C = 0.75). It is assumed that the initial crack width is 0.5 mm, and that the maximum shear level is 1.25 N/mm^2. Fig. 9 shows the influence of the restraint stiffness (N/mm^3) on the development of the maximimum shear displacement and maximum crack opening during cycling.

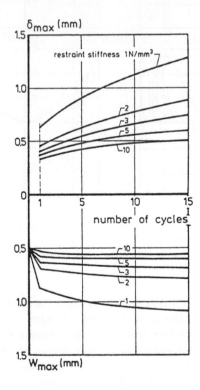

Figure 9.   Influence of restraint stiffness on maximum shear displacement and crack width during cyclic shear loading (calculated).

CONCLUSIONS

1) The behaviour of cracks subjected to cyclic shear loading can be described by a two-parameter model. The two parameters needed for a realistic description of the behaviour are the coefficient of friction between the matrix and the aggregate particles and a parameter involving the matrix strength and the relation matrix to particle strength (fracture index).

2) To describe the behaviour, it is sufficient to simplify the aggregate particles to spheres with an infinite stiffness and the matrix to a material with a rigid-brittle behaviour.

3) The load-history effect is caused by crushing of the matrix. Due to this crushing the matrix is irreversibly damaged during cycling, which influences the behaviour during the next cycles of loading.

4) The influence of the damage on the geometrical shape of the crack faces can be followed cycle by cycle, which could be used if questions with regard to the leakage behaviour have to be dealt with.

REFERENCES

1. Laible, J.F., White, R.N., Gergely, P., Experimental investigation of seismic shear transfer across cracks in nuclear containment vessels, ACI Special Publication 53, pp. 203-226, 1977.

2. Briseghella, L., Gori, R., Aggregate interlock cyclic response of R.C. critical section. In Proceedings of the San Francisco Earthquake Conference, 1984.

3. Walraven, J.C., Fundamental analysis of aggregate interlock. Journal of the Structural Division, ASCE, Vol. 107, No. 11, Nov. 1981, pp. 2245-2270.

4. Mattock, A.H., Hawkins, N.M., Research on shear transfer in reinforced concrete - recent research. PCI-Journal, V. 17, No. 2, March/April 1972, pp. 55-75.

5. Walraven, J.C., Reinhardt, H.W., Cracks in concrete subject to shear. Journal of the Structural Division, ASCE, Vol. 108, No. 1, Jan. 1982, pp. 225-244.

6. Walraven, J.C., Reinhardt, H.W., Theory and experiments of the mechanical behaviour of cracks in plain and reinforced concrete subjected to shear loading. Heron, Vol. 26, No. 1a, 1981.

7. Laible, J.F., An experimental investigation of interface shear transfer, PhD-Thesis, Cornell University, Ithaca, N.Y., 1973.

8. Tassios, T.P., Vintzeleou, E.N., Concrete-to-concrete friction. Journal of the Structural Division, ASCE, Vol. 113, No. 4, April 1987, pp. 832-849.

OPTIMUM DESIGN OF STIRRUPS IN R/C AND P/C THIN-WEBBED BEAMS FAILING IN SHEAR

MARCO DI PRISCO and PIETRO G. GAMBAROVA
Department of Structural Engineering
Politecnico di Milano
Piazza Leonardo Da Vinci 32, Milan 20133, Italy

ABSTRACT

A recently developed model for the description of the ultimate behavior of R/C and P/C thin-webbed beams failing in shear is recalled here in order to introduce two new aspects (dowel action and non-uniform distribution of the interface displacements along the shear cracks), and to work out a set of charts aimed at the optimum design of the stirrups. The model is based on the truss analogy, with the ties (i.e. stirrups) at the onset of yielding and the end-sections of the concrete struts (between the shear cracks) at the onset of crushing in shear and compression. As assumed by several wellknown scholars, the stress field in the web is represented by means of diagonal compression. Finally, the model is shown to fit some recent test results by Regan and Rezai-Jorabi fairly well.

INTRODUCTION AND NATURE OF PROBLEM

As is well known, several shear transfer mechanisms are mobilized in RC and PC beams in the ultimate load situation, particularly in the zones where large shear forces have to be transferred. The simultaneous activation and/ or overlapping of the different mechanisms make it difficult to analyze the role of each of them; consequently, the modelization by means of relatively simple schemes, such as the truss-and-tie model for transversely reinforced beams, is instrumental in the study of each mechanism's role and of their interaction. On this point, the truss-and-tie model is particularly well suited to thin-webbed beams, because of the fairly regular crack pattern in the web and well-defined zone where dowel action is mobilized (bottom flange, if subjected to cracking).

In fact truss analogy with variable strut inclination has been -and still is- widely adopted, either within a plasticity-oriented approach (for instance, see Nielsen et al. [1]) or according to a more straightforward understanding of RC behavior (for instance, see the Diagonal Compression Field developed by Collins [2], and Vecchio and Collins [3], and the cracked-web model with rough shear cracks, developed by Kupfer et al. [4]).

The cracked-web model is the starting point of the model recalled here

[5,6], which is still based on truss analogy, with the following assumptions: (a) the shear cracks are already formed and either limited to the web (web-shear cracks, Case I typical of PC beams) or extended to the bottom flange (flexure-shear cracks, Case II typical of RC beams); (b) the stirrups are at or beyond the yield threshold; (c) the partially cracked end-sections of the inclined concrete struts are at the onset of failure in shear-compression; (d) the average behavior of the concrete is represented by a Diagonal Compression Field.

While in two previous papers (Dei Poli et al. [5,6]) the distribution of the displacements along the shear cracks was uniform and dowel action (in Case II) was neglected, here (as in [7]) crack opening and slip are given suitable non-uniform and realistic distributions, and dowel action is "added" to the other shear-resistant mechanisms (aggregate interlock, shear and bending of concrete struts, tension stiffening due to stirrup/concrete bond); aggregate interlock is described by means of the rough crack model (Bazant, Gambarova and Karakoç [8,9]).

Since the truss-and-tie system is studied by writing a set of equilibrium equations, respecting the ultimate strength everywhere, both "as a whole" (Diagonal Compression Field) and locally, and the ultimate strength is only reached in a few regions (along the shear cracks, where the stirrups are plasticized, and at the end-sections of the struts, where the concrete is at the onset of collapse in shear-compression), the proposed approach falls within the static theorem of limit analysis. Nevertheless, the compatibility equations (necessary in order to evaluate the interlock forces) and the approximate introduction of the dowel action are beyond the assumptions of the aforementioned theorem and to what extent the results are on the safe side is still an open question.

After a brief description of how crack displacements and dowel action are introduced, 8 beams tested by Regan and Rezai-Jorabi [10] are analysed, and the theoretical and experimental results are shown to agree satisfactorily. Finally, two charts for the optimum design of the stirrups are presented (see also [11]), and some conclusions are drawn.

LIMIT ANALYSIS MODEL

As in [5] and [6] the beam is considered as a truss consisting of two parallel chords (the flanges with the main reinforcement) connected by inclined compression struts (the concrete struts bound by the shear cracks) and vertical rods acting in tension (the stirrups).

Let us assume that (a) the crack pattern in the web is regular with closely spaced inclined cracks, Fig.1a; (b) the faces of the inclined cracks are rough and interlocked, Fig.1b; (c) the axial force in the struts has a certain excentricity because of the initial cantilever-type behavior of the struts, Figs.1c,d; (d) the shear failure is governed by the yielding of the stirrups and is accompanied by the failure in shear and compression of the end sections of the struts (Case I, Fig.1c) or of the top section (Case II, Fig.1d); (e) a uniaxial and diagonal compression field prevails in the web (σ_2, α, Fig.1e). A set of equilibrium and compatibility equations can be written, as well as a set of constitutive relationships regarding aggregate interlock, solid concrete, crack spacing and bond (between the stirrups and the concrete).

Here (as in [7]) the displacements at the interface of the inclined cracks are not considered uniform, but realistic laws are adopted for crack opening and slip (Fig.2). Two cases are always referred to: Case I with web-shear cracks and Case II with flexure-shear cracks. Crack behavior is assumed to be governed by Mode I close to the tip (Case II) or at both tips

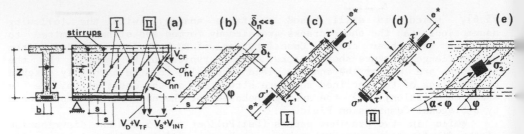

Figure 1. Cracked web: (a) web-shear cracks-Case I, and flexure-shear cracks -Case II; (b) average crack opening and slip; (c,d) forces acting on a concrete strut; (e) Diagonal Compression Field: α, φ = DCF and crack angles.

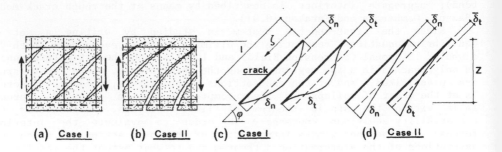

Figure 2. Different distributions adopted for crack opening and slip (δ_n, δ_t): (a,c) web shear cracks and (b,d) flexure-shear cracks.

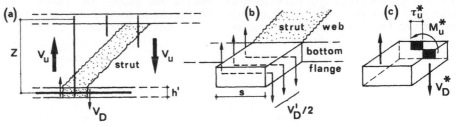

Figure 3. (a) Dowel action developed by longitudinal bars (main reinforcement); (b) dowel action developed by the bars embedded in the outer parts of the bottom flange; (c) limit state in torsion.

TABLE 1

		Characteristics of beams	[10]	LA	[10]	LA = Limit Analysis		
Beam	ρ_{st}	f_c' (MPa)	τ_u/f_c'	τ_u/f_c'	θ_u^σ	α	θ_u^σ	σ_2 (MPa)
R1	0.706	43	0.215	0.234	30	32	31	18
R2	0.706	34	0.245	0.275	33	34	33	18
R3	0.706	54	0.211	0.201	25	32	29	23
R4	0.404	60	0.158	0.139	19	27	36	15
R5	0.404	26	0.235	0.216	27	28	34	13
R6	0.404	65	0.136	0.129	19	26	30	16
R7	1.010	41	0.248	0.300	32	34	31	24
R9	1.010	63	0.216	0.219	30	32	27	25

I, II and θ_u^σ = directions and angle of the principal stresses.

(Case I), whilst a Mixed Mode behavior is introduced in the regions far from the tips. In order to model the above-mentioned behaviors, crack opening and slip are given suitable distributions along an inclined crack (Figs.2c,d):

- <u>Case I (web-shear cracks, Figs.2a,c)</u>

$$\delta_n = 6 \, \bar{\delta}_n (\xi - \xi^2), \qquad \delta_t = \bar{\delta}_t [1 - \cos(2\pi\xi)] \qquad (1)$$

- <u>Case II (flexure-shear cracks, Figs.2b,d)</u>

$$\delta_n = 2 \, \bar{\delta}_n \xi, \qquad \delta_t = \frac{\pi}{\pi - 2} \bar{\delta}_t [1 - \cos(\pi\xi/2)] \qquad (2)$$

where $\xi = \zeta/1$, $1 = z/\sin\phi$; $\bar{\delta}_n$, $\bar{\delta}_t$ = average values of crack displacements as introduced in [5] and [6].

At increasing values of τ_u / f_c' (τ_u = ultimate shear stress = V_u/bz with V_u = ultimate shear force, z = internal moment lever arm, b = web thickness), the highly nonlinear system of equilibrium, compatibility and constitutive equations is solved iteratively in order to evaluate $\Delta\tau_{INT}$ and $\Delta\tau_{BND}$ (contributions of aggregate interlock and strut bending stiffness to web capacity in shear); s (crack spacing); ρ_{st} (stirrup ratio); α, σ_2 and ε_2 (orientation, stress and strain of the diagonal compression field); α_{ts} (= $\varepsilon_s / \varepsilon_{sy}$ = ratio of the average strain in the stirrup to the yield strain of the steel); $\bar{\sigma}_{nt}^c$, $\bar{\sigma}_{nn}^c$, $\bar{\delta}_t$ and $\bar{\delta}_n$ (average values of the stresses and displacements along a rough inclined crack). At the end of each iteration, δ_t and δ_n are evaluated in a set of points along an inclined crack, according to Eqs.1 and 2, in order to update the values of $\bar{\sigma}_{nt}^c$ and $\bar{\sigma}_{nn}^c$, and start a new iteration.

The model in its actual form does not consider dowel action (which is present only in Case II, bottom flange in Figs.1a, 2b), as well as shear transfer in the uncracked flanges (both flanges in Case I, top flange in Case II). However, as a first and probably fairly good approximation (at least in a Limit Analysis approach) dowel action is here evaluated separately and then added to the other resistant contributions. Of course the Authors are aware that dowel action should be directly introduced into the equilibrium equations of a concrete strut, and by the time this paper is published dowel action will have been introduced in a more consistent way. As for dowel action effectiveness (Fig.3a), only the bars placed close to the web mean plane, i.e. inside the joint between the web and the lower cracked flange (see for instance Ø16 bars in Fig.4) develop large dowel forces, since they push against concrete cover or core (depending on which face of the crack is considered), whilst the bars placed in the outer parts of the cracked flange (i.e. far from the longitudinal mean plane, see for instance Ø20 bars in Fig.4) are far less effective, since dowel forces are limited by the amount of torque that concrete can transfer from the projecting parts of the flange to the inner part, where the flange and the web join together (Figs.3b,c).

As a rule, the contribution of dowel action has to be evaluated on a case-by-case basis, from the actual values of crack opening and slip in the centroid of the longitudinal steel in tension or from the torque capacity of the concrete cantilevers (Figs.3b,c). Dowel action contribution is then added to the ultimate web capacity τ_u. To this end, the following equations are used:

- bars placed outside the web-flange joint (Fig.3b):

$$V_D' \leq 2V_D^* = 2(M_u^*/s) \text{ where } M_u^* = k\tau_u^* ab^2, \quad \tau_u^* = f_{ct} = 0.27f_c'^{0.66},$$

$$a = \text{Max}(h',s), \quad b = \text{Min}(h',s), \quad k = 0.47 \text{ (plastic distribution)}$$

- bars placed inside the web-flange joint

$$V_D'' = (C_1\phi - C_2) \sqrt{f_c'/30} \text{ with } C_1 = 1.6, \ C_2 = 15 \text{ (see the tests by Dei Poli et al. in "Studi e Ricerche", Vol.10/88, Politecnico di Milano, 1989, pp.217-303).}$$

When fitting test results, the spacing of the stirrups (which is always assumed to be equal to crack spacing, Figs.1a,b) is an input data and the constitutive equation governing crack spacing is ignored. When using the proposed model for designing a double-tee beam, the optimal stirrup ratio is found by solving the afore-mentioned system of equations, including the constitutive equation of crack spacing.

REFERENCE TESTS

Eight of the nine double-tee RC beams tested by Regan and Rezai-Jorabi [10] were analysed by truss analogy (limit analysis), namely the beams R1-R7 and the beam R9; the beam R8 was not included because the stirrups did not yield, so the authors said.

The geometry of the beams and the most relevant details of the reinforcement (stirrups and main bars) are shown in Fig.4. The variables were the strength of the concrete (cylindrical strength $f_c' = 26-65$ MPa) and the stirrup ratio ($\rho_{st} = 0.40\%$, 0.71%, 1.01%), as reported in Table 1.

According to the crack patterns shown in [10] (beams R1 and R4) and in a report published in 1987 (beams R1,R2,R3,R4,R5 and R6, "Shear resistance of I-sectioned reinforced concrete beams", Structure Research Group, Polytechnic of Central London), two beams were subjected to severe cracking in the bottom flange (R1 and R2), three beams to light cracking (R3,R4 and R6), R5 did not show any cracking and the crack patterns of R7 and R9 were not shown. Consequently, the beams R1-R4 and R6 fall into Case II of limit analysis (flexure-shear cracks), the beams R5 falls into Case I (web-shear cracks); as for the beams R7 and R9, both cases have to be considered. Eventually, all the beams but R1,R2 and R5 were analysed in both ways in order to ascertain the differences between the predictions based on the two above behaviors.

The beams R1,R4 and R7 were also analysed [7] by means of a NLFE code (essentially an improved version of ADINA), where concrete subjected to a normal tensile strain is modeled via a simplified bilateral stress-strain law in order to introduce strain softening, and then -beyond a prefixed strain value- via aggregate interlock laws, once shear develops at the crack interface.

In the limit analysis, the 2 $\emptyset$16 bars in the bottom flange were lumped into a single $\emptyset$22 bar for the evaluation of dowel action, since the 2 $\emptyset$16 bars are placed side-by-side and tend to behave as a single bar.

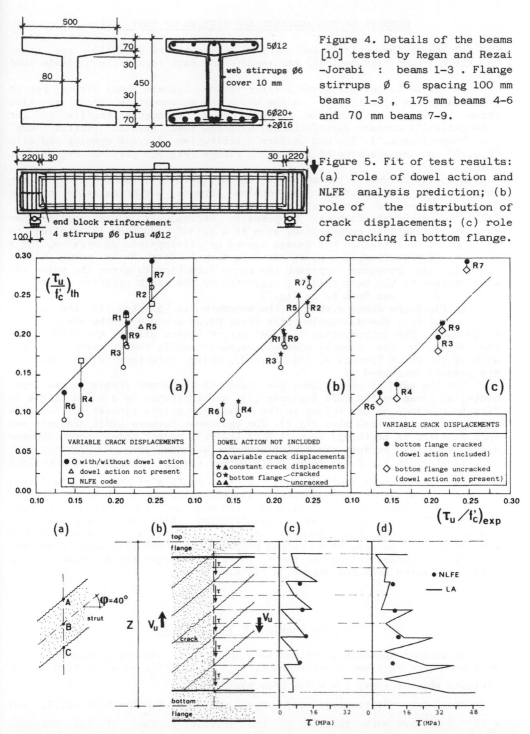

Figure 4. Details of the beams [10] tested by Regan and Rezai –Jorabi : beams 1-3 . Flange stirrups Ø 6 spacing 100 mm beams 1-3 , 175 mm beams 4-6 and 70 mm beams 7-9.

Figure 5. Fit of test results: (a) role of dowel action and NLFE analysis prediction; (b) role of the distribution of crack displacements; (c) role of cracking in bottom flange.

Figure 6. Beam R7 [10]: diagrams of the shear stress in the concrete, in the cross-section at 1/4 (a,b), dowel action not included (c) and included (d).

RESULTS OF THE ANALYSIS AND FITTING OF TEST DATA

The results of the analysis and the most relevant test data [10] are shown in Table 1 and Figs.5-8. All the results and data regard the ultimate load situation.

In Table 1, columns 4 and 5 refer to the ultimate shear stress, put in a dimensionless form (τ_u/f_c'): on the whole the agreement is good, as also shown in Fig.5a, where the computed shear strengths (suffix "th" for "theoretical") compare satisfactorily with the test results (suffix "exp" for "experimental"). The non-uniform distribution of crack opening and slip (Fig.5b) and the cracking of the bottom flange (Fig.5c) play a significant role.

Also the values of the angles θ_u^σ and α (Table 1, columns 6-8, $\theta_u^\sigma=$ orientation of the principal compressive stress, α = orientation of the Diagonal Compression Field) are generally in good agreement. Moreover, σ_2 (Table 1, column 9 Diagonal Compression stress) is always smaller than concrete strength in compression, even if a 20-30% reduction is applied to f_c' in order to introduce the damage caused by stirrup-induced strains, and a further 15% reduction is applied for long-term loading. It is worth noting that in the proposed approach the crack direction is given the value 40° with respect to the beam axis, as suggested by the crack patterns shown in [10] (ϕ=38-43°, see Beam R1 in Fig.7).

In Fig.6 the shear stress in the concrete is plotted in the cross-section, at 1/4: the discontinuities (Figs.6c,d) are due to the shear cracks (Figs.6a,b). The introduction of dowel action (which affects the bending and the shear in the struts) does not markedly modify the stresses along the axis of the strut (points B, Figs.6a,c,d), while interface discontinuities are greatly increased.

For the sake of comparison, the values of the shear stress in the concrete, as given by the NLFE Analysis [7], are indicated by a solid circle in Figs.6c,d: these values belong to the 4 Gaussian points closest to the beam section at 1/4 (see also Fig.7). The agreement between Limit Analysis and NLFE Analysis is striking, even if NLFE Analysis cannot model crack-induced discontinuities, because crack displacements are smeared.

Finally, in Fig.8 the local contribution ($\Delta\tau_{INT}$) of aggregate interlock to shear transfer is plotted along the cross section of the web. Fig.8 does not refer to any specific beam tested in [10]: only the depth of the section (H=450 mm), the stirrup diameter ($\emptyset$6 mm) and the stirrup strength (f_{sy} =745 MPa) are the same as in Regan's beams, while the stirrup ratio and spacing are among the unknowns and are optimized, according to the most general use of the code based on the proposed Limit Analysis model (see also [5,6]): of course, the larger the τ_u value, the larger the required stirrup ratio ρ_{st}, which does not appear in Fig.8.

DESIGN CHARTS FOR THE STIRRUPS

The theoretical approach above has been applied to the evaluation of the optimum degree of shear reinforcement, $\omega_{st}= \rho_{st} f_{sy}/f_c'$, as a function of the ultimate shear stress, $\tau_u= V_u/bz$ (Fig.9).

The two limitations of the diagrams (τ_u/f_c'= 0.075-0.10, and $\tau_u/f_c'\approx$ 0.35) are both related to the brittle failure of the Diagonal Compression Field ($\sigma_2 = f_c'$= 0.75x0.85 f_c', where 0.75 introduces the damage due to the stirrup-induced tensile strains, and 0.85 stands for long-term

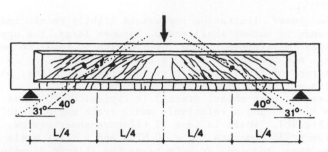

Figure 7. Example of crack pattern, beam R1 [10]: average crack angle at 1/4 = φ = 40° (input data in the proposed approach); angle of the Diagonal Compression Field = α = 31° (output data in the proposed approach).

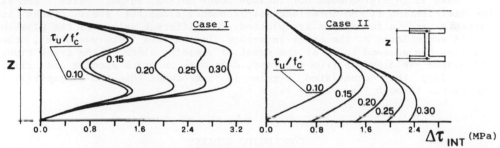

Figure 8. Limit analysis : diagrams of the local contribution of aggregate interlock to shear transfer in the web. φ = 40°, f'_c = 45 MPa, f_{sy} = 745 MPa, z = 380 mm, same section as in [10].

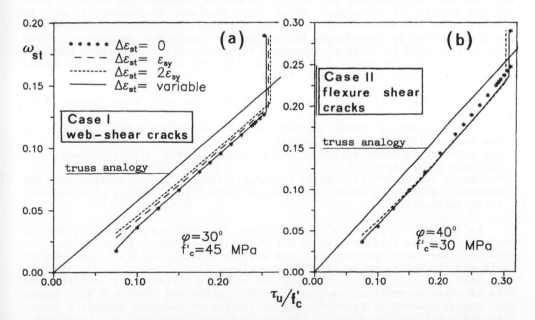

Figure 9. Charts for the optimum design of shear reinforcement in thin-webbed beams : z = 1000 mm, f_{sy} = 440 MPa, d_{st} = 8 mm.

loading). The lower limitation represents lightly reinforced beams, where the angle α tends to become small and σ_2 becomes large. The upper limitation regards heavily reinforced beams, for which σ_2 tends to become a linear function of τ_u, and the angles α and φ tend to coincide.

The diagrams of Fig.9 are more or less straight lines, parallel to the full heavy line which represents truss analogy.

Since Case I (web-shear cracks) is typical of PC beams, a high-grade concrete ($f_c' = 45$ MPa) and a relatively small crack angle (φ=30°) have been introduced (Fig.9b), while in Case II (flexure-shear cracks, typical of RC beams) a medium-grade concrete ($f_c' = 30$ MPa) and a crack angle close to π/4 (φ=40°) have been adopted. In both cases the stirrup diameter is the same ($d_{st}= 8$ mm, # 3), but the role of this parameter is not relevant (see [5] and [11]), except for the evaluation of the optimum web thickness ($b≈d_{st}tg\phi$ [6, 11].)

Case II in Fig.9b does not include dowel action. In both cases I and II the distribution of the interface displacements along the shear cracks is variable and a few curves are plotted, for different values of the plastic strain accumulated at the crack interface. In the case "$\Delta\varepsilon_{st}$ variable", the stirrups are assumed to be at the onset of yielding close to the crack tips, and beyond yielding far from the tips, depending on crack opening and slip. Eventually, the accumulated strain is introduced into the compatibility equations, through its average value $\Delta\varepsilon_{st}$.

CONCLUDING REMARKS

The limit analysis of a thin-webbed beam failing in shear due to stirrup yielding and to the local crushing of the concrete struts in shear-compression is performed here by improving and extending a previous model based on the truss analogy and on the concept of Diagonal Compression Field. Two new aspects are considered here: the relative displacements at the interface of the shear cracks are given suitable, non-uniform distributions and dowel action is introduced as a further resistant mechanism, whose effects are added to those of aggregate interlock, stirrup/concrete bond and shear/bending in the concrete struts, the last three having been introduced in a previous phase of the researchwork.

The results of the analysis and the comparison with some recent test data show that:
- the contribution of dowel action to the ultimate shear capacity is substantial (flexure-shear cracks);
- the non-uniform distribution of crack displacements always reduces the ultimate shear capacity;
- values as low as 20°-25° may be reasonably adopted in the design, for the orientation of the Diagonal Compression Field (truss analogy with variable strut inclination).

As regards the practical implications, the proposed approach allows the evaluation of the optimum stirrup ratio and web thickness, provided that the material and bond properties, the average crack orientation, the section depth and the design shear capacity are given.

Last but not least, the proposed model is suitable for analyzing the role of other parameters, such as the fiber amount and type in F.R.C. beams, the damage caused by previous load cycles (which decrease aggregate inter-lock and bond) and the share of the shear strength developed by the chords.

255

ACKNOWLEDGEMENTS

The financial support of the Ministry of University, Research and Technological Development (MURST, funds 40%, 1989-90) is gratefully acknowledged.

REFERENCES

1. Nielsen, M.P., Braestrup, M.W. and Bach, F., Rational Analysis of Shear
   in Reinforced Concrete Beams. IABSE Proceedings P15/78, 1978.

2. Collins, M.P., Towards a Rational Theory for R/C Members in Shear,
   Journal of the Structural Division. ASCE, 104(4), 1978, pp.649-666.

3. Vecchio, F.J. and Collins, M.P., The Modified Compression Field Theory
   for Reinforced Concrete Elements Subjected to Shear. ACI Journal, Technical Paper No.83-22, March-April 1986, pp.219-231.

4. Kupfer, H., Mang, R. and Karavesyroglou, M., Failure of the Shear Zone
   in R/C and P/C Beams - Theoretical Analysis Including Aggregate Interlock (in German). Bauingenieur Bulletin No.58, 1983, pp.143-149.

5. Dei Poli, S., Gambarova, P.G. and Karakoç, C., Aggregate Interlock Role
   in R/C Thin-Webbed Beams in Shear. Journal of Structural Engineering,
   ASCE, 113(1), 1987, pp.1-19.

6. Dei Poli, S., Di Prisco, M. and Gambarova, P.G., Stress Field in Web of
   RC Thin-Webbed Beams Failing in Shear. Journal of Structural Engineering, ASCE, 116(9), 1990, pp.2496-2515.

7. Di Prisco, M., Gambarova, P.G. and Valente, G.F., Evolutive versus Limit
   Analysis in Modelling RC Thin-Webbed Beams Failing in Shear. Proceedings
   of SCI-C 1990, Second International Conference on Computer Aided
   Analysis and Design of Concrete Structures, Zell am See (Austria), April
   1990, pp.21-32.

8. Bazant, Z.P. and Gambarova, P.G., Rough Cracks in Reinforced Concrete.
   Journal of the Structural Division, ASCE, 106(4), 1980, pp.819-842.

9. Gambarova, P.G. and Karakoç, C., A New Approach to the Analysis of the
   Confinement Role in Regularly Cracked Concrete Elements. Transactions of
   the 7th SMiRT Conference, Vol.H, Paper H5/7, Chicago, August 1983,
   pp.251 261.

10. Regan, P.E. and Rezai-Jorabi, H., The Shear Resistance of Reinforced
    Concrete I-Beams. Studi e Ricerche, Vol.9-87, School for the Design of
    RC Structures, Politecnico di Milano, Milan (Italy), 1988, pp.305-321.

11. Di Prisco, M. and Gambarova, P.G., On the ultimate shear behavior of RC
    and PC thin-webbed beams (in Italian). Studi e Ricerche, Vol. 8-86,
    School for the Design of RC Structures, Politecnico di Milano, Milan
    (Italy), 1987, pp.199-268.

A CONSISTENT MODEL FOR THE DESIGN OF SHEAR REINFORCEMENT IN SLENDER BEAMS WITH I- OR BOX-SHAPED CROSS-SECTION

H. KUPFER
Professor for Structural Engineering

H. BULICEK
Research Assistant

Technische Universität München, Germany

ABSTRACT

This paper presents a theoretical model for calculating the shear capacity in ultimate limit state (ULS) of slender RC or PC beams with vertical shear reinforcement. The strain state of the web is taken into account being compatible to the longitudinal strain in the chords and the average vertical strain in the stirrups. The compatibility of strains and the equilibrium of stresses in the web are linked together by the use of applicable constitutive laws including aggregate interlock behavior.

The results of the numerical evaluation show the substantial influence of axial strains of the beam on the shear capacity. The low amount of shear reinforcement according to the theory of plasticity can only be confirmed if axial compressive strains prevail in the beam. Axial tensile strains cause a significant decreasing effect on the shear capacity due to the reduced action of aggregate interlock.

NOTATION LIST

φ inclination of shear cracks in the web
θ inclination of the principal compressive stress σ_2 in the web
δ inclination between crack inclination φ and inclination θ of the principal compressive stress σ_2 in the web, $\delta = \varphi - \theta$
V_s shear force carried by the web concrete in ultimate limit state under consideration of inclined tendons and inclined chords
E_s Modulus of elasticity of steel, $E_s = 210000$ MPa
G_s secant shear modulus of concrete
K_s secant compressive modulus of concrete
f_y yield strength of steel
f_{ct} tensile strength of concrete
f_c cylinder compressive strength of concrete
f_{cube} cube compressive strength of concrete, $f_{cube} \approx 1.1 \ f_c$

f_{c2}	uniaxial compressive strength of the concrete in the web (s. equ. 13)
τ	acting shear stress in the web, $\tau = V_s/(b \cdot z)$
τ_f	shear stress due to load effect at the onset of cracking
τ_p	shear stress due to the effect of inclined tendons or prestressing shear reinforcement (negative if contrary to τ)
τ_c	shear stress parallel to the crack
τ_0	octaeder shear stress
σ_c	normal stress rectangular to the crack
σ_d	normal stress in the struts parallel to the shear cracks
σ_y	vertical normal stress in the web concrete produced by the forces of the stirrups, $\sigma_y = -\mu \cdot f_y$
σ_x	axial normal stress in the web concrete
σ_{xp}	axial stress due to prestressing forces, derived from the uncracked state of the concrete
σ_{xf}	axial stress due to load effect at the onset of cracking
σ_2	principal stress in the web directed parallel to the inclination θ
σ_1	principal stress in the web directed rectangular to the inclination θ
ϵ_{20}	principal strain of the web concrete, directed associated to σ_2
ϵ_{10}	principal strain of the web concrete, directed associated to σ_1
ϵ_x	overall smeared axial strain of the web (including smeared v and w)
ϵ_y	overall smeared vertical strain of the web (incl. smeared v and w)
ϵ_s	strain of the shear reinforcement at the shear cracks
$\Delta\epsilon_s$	additional smeared vertical strain due to anchorage slip of the shear reinforcement
ϵ_{cs}	shrinkage strain of the concrete ($\epsilon_{cs} < 0$)
k	value taking account of the tension stiffening effect referring to the shear reinforcement
A_{sw}	cross-section of the stirrup legs within the area s·b
a_{sw}	cross-section of the shear reinforcement per unit length, $a_{sw} = A_{sw}/s$
s	stirrup spacing in axial direction
b	width of the web under consideration of a diminution due to ducts in the web
d_s	diameter of the steel used for shear reinforcement
v	shear crack displacement parallel to the shear crack
w	shear crack opening rectangular to the shear crack
a	average crack spacing -measured rectangular to the crack
z	inner lever arm
n	ratio between σ_{xf} and τ_f, $n = \sigma_{xf}/\tau_f$
ν	acting shear stress ratio, $\nu = 2\tau/f_{c2}$
μ	geometrical shear reinforcement ratio, $\mu = a_{sw}/b$
ω	mechanical shear reinforcement ratio, $\omega = 2\mu \cdot f_y/f_{c2}$

tensile stresses and tensile strains are defined as being positive

INTRODUCTION

The stress state in cracked webs of slender RC or PC beams is characterized by a biaxial stress field in the concrete and a uniaxial tension field represented by the shear reinforcement [1-3]. In general the inclination θ of the principal compressive stress σ_2 in the concrete differs from the crack inclination φ, so that mostly flatter angles adjust for θ than for φ (for notations see figure 1). This deviation between θ and φ can predominantly be founded by the action of aggregate interlock in the crack plane. The crack stresses due to aggregate interlock depend on the magnitude of the crack displacement v (parallel to φ) and the crack opening w (rectangular to φ). On the other hand the values of v and w are determined by the kinematic state in the beam: The strain state of the web concrete (including the smeared strain state produced by v and w) has to be compatible with the strain in the stirrups and the axial strain in the beam. The strains of the web concrete itself are obtained from the concrete stresses -fulfilling equilibrium conditions- by the use of constitutive laws.

The following evaluation only deals with slender beams with I- or box-shaped cross-section and vertical shear reinforcement neglecting the influences due to bending of the struts as well as due to dowel action of the longitudinal reinforcement.

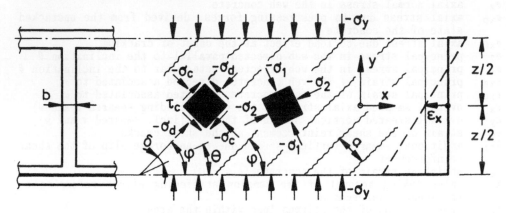

Figure 1. Notations

THEORETICAL   FOUNDATION

Calculation of the Crack Inclination φ

Shear cracks form in the direction of the principal compressive stresses when principal tensile stresses exceed the concrete tensile strength f_{ct}. For calculating the crack inclination φ the principal stresses have to be derived from the uncracked state of the concrete.

Hence the following equation for φ can be derived in case of acting axial stresses σ_{xp} due to prestressing forces only:

$$\cot \varphi = \sqrt{1 - \frac{\sigma_{xp}}{f_{ct}}} \tag{1a}$$

If the stresses are caused only by the load effects σ_{xf} and $\tau_f = \sigma_{xf}/n$ acting at the onset of cracking the following condition is obtained:

$$\cot \varphi = -\frac{n}{2} + \sqrt{1 + \left(\frac{n}{2}\right)^2} \tag{1b}$$

If the stresses are produced by load effects as well as by prestressing forces the crack inclination is given by:

$$\cot \varphi = -\frac{n}{2} + \sqrt{1 + \left(\frac{n}{2}\right)^2 - \frac{\sigma_{xp} - n \cdot \tau_p}{f_{ct}}} \tag{1c}$$

259

Equilibrium Conditions of Stresses

The stress state in the inclined struts beyond two neighbouring shear cracks is given by the normal stress σ_d parallel to the crack inclination φ and the acting crack stresses σ_c and τ_c due to aggregate interlock, neglecting the tensile stresses due to tension stiffening of the stirrups (see figure 2).

Along a horizontal section the shear stress τ and the compressive stress $-\sigma_y$, which is produced by the forces of the stirrups, are acting.

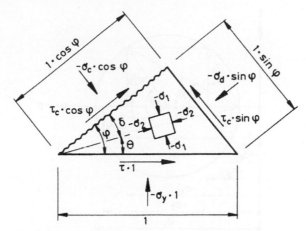

Figure 2. Stress state in the web concrete

The following equations have to be fulfilled as equilibrium conditions in horizontal (equ. (2)) and vertical (equ. (3a)) direction:

$$- \sigma_d = \frac{2\,\tau}{\sin 2\varphi} + 2\,\tau_c \cdot \cot 2\varphi - \sigma_c \tag{2}$$

$$- \sigma_y = (\tau - \tau_c) \cdot \tan \varphi - \sigma_c \tag{3a}$$

or
$$- \sigma_y = \tau \cdot \tan \theta - \sigma_1 \tag{3b}$$

The principal stresses σ_1 and σ_2 in the web concrete can be derived by equation (4):

$$\sigma_{1,2} = \frac{1}{2} \cdot \left[\sigma_d + \sigma_c \pm \sqrt{(\sigma_d - \sigma_c)^2 + 4\,\tau_c{}^2} \right] \tag{4}$$

The following equation gives the inclination δ between the principal compressive stress σ_2 and the crack inclination φ:

$$\tan 2\delta = \frac{2\,\tau_c}{-\sigma_d + \sigma_c} \tag{5a}$$

The inclination θ of the principal compressive stress σ_2 can be derived by

$$\theta = \varphi - \delta \tag{5b}$$

or by
$$\cot 2\theta = \cot 2\varphi + \frac{\tau_c}{\tau} \cdot \frac{1}{\sin 2\varphi} \tag{5c}$$

The relations between the stresses and their inclinations in the web concrete can also be illustrated by Mohr's stress circle which is shown in figure 3:

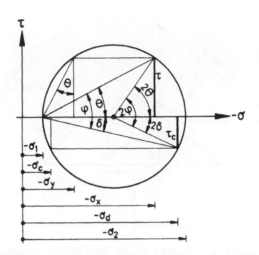

Figure 3. Mohr's stress circle for the stress state in the web concrete

Kinematic Conditions of Strains

In any considered direction the overall smeared strain of the web results from the pure concrete strain ϵ_0 and the smeared average strains ϵ_w and ϵ_v due to crack opening w and crack displacement v:

$$\epsilon = \epsilon_0 + \epsilon_w + \epsilon_v$$

Thus the overall smeared strains ϵ_x and ϵ_y in the considered directions x and y can be expressed as:

$$\epsilon_x = \epsilon_{xo} + \epsilon_{xw} + \epsilon_{xv} \tag{6a}$$

$$\epsilon_y = \epsilon_{yo} + \epsilon_{yw} + \epsilon_{yv} \tag{6b}$$

The pure concrete strains ϵ_{xo} and ϵ_{yo} can be derived from the principal strains ϵ_{10} and ϵ_{20} of the web concrete (see figure 4a):

$$\epsilon_{xo} = \epsilon_{10} \cdot \sin^2 \theta + \epsilon_{20} \cdot \cos^2 \theta \tag{7a}$$

$$\epsilon_{yo} = \epsilon_{10} \cdot \cos^2 \theta + \epsilon_{20} \cdot \sin^2 \theta \tag{7b}$$

The smeared uniaxial strain state w/a rectangular to the crack inclination φ has components in x and y direction (see figure 4b):

$$\epsilon_{xw} = w/a \cdot \sin^2 \varphi \qquad (8a)$$

$$\epsilon_{yw} = w/a \cdot \cos^2 \varphi \qquad (8b)$$

The smeared shear strain v/a parallel to the crack inclination φ has components in x and y direction (see figure 4c):

$$\epsilon_{xv} = -v/a \cdot \sin \varphi \cos \varphi \qquad (9a)$$

$$\epsilon_{yv} = v/a \cdot \sin \varphi \cos \varphi \qquad (9b)$$

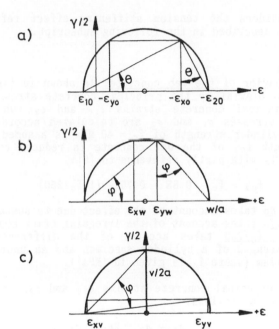

Figure 4. Mohr's strain circles for
a) pure concrete strains
b) smeared strains due to crack opening w
c) smeared strains due to crack displacement v

By the use of the preceding equations 6 - 9 the smeared strains w/a and v/a follow as:

$$w/a = \epsilon_x + \epsilon_y - \epsilon_{10} - \epsilon_{20} \qquad (10a)$$

$$v/a = - \epsilon_x \cot\varphi + \epsilon_y \tan\varphi - \epsilon_{10}\frac{\sin^2\varphi - \sin^2\theta}{\sin\varphi\cos\varphi} + \epsilon_{20}\frac{\cos^2\varphi - \sin^2\theta}{\sin\varphi\cos\varphi} \qquad (10b)$$

In the presented model the average axial strain ϵ_x of the web is taken for granted or can be derived from the longitudinal strain of the chords of an I- or box-shaped beam according to the following equation:

$$\epsilon_x = \frac{\epsilon_{x,top\ chord} + \epsilon_{x,bottom\ chord}}{2} \tag{11}$$

where the longitudinal strains of the chords then have to be calculated by taking into account the influence of the axial stresses σ_x in the web (see figure 3). The influence of the axial shrinkage of the chords and the web cancel each other out.

The overall vertical strain ϵ_y of the web is given by the strain ϵ_s of the stirrups and an additional strain for anchorage slip $\Delta\epsilon_s$ of the shear reinforcement. The shrinkage $\epsilon_{cs} < 0$ of the concrete is (referring to the results) equivalent to an elongation of the stirrups:

$$\epsilon_y = (\epsilon_s + \Delta\epsilon_s - \epsilon_{cs}) \cdot k \tag{12}$$

The factor k considers the tension stiffening effect referring to the stirrup strains as described in the following subscript.

Constitutive Laws

Nonlinear behavior of the web concrete: As shown in figure 1 the web concrete is loaded biaxial by the principal concrete stresses σ_1 and σ_2. The resulting principal concrete strains ϵ_{10} and ϵ_{20} which correspond nonlinear to the stresses σ_1 and σ_2 are calculated according to [4] as follows where a cylinder strength of $f_c = 40$ MPa is assumed. The uniaxial compressive strength f_{c2} of the web concrete is reduced compared to the cylinder strength f_c with partly reference to [5]:

$$f_{c2} = f_c \cdot 0.85 \cdot 0.75 \cdot (1 - f_c/250) \tag{13}$$

where   0.85 takes account of the effect due to sustained load
0.75 takes account of the irregular crack course
$1 - f_c/250$ takes account of the difference between the strength of a cylinder specimen and an uncracked concrete prism (where f_c is given in [MPa]).

The resulting principal concrete strains ϵ_{10} and ϵ_{20} are given by the use of reference [4] as follows:

$$\epsilon_{10} = \frac{\sigma_1 + \sigma_2}{9\ K_s} + \frac{2\sigma_1 - \sigma_2}{6\ G_s} \tag{14a}$$

$$\epsilon_{20} = \frac{\sigma_1 + \sigma_2}{9\ K_s} + \frac{2\sigma_2 - \sigma_1}{6\ G_s} \tag{14b}$$

where the secant shear modulus G_s is

$$G_s = 13000 \cdot \left[1 - 3.5\ \left(\frac{\tau_0}{f_{c2}}\right)^{2.5}\right] \qquad [\text{MPa}] \tag{15a}$$

the secant compressive modulus K_s is

$$K_s = 17000 \cdot \left[1 - 1.6\ \left(\frac{\tau_0}{f_{c2}}\right)^{1.8}\right] \qquad [\text{MPa}] \tag{15b}$$

and the octaeder shear stress τ_0 is

$$\tau_0 = \sqrt{2}/3 \cdot \sqrt{\sigma_1{}^2 + \sigma_2{}^2 - \sigma_1\sigma_2} \qquad (15c)$$

Stress and strain of shear reinforcement: In the area where the shear cracks cross the stirrups the stirrup stresses are fixed to $f_y = 500$ MPa; the associated strains $\epsilon_s \geq f_y/E_s$, however, are assumed to reach a magnitude leading to the optimal shear capacity [6]. The valuation of the tension stiffening effect after repeated service load is derived assuming an acting average concrete tensile stress of $0.20\, f_{ct}$ along the stirrup axis so that the factor k in equation (12) is:

$$k = 1 - 0.20\, \frac{f_{ct}}{\mu \cdot f_y} = 1 - 0.40\, \frac{f_{ct}}{f_c{}^2} \cdot \frac{1}{\omega}$$

assuming $\frac{f_{ct}}{f_c{}^2} \approx 0.15 \rightarrow \qquad k = 1 - 0.06\, \frac{1}{\omega} \qquad (16)$

Aggregate interlock: To describe the correlations between the crack stresses σ_c, τ_c and the obtained magnitudes of crack opening w and crack displacement v the following equations - valid for the range of application within this model- are given in reference [7]:

$$\tau_c = -\,f_{cube}/30 + \Big[1.8w^{-0.8} + (0.234w^{-0.707}-0.20)\cdot f_{cube}\Big]\cdot v \qquad (17a)$$
$$\geq 0$$

$$\sigma_c = f_{cube}/20 - \Big[1.35w^{-0.63} + (0.191w^{-0.552}-0.15)\cdot f_{cube}\Big]\cdot v \qquad (17b)$$
$$\leq 0$$

where f_{cube}, τ_c, σ_c are in [MPa] and v, w in [mm]

Shear crack spacing: The average spacing of shear cracks in slender beams is mainly influenced by the mechanical shear reinforcement ratio ω, the diameter d_s of the shear reinforcement and the inner lever arm z. Therefore the following semi-empirical formula is derived for the crack spacing "a" based on an evaluation of series of test specimens under shear load which are published in reference [8] and reference [9]:

$$\frac{1}{a} = \frac{5\mu}{d_s} + \frac{2}{z} = \frac{2.5\omega f_{c2}}{d_s f_y} + \frac{2}{z} \qquad (18)$$

where "a" is measured perpendicular to the crack inclination φ.

NUMERICAL EVALUATION AND RESULTING SHEAR CAPACITY

The numerical evaluation is made for a crack inclination of $35^0 \leq \varphi \leq 55^0$ where the average axial strain ϵ_x is varied in the range of $-0.001 \leq \epsilon_x \leq +0.001$.

The obtained design curves (see figure 5) point out that the results according to the plastic method (assuming an optimal utilized uniaxial compression field $(\sigma_2 = -f_{c2}, \sigma_1 = 0)$) represent an "upper bound" for the shear capacity ν of slender RC- or PC-beams. This is supported by the results of an evaluation in reference [10] based on a model with a uniaxial compressive field. There the action of aggregate interlock is considered approximately by the assumption that the resulting crack opening adjusts rectangular to the compression field.

264

Furthermore the numerical evaluation of the presented model shows that a biaxial stress state (characterized by a biaxial compression field or a biaxial compression/tension field) prevails in the web concrete. It turns out, however, that the stress σ_1 is small so that its neglection seems to be allowable for design.

The curvatures in figure 5 prove the decreasing influence of axial tensile strains as well as the increasing influence of axial compressive strains upon the shear capacity as also confirmed in reference [11] by the results of an extensive test program.

At least the curves for variable crack inclinations φ represent its influence upon the shear capacity. It turns out that the inclination of cracks has a small influence upon the shear capacity in the range of moderate acting shear stresses but a clear one in the range of small and high acting shear stresses.

Taking account of axial strains ϵ_x ($-2\text{‰} \leq \epsilon_x \leq +2\text{‰}$) the range of application for practical design is located between the "Ritter-Mörsch-Line" and the limit line according to the theory of plasticity.

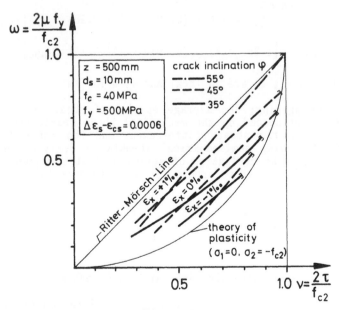

Figure 5.
Design-curves for the mechanical shear reinforcement ratio ω in terms of the acting shear stress ratio ν in ultimate limit state

SUMMARY AND CONCLUSIONS

In this paper a consistent model has been developed for evaluating the shear capacity in ultimate limit state of slender RC oder PC beams with vertical shear reinforcement. The equilibrium in the web as well as the compatibility of strains and aggregate interlock conditions have been fulfilled. It has been found that a biaxial stress state prevails in the web concrete characterized by a predominating principal compressive stress and a perpendicular acting subordinated principal compressive or tensile stress.

Furthermore the favourable effect of axial compressive strains and the diminishing influence of axial tensile strains in the beam upon the shear capacity are pointed out taking account of the shear crack inclination.

As a final conclusion it can be said that the shear capacity ν according to the theory of plasticity has turned out as an "upper bound" with respect to ν so that a significant overestimation is committed if axial tensile strains prevail in the beam.

REFERENCES

1. Kupfer, H., Mang R. and Karavesyroglou M., Bruchzustand der Schubzone von Stahlbeton- und Spannbetonträgern- Eine Analyse unter Berücksichtigung der Rißverzahnung. In Bauingenieur, Ernst und Sohn, Berlin ,1983, 58, pp. 143-9.

2. Kirmair, H., Das Schubtragverhalten schlanker Stahlbetonbalken. In Heft 385 DAfStb, Ernst und Sohn, Berlin, 1987.

3. Kirmair, H., Mang R., Das Tragverhalten der Schubzone schlanker Stahlbeton- und Spannbetonträger bei Biegung mit Längskraft. In Bauingenieur, Ernst und Sohn, Berlin, 1987, 62, pp.165-70.

4. Kupfer, H. B., Gerstle K. H., Behavior of Concrete under Biaxial Stresses. In Journal of the Engineering Mechanics Division, ASCE, Cleveland, August 1973, pp. 853-66.

5. CEB-FIP Model Code 1990, First Draft, Bulletin d'Information N 196, Comite Euro-International du Beton, Lausanne, March 1990.

6. Cerruti L, Marti P., Staggered Shear Design of Concrete Beams: Large Scale Tests. In Can. J. Civ. Eng. 14, 1987, pp. 257-68.

7. Walraven, J.C., Experiments on shear transfer in cracks in concrete - Part 2: Analysis of results. Delft University of Technology, 1979, Dept. of Civil Engineering, Rpt. 5-79-10.

8. Kupfer, H., Baumann, T., Versuche zur Schubsicherung und Momentendeckung von profilierten Stahlbetonbalken. In Heft 218 DAfStb, Ernst und Sohn, Berlin, 1972.

9. Kupfer, H., Guckenberger K., Versuche zur Schubtragwirkung von profilierten Stahlbeton- und Spannbetonträgern mit überdrückten Gurtplatten. In Heft 377 DAfStb, Ernst und Sohn, Berlin, 1986.

10. Reineck K.-H., Hardjasaputra H., Zum Dehnungszustand bei der Querkraftbemessung profilierter Stahlbeton- und Spannbetonträger. In Bauingenieur, Ernst & Sohn, Berlin, 1990, pp. 73-82.

11. Bhide S. B., Collins M.P., Reinforced Concrete Elements in Shear and Tension. Publication No. 87-02, University of Toronto, January 1987.

INTERACTIVE PLASTIC DESIGN OF WALLS

JOHANNES MAIER*
Wettingen, Switzerland

ABSTRACT

The strength design of reinforced concrete walls may be based on the lower bound theorem of the theory of plasticity. The theorem requires the finding of a stress field which meets the equilibrium and static boundary conditions everywhere in the structure and which does not violate the yield condition anywhere. The introduction of stress discontinuity lines allows a stress field which consists of polygonal parts with a constant state of stress.

The article presents the results of a stress analysis for a six-storey-high wall system with openings, loaded with horizontal and vertical forces. The analysis was facilitated by the use of a recently developed interactive computer program.

INTRODUCTION

Strength, functionality and durability are the basic demands for the structural design. One of the first steps in the design process of reinforced concrete structures is the determination of the necessary concrete dimensions and the identification of those parts, which need reinforcement by steel bars or by prestressing. In a following step a layout of this reinforcement has to be found. Hitherto the design of walls was based on empirical rules or on linear-elastic analysis without consideration of stress redistribution.

The description of the actual mechanical condition of a solid body requires the knowledge of the stresses and of the deformations in every point of its volume. As soon as material behaviour other than elastic material behaviour is considered, even the load history plays an important role. For real structures the ever-present—but unknown—initial residual stresses must be added to the results of the stress analysis.

Extensive tests and many years of practical application have confirmed that the methods based on the limit theorems provided by the theory of plasticity can be very useful for the strength design of beams and slabs. The research project *Reinforced Concrete Wall Structures*, carried out at the Swiss Federal Institute of Technology in Zürich, investigated the applicability of these methods on wall structures [1, 2]. The results of the experimental part of this research project are presented in the article *Shear Wall Tests*.

*Former research engineer at the Institute of Structural Engineering, ETH-Hönggerberg, CH-8093 Zürich, Switzerland.

THE LOWER BOUND THEOREM OF THE THEORY OF PLASTICITY

The following lower bound theorem of the theory of plasticity, which was formulated forty years ago, is valid if the material behaviour is elastic–ideal-plastic or rigid–ideal-plastic, and if equilibrium can be formulated for the undeformed system:

> *Every load for which a stable and statically admissible stress field exists is not higher than the ultimate load of the structure.*

A stress field is statically admissible if it meets the *equilibrium* and *static boundary conditions* everywhere in the structure. It is stable as long as the *yield condition* is not violated anywhere.

The lower bound theorem raises no requirements on the strain field. It allows the safe design of structures for ultimate load without consideration of either initial residual stresses or load history. However, contrary to the linear theory of elasticity, independent load cases cannot be superimposed.

THE STRESS FIELD

The equilibrium condition for the stress field of a plane structure loaded only by membrane forces is given by two partial differential equations between the three stress components σ_x, σ_y, τ_{xy} and the two components f_x, f_y of the load vector:

$$\frac{\partial \sigma_x}{\partial x} + \frac{\partial \tau_{xy}}{\partial y} + f_x = 0 \tag{1}$$

$$\frac{\partial \tau_{xy}}{\partial x} + \frac{\partial \sigma_y}{\partial y} + f_y = 0 \tag{2}$$

A continuous analytic solution for this system of coupled differential equations and the static boundary conditions of the structure is rarely found. Most solutions presented so far assume in addition a linear-elastic material behaviour. Therefore the consideration of a plastic stress redistribution is impossible. But the load associated with such a stress field is below the ultimate load as long as the yield condition is not violated.

The subdivision of the structure in smaller parts by the introduction of stress discontinuity lines presents new possibilities for the development of a stress field. The equilibrium condition for an element on the discontinuity line requires that both sides have the same normal stress perpendicular to the discontinuity line, and the same shear stress:

$$\sigma_{nA} = \sigma_{nB} \qquad \text{and} \qquad \tau_{ntA} = \tau_{ntB} \tag{3}$$

Only the normal stresses σ_{tA} and σ_{tB} parallel to the discontinuity line may be different. The assumption of a constant state of stress in every part of the subdivided structure and of straight discontinuity lines facilitates the finding of the stress field. With the notations of Figure 1 the following relations can be formulated between the stresses σ_{rk}, τ_{rk} at the edge k, and the state of stress σ_x, σ_y, τ_{xy} in the interior of the subarea:

$$\sigma_{rk} = \sigma_x \cdot \cos^2\alpha_k + \sigma_y \cdot \sin^2\alpha_k + \tau_{xy} \cdot \sin 2\alpha_k \qquad k = 1, 2, \ldots, n \tag{4}$$

$$\tau_{rk} = 0.5 \cdot (\sigma_y - \sigma_x) \cdot \sin 2\alpha_k + \tau_{xy} \cdot \cos 2\alpha_k \qquad k = 1, 2, \ldots, n \tag{5}$$

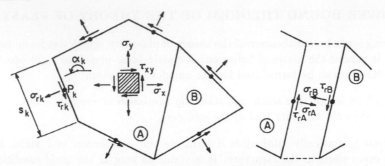

Figure 1: Polygonal areas with constant state of stress.

The two equations given by 3 claim continuity of the normal and shear stresses at the common edge of the subareas A and B in Figure 1. Two of the four equations given by 4 and 5 for this edge are therefore linearly dependent. With

n_g = number of internal discontinuity lines and external edges,

n_f = number of subareas with three unknown stress components,

n_r = number of unknown external edge stresses and

n_p = number of indeterminate coordinates of the subarea corners

the degree of static determination n_s of the stress field is given by the formula

$$n_s = 2 \cdot n_g - 3 \cdot n_f - n_r - n_p \qquad \begin{cases} n_s > 0: & \text{statically overdeterminate} \\ n_s = 0: & \text{statically determinate} \\ n_s < 0: & \text{statically underdeterminate} \end{cases} \qquad (6)$$

This equation is also valid for every subarea. The fulfillment of the requirement for a statically determinate stress field is a necessary condition for a solution of the stress analysis problem based only on equilibrium.

INTERACTIVE DESIGN OF THE STRESS FIELD

Hitherto the determination of a stress field with discontinuity lines has been a step-by-step graphic-analytic procedure: Starting from loaded, external edges of the structure a first subarea is constructed; in the following stage a new subarea is introduced, which is in equilibrium with the already determined part of the stress field. This procedure is continued until the whole structure is covered with subareas. An intuitively chosen load-bearing model serves as a guideline. The stresses of every subarea must be checked with the yield condition. The step-by-step development of the stress field is very similar to a jigsaw puzzle: The edges of adjoining parts must fit into one another in both cases and the state of stress can be compared with the completed puzzle.

The graphic-analytic procedure uses not only polygonal subareas with constant state of stress but also fan and arch shaped subareas with curved discontinuity lines and a more complex state of stress [3]. This may lead to analytic difficulties. However, quite often a qualitative knowledge of the stresses and of the geometry is sufficient.

The translation of this graphic-analytic step-by-step procedure to a computer program requires a reduction of the number of different subarea types. The state of stress in the subareas should be as simple as possible. But a limited freedom in the choice of load-bearing model is not acceptable. The program SFS (Stress Field Solver) presented in [4] uses a triangle with a constant state of stress as the only subarea type to meet all these requirements. The program builds the stress field by introducing blocks of two, three or four stress triangles at the same time instead of only one new stress triangle at every step. Figure 2 shows these three different types of stress triangle blocks.

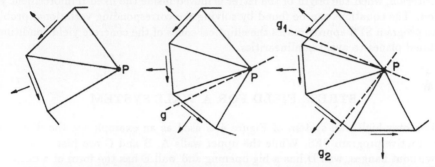

Figure 2: Blocks with two, three and four stress triangles used by the program SFS.

Condition 6 of the preceding section allows for the block of two triangles a free choice of the new point P. For the block of three triangles, equilibrium is only possible if the point P is on the line g. In the last case, the block with four new stress triangles, there is no freedom at all. The point P must be on the intersection of the two lines g_1 and g_2. The exclusive use of the block type with two stress triangles keeps the number of subarea edges constant and therefore prevents a triangularisation of the whole structure. The two other types give the necessary reduction of the number of subarea edges.

The yield condition limits the freedom of choice for the geometry of the new triangles. This leads to an interactive process, which the following flow diagram illustrates:

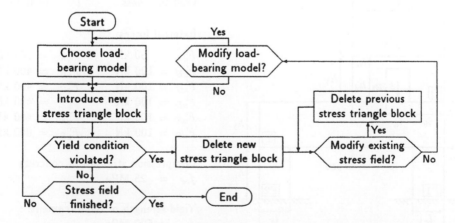

Figure 3: Flow diagram for the interactive stress field program SFS.

YIELD CONDITION

The yield condition of an ideal-plastic, reinforced concrete wall element can be composed from the separate yield conditions of the two components concrete and steel. Ignoring the tensional and the increased biaxial compression strength, the well known square yield criterion results for the concrete. In the σ_x-σ_y-τ_{xy}-space this is equivalent to two elliptical cones. The uniaxial steel bars can reach the yield strength in tension and in compression. The combined yield criterion is given by the cover of all possible positions of the concrete yield criterion, when the origin of the latter is moved inside the fixed reinforcement yield criterion. The equations can be found by solving the corresponding variational problem.

The program SFS approximates the elliptical cones of the concrete yield condition by 16 inclined planes to avoid nonlinearities.

STRESS FIELD FOR A WALL SYSTEM

The six-storey-high wall system of Figure 4 is used as an example for the design with the interactive program SFS. While the upper walls A, B and C are just normal shear walls without flanges, wall D has a big opening and wall E has the form of a deep beam supported by the more column like walls F1 and F2. The horizontal and vertical forces acting on every storey are applied through the slabs. This permits a free choice of the point of application for the horizontal force and thus allows a redistribution of the shear

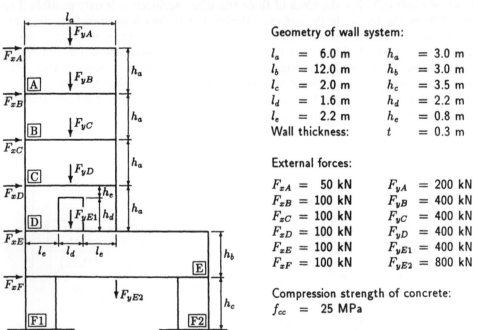

Geometry of wall system:

l_a	= 6.0 m	h_a	= 3.0 m
l_b	= 12.0 m	h_b	= 3.0 m
l_c	= 2.0 m	h_c	= 3.5 m
l_d	= 1.6 m	h_d	= 2.2 m
l_e	= 2.2 m	h_e	= 0.8 m
Wall thickness:		t	= 0.3 m

External forces:

F_{xA} =	50 kN	F_{yA}	= 200 kN
F_{xB} =	100 kN	F_{yB}	= 400 kN
F_{xC} =	100 kN	F_{yC}	= 400 kN
F_{xD} =	100 kN	F_{yD}	= 400 kN
F_{xE} =	100 kN	F_{yE1}	= 400 kN
F_{xF} =	100 kN	F_{yE2}	= 800 kN

Compression strength of concrete:
f_{cc} = 25 MPa

Yield strength of reinforcement:
f_{sy} = 500 MPa

Figure 4: Geometry of wall system and external forces.

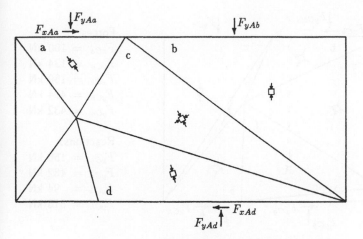

Forces:
$$F_{xAa} = 50 \text{ kN}$$
$$F_{yAa} = 67 \text{ kN}$$
$$F_{yAb} = 133 \text{ kN}$$

Reactions:
$$F_{xAd} = 50 \text{ kN}$$
$$F_{yAd} = 200 \text{ kN}$$

Figure 5: Stress field for wall A.

stress between the walls. The vertical forces are modelled by distributed, constant normal forces. As the program SFS does not accept external forces on internal discontinuity lines, every wall must be analysed separately.

The load path for the resultant of horizontal and vertical force indicates the possibility of a direct support load-bearing model for the walls A, B, C, F1 and F2. A triangular truss with two inclined compression struts and a horizontal tension tie allows the force transfer above the opening of wall D. This truss is supported by one inclined strut on each side of the opening. The load-bearing model chosen for the deep beam like wall E is a combination of direct and indirect support. The suspension of the vertical force acting on the lower edge requires additional vertical reinforcement of this wall.

The result of the computer analysis is presented in Figures 5 to 10. The state of stress of each subarea is indicated by the principal stresses. The length of the arrows gives a rough estimate of the corresponding stress values. Stress triangles without stress arrows

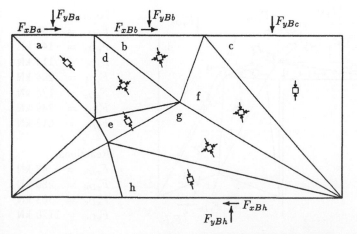

Forces:
$$F_{xBa} = 100 \text{ kN}$$
$$F_{yBa} = 100 \text{ kN}$$
$$F_{xBb} = 50 \text{ kN}$$
$$F_{yBb} = 222 \text{ kN}$$
$$F_{yBc} = 278 \text{ kN}$$

Reactions:
$$F_{xBh} = 150 \text{ kN}$$
$$F_{yBh} = 600 \text{ kN}$$

Figure 6: Stress field for wall B.

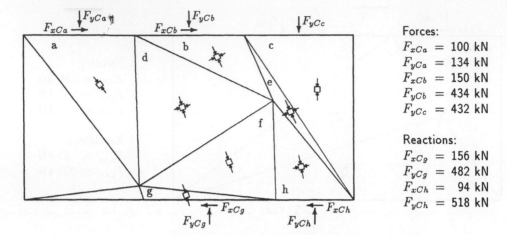

Forces:

$F_{xCa} = 100$ kN
$F_{yCa} = 134$ kN
$F_{xCb} = 150$ kN
$F_{yCb} = 434$ kN
$F_{yCc} = 432$ kN

Reactions:

$F_{xCg} = 156$ kN
$F_{yCg} = 482$ kN
$F_{xCh} = 94$ kN
$F_{yCh} = 518$ kN

Figure 7: Stress field for wall C.

are not loaded. The loaded or supported edges are marked by force arrows.

The analysis shows that the walls A, B, C, F1 and F2 do not need any reinforcement. The four stress triangles i to l above the opening of wall D require a horizontal reinforcement with $\varrho_x = 0.76\%$. The deep beam E needs both vertical and horizontal reinforcement: The stress triangles j to l, n to p and s to y should have a minimum vertical reinforcement of $\varrho_y = 0.16\%$; the necessary horizontal reinforcement of the stress triangles Φ and r to z is $\varrho_x = 5.22\%$. The reinforcement must always be anchored outside of the stress triangles.

The load-bearing capacity of the other stress triangles is limited by the concrete yield condition alone. The absolute value of all stresses in the walls A and B are below 1.0 MPa. The same is valid for wall C except the stress triangle e with $\sigma_2 = -2.2$ MPa. The concrete stresses above the opening of wall D reach considerably higher values. For the given loads an extreme stress $\sigma_2 = -10.7$ MPa results for the triangle h. But the

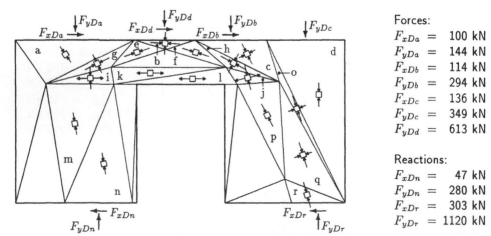

Forces:

$F_{xDa} = 100$ kN
$F_{yDa} = 144$ kN
$F_{xDb} = 114$ kN
$F_{yDb} = 294$ kN
$F_{xDc} = 136$ kN
$F_{yDc} = 349$ kN
$F_{yDd} = 613$ kN

Reactions:

$F_{xDn} = 47$ kN
$F_{yDn} = 280$ kN
$F_{xDr} = 303$ kN
$F_{yDr} = 1120$ kN

Figure 8: Stress field for wall D.

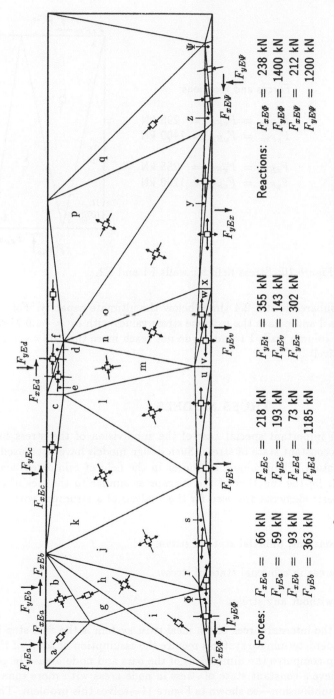

Forces:

$F_{xEa} = 66$ kN
$F_{yEa} = 59$ kN
$F_{xEb} = 93$ kN
$F_{yEb} = 363$ kN

$F_{xEc} = 218$ kN
$F_{yEc} = 193$ kN
$F_{xEd} = 73$ kN
$F_{yEd} = 1185$ kN

$F_{xEt} = 355$ kN
$F_{yEt} = 143$ kN
$F_{yEv} = 302$ kN

Reactions:
$F_{xE\Phi} = 238$ kN
$F_{yE\Phi} = 1400$ kN
$F_{xE\Psi} = 212$ kN
$F_{yE\Psi} = 1200$ kN

Stress triangels with vertical reinforcement: j...l, n...p and s...y.
Stress triangels with horizontal reinforcement: Φ and r...z.

Figure 9: Stress field for wall E.

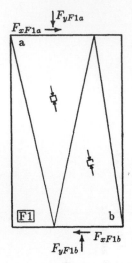

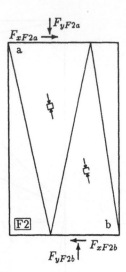

Forces and reactions:

$$F_{xF1a} = F_{xF1b} = \quad 295 \text{ kN}$$
$$F_{yF1a} = F_{yF1b} = 1400 \text{ kN}$$

$$F_{xF2a} = F_{xF2b} = \quad 255 \text{ kN}$$
$$F_{yF2a} = F_{yF2b} = 1200 \text{ kN}$$

Figure 10: Stress field for walls F1 and F2.

state of stress in this subarea is still 2.4 times below the ultimate capacity. For wall E the two stress triangles d and e have the extreme stress values with $\sigma_1 = -4.0$ MPa and $\sigma_2 = -14.0$ MPa. The column walls F1 and F2 do not reach more than $\sigma_2 = -3.2$ MPa and -2.8 MPa respectively.

TRUSS MODELS

The truss model is an important special case of the subdivision of the stress field in polygonal areas with a constant state of stress. Such design models have been used since the end of the last century. Hennebique, a pioneer in the field of reinforced concrete, described in his patent [5] the function of the stirrups as similar to the ties of a steel truss. The following basic elements are used for the analysis of a structure with a truss model:

- Truss bars in a constant, uniaxial state of stress.

- Polygonal node areas in a biaxial state of stress.

- Polygonal areas without any stress.

The calculation of the internal forces on an ideal truss system is the first step in the analysis. Statically underdeterminate systems require the assumption of some of the bar forces. The second step computes the dimensions of the bars and node areas. In general it is impossible to achieve a constant state of stress in node areas with more than three meeting truss bars. A subdivision—as shown in Figure 11—solves this problem. The last step checks the dimensions with the outline of the structure and against overlapping of the elements. This design procedure does not demand the use of a computer, especially if the stresses of all truss bars meeting in a node area are the same. However, the use

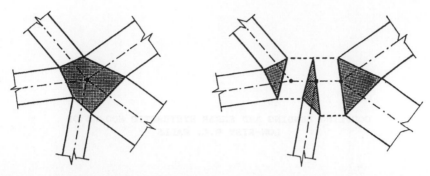

Figure 11: Subdivision of a truss node in stress triangles with constant state of stress.

of a special, interactive computer program [6] accelerates the analysis and permits the testing of different load-bearing models.

CONCLUSIONS

The lower bound theorem of the theory of plasticity allows the safe strength design of reinforced concrete wall structures. The construction of a possible stress field for a given load combination is facilitated by the use of discontinuity lines and subareas with a constant state of stress. The sometimes tedious process of calculation can be accelerated by interactive computer programs. Such a program leaves the choice of an appropriate load-bearing model to the designer and prevents thereby unsuitable solutions.

REFERENCES

[1] J. Maier, B. Thürlimann: *Bruchversuche an Stahlbetonscheiben,* Institut für Baustatik und Konstruktion, ETH-Zürich, Report 8003-1, Birkhäuser Verlag, Basel, 1985.

[2] J. Maier: *Tragfähigkeit von Stahlbetonscheiben,* Institut für Baustatik und Konstruktion, ETH-Zürich, Report 169, Birkhäuser Verlag, Basel, 1988.

[3] A. Muttoni, J. Schwartz, B. Thürlimann: *Design and Detailing of Reinforced Concrete Structures Using Stress Fields,* Swiss Federal Institute of Technology, Zürich, 1989.

[4] R. Hajdin: *Computerunterstützte Berechnung von Stahlbetonscheiben mit Spannungsfeldern,* Institut für Baustatik und Konstruktion, ETH-Zürich, Report 175, Birkhäuser Verlag, Basel, 1990.

[5] F. Hennebique: *Poutre légère et de grande résistance en béton de ciment avec barres de fer noyées,* Swiss Patent No. 6533 Class 5, Bundesamt für geistiges Eigentum, Bern, 1893.

[6] M. Schlaich: *Computerunterstützte Bemessung von Stahlbetonscheiben mit Fachwerkmodellen,* Professur für Informatik, ETH-Zürich, 1989.

COUPLING BENDING AND SHEAR HYSTERETIC MODELS OF
LOW-RISE R.C. WALLS

FRANKLIN Y. CHENG
Curators' Professor of Civil Engineering
University of Missouri-Rolla, Rolla, MO 65401 USA

ABSTRACT

This paper presents an analytical technique for calculating static and
seismic response of low-rise shear wall structures. The displacement
response is resulting from bending, shear, and bond deformations. The
bending and shear deformations are coupled and can be determined separately.
The hysteresis rules are developed for inelastic structural analysis and are
also modelled separately for bending and shear. The ductilities are
expressed in terms of bending and shear displacements and the damage index is
in terms of ductility and strain energy dissipation. Several numerical
examples are provided from which one may observe that the shear deformation
is significant and that the analytical results are compared favorably with
experimental data.

INTRODUCTION

Low-rise shear walls are commonly used in both industrial buildings, such as
auxiliary buildings at nuclear power plants, and low-rise civil structures.
Because these walls have strong coupling behavior of bending and shear
deformation, and the shear deformation can sometimes be larger than the
bending deformation. It is necessary to perform experimental and analytical
work to derive mathematical models for inelastic analysis of buildings
subjected to strong earthquake motions. In this paper an analytical method
is presented for calculating the shear and bending deformation and generating
hysteresis loops for analysis of general low-rise shear wall structural
systems.

MATHEMATICAL MODELING

Stresses and Strains in Concrete and Steel

For a given wall configuration there exists a unique moment (M) shear (V)
curvature (ϕ) shear strain (γ) relationship as $M - V - \phi - \gamma$. Let the
bending strain be linear with respect to the width of the wall as shown in

Fig. 1a, the strain distribution is a function of the strain on the tension side ϵ_1 and compression side ϵ_2 of the wall. Curvature ϕ is the slope of the linear strain distribution. Shear strain γ is assumed to be constant across the width of the wall. The horizontal expansion of the wall is restrained by the boundary elements at the base and at each floor, the transverse strain is consequently assumed to be zero. The combined bending and shear strain is sketched in Fig. 1b.

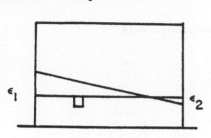

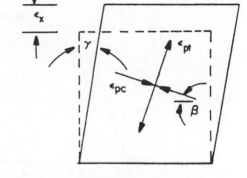

Figure 1a. Bending strain

Figure 1b. Combined bending and shear strain

Since both normal and shear strains act on a concrete element, Mohr's circle is used to determine the principal concrete compressive strain ϵ_{pc}, the principal concrete tensile strain ϵ_{pt}, and the angle of orientation, β. The principal concrete compressive stress σ_{pc} is softened to account for tensile strain that exists perpendicularly to the principal compressive strain. The softening term $1/\lambda$ controls the amount of softening as given in Fig. 2a. The parameter λ is modified from the previous work (1, 2) for the present study. The concrete tensile-stress strain has several models used in the past. The graded model (III) sketched in Fig. 2b is used to determine the principal tensile stress σ_{pt}. The graded tensile model assumes that 25% of the concrete per unit area cracks at 25% of f'_{cr}, 50% of the concrete is cracked at 50% of f'_{cr}, 75% of the concrete is cracked at 75% of f'_{cr}, and the balance of the unit concrete area cracks at f'_{cr}. This model provides a smooth transition between uncracked and cracked concrete in order to generate continuous $M - V - \phi - \gamma$ relationships. The concrete principal compressive and tensile stresses have the same orientation as their respective strains. Mohr's circle for stresses is used to determine the longitudinal concrete stress, σ_{d}, and shear stress, τ_c.

Interaction Surfaces
Axial load, moment, and shear are determined by integrating numerically over the wall cross section. Typically the maximum longitudinal compressive strain ϵ_2 is fixed, and the maximum longitudinal tensile strain ϵ_1 is varied until the sum of axial loads are approximately zero. Several iterations are usually required before a strain distribution is acceptable. Repeating these calculations for different combinations of ϵ_2 and γ defines the moment-shear-curvature $M - V - \phi$ interaction surface, and the moment-shear-shear strain $M - V - \gamma$ interaction surface are as shown in Figs. 3a and 3b.

Backbone Curves for Bending and Shear
For a given moment-to-shear ratio, the curvature and base rotation are combined to yield the rotation of a unit length member, θ_u. Plotting M versus

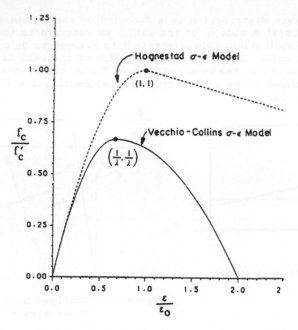

Figure 2a. Concrete principal compressive stress-strain model

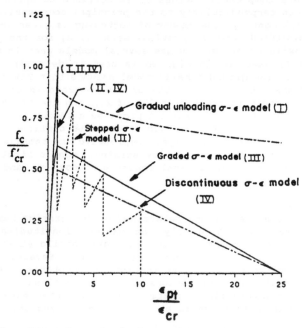

Figure 2b. Concrete principal tensile stress-strain model

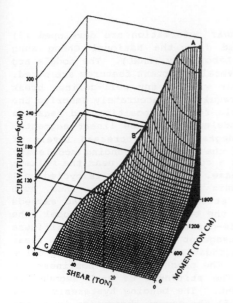

Figure 3a. Moment-shear-curvature interaction surface

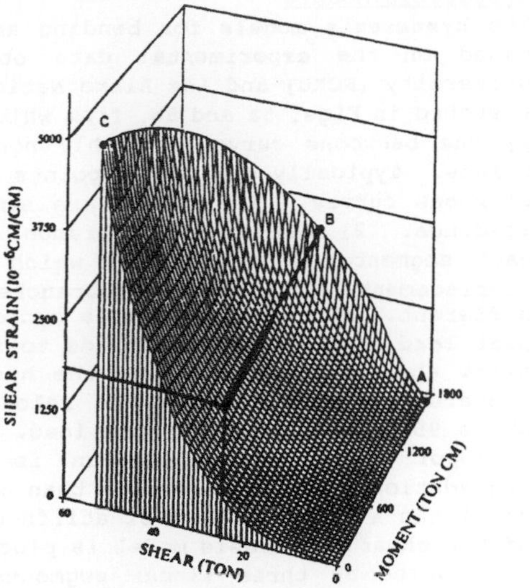

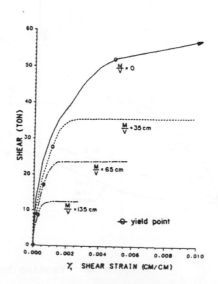

Figure 3b. Moment-shear-shear strain interaction surface

θ_u and V versus γ (or γ_u) yields the bending and shear backbone curves as shown in Figs. 4a, and 4b, respectively. Examining both bending and shear backbone curves reveals that decreasing the moment to shear ratio decreases the bending capacity and bending ductility while increasing the shear capacity and shear ductility.

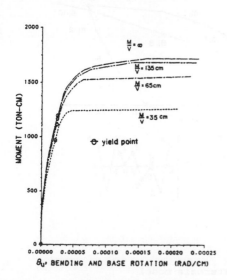

Figure 4a. Moment-curvature relationship at various moment to shear ratios

Figure 4b. Shear-shear strain relationship at various moment to shear ratios

Hysteresis Models

The hysteresis models for bending and shear deformation are developed (3) based on the experimental data obtained from the National Cheng Kung University (NCKU) and Los Alamo National Laboratory (LANL). The models are sketched in Figs. 5a and 5b, from which several important features are noted: 1) The backbone curve is highly nonlinear without any well-defined break points. Typically six to ten points are required to accurately describe the backbone curves. The curves are symmetric for both positive and negative loadings. 2) The unloading branch is modeled with three linear segments; each segment has a stiffness which is degraded with increased levels of displacement. The unloading branches of the shear and bending models are different. 3) When the wall is cycled at a constant load, near the maximum past load, the deformation tends to increase. When the wall is cycled well below the maximum past load, the hysteresis loops are stable. Thus, both hysteresis models degrade the reloading stiffness, when the load exceeds about 95% of the maximum past load. For the bending hysteresis model, the rate of stiffness degradation is cycle-dependent with more stiffness degradation on the first cycle than on subsequent cycles. Whereas the shear model has a constant rate of stiffness degradation. 4) The reloading curve of the shear hysteresis model is pinched. The pinch is highly nonlinear and is modeled by three linear segments. The stiffness of each segment is dependent on the maximum past displacement. The bending hysteresis model doesn't have this pinching behavior. And 5) the energy absorption increases with increasing peak displacements. This is accomplished in both hysteresis models by varying the unloading and loading stiffness. However, the energy absorption of the shear hysteresis model is less than the bending hysteresis model because of pinching.

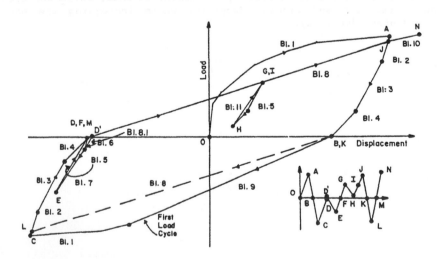

Figure 5a.   Bending hysteresis model

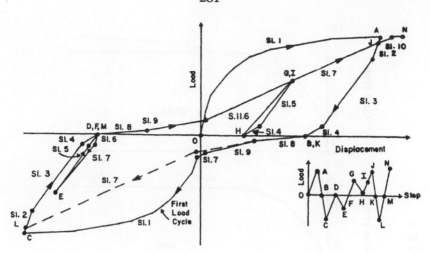

Figure 5b. Shear hysteresis model

Ductility and Damage Index

The ductility for individual walls is expressed as

$$\mu_T = \frac{\delta_{bm} + \delta_{sm}}{\delta_{yb} + \delta_{ys}}, \qquad \mu_b = \frac{\delta_{bm}}{\delta_{yb}}, \qquad \mu_s = \frac{\delta_{sm}}{\delta_{ys}} \qquad (1a,b,c)$$

in which δ_{bm} and δ_{sm} are the maximum displacements due to bending and shear, respectively; δ_{yb} and δ_{ys} are the yielding displacements due to bending and shear, respectively; and μ_T, μ_b, and μ_s are the total, bending, and shear ductilities, respectively. The damage index is defined as

$$DI = \frac{\delta_{max}}{\delta_{ult}} + \frac{\beta}{F_y \, \delta_{ult}} \int_0^t d(PSE) \qquad (2)$$

where δ_{max} is the maximum displacement under seismic response, δ_{ult} is the failure displacement under monotonic loading, F_y is the yielding force, the integration term of PSE is related to the plastic strain energy for the time duration t of the seismic response, and $\beta = 0.2$ represents the hysteretic energy coefficient (3).

NUMERICAL COMPARISONS

Comparison of Analytical and Experimental Monotonic Displacements Due to Bending and Shear

The total deformation of bending and shear of NCKU wall SW6 (4, 5) is compared with the analytical model in Fig. 6. This wall is 100 cm wide, 50 cm tall, and 10 cm thick. The concrete strength is 288 kg/cm^2 and the steel yield point is 4900 kg/cm^2. Steel reinforcing ratios are 0.774% vertically, and 1.03% horizontally. The reinforcing steel is uniformally distributed across the width of the wall. Comparisons between the calculated and experimental backbone curves are good.

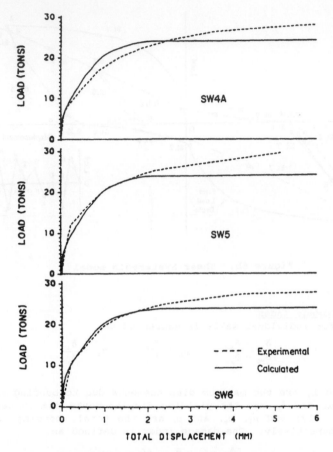

Figure 6.   Calculated and experimental monotonic response

Comparison of Analytical and Experimental Response for Earthquake Excitations

A two-story auxiliary building model called LANL 3D-11 shown in Fig. 7a was
subjected to a scaled version of the 1940 El Centro N-S earthquake (6, 7).
The response was measured by accelerometers; the experimental displacements
are derived from the acceleration data. Damping is assumed to be 5.5%. The
dynamic equation is solved by the linear acceleration method with the
stiffness being updated at each time step (3).

The acceleration power spectral density (PSD) for the roof is given in
Fig. 7b and the comparison of the displacement response at the floor is
sketched in Fig. 7c. Calculated displacement are somewhat higher than the
experimental peak displacement. Observing the acceleration power spectral
density indicates that both calculated and experimental first modes are at 83
hz, and the experimental second mode is slightly lower than the calculated
second mode. This example is used to demonstrate the applicability of the
shear wall panel in analysis.

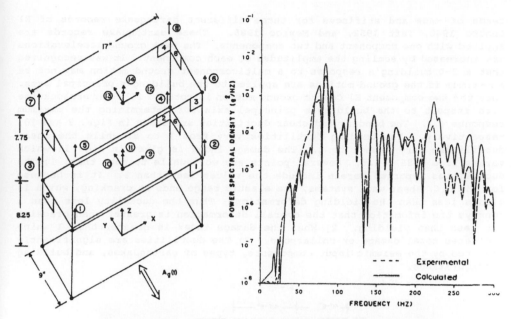

Figure 7a.　LANL box 3D11,
　　　　　degrees of freedom

Figure 7b. Comparison of calculated and
　　　　　experimental root acceleration PSD

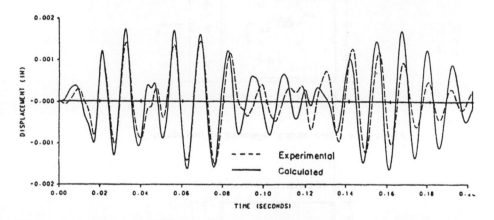

Figure 7c.　Comparison of calculated and experimental roof displacements

Ductility and Damage Studies

The two-story buildings of symmetric and unsymmetric plan shown in Figs. 8a
and 8b are used in this study. The first floor is 15 ft. and the 2nd floor
is 12 ft. high. The shear walls are 30 ft. wide which have the height to
width ratios at the first and second floor of 0.50 and 0.40, respectively.
These walls are 8 in. thick and are made of 4000 psi (f_c') concrete. The
walls are reinforced with 2 #3 at 11 in. each way with vertical and
horizontal reinforcement ratios of ρ = 0.25%. The steel has the yielding
stress of 60 ksi. The buildings are analyzed with 5% damping expressed in

284

terms of mass and stiffness for three different earthquake records of El
Centro 1940, Taft 1952, and Mexico 1985. These earthquake records are
applied with one component and two components. The peak ground accelerations
are increased by scaling the amplitudes of each component. It was recognized
that a 3-D building's response to a multicomponent ground motion may not be
a maximum if the ground motions are applied to the building's principal axes.
Thus the two component El Centro ground motion is rotated 26° counterclockwise
with respect to the building's principal axis for determining the maximum
response (3). The bending and shear ductilities are shown in Figs. 9 and 10,
respectively. The bending ductilities vary from 1 to 8, while the shear
ductilities vary from 1 to 4. The damage index is given in Fig. 11 which
varies from 0.25 ~ 1.5. Several points are worthwhile to note that: 1) The
ductilities reported herein include the values less than 1. It is because
for the R.C. shear wall systems, the elastic range ends at cracking, which is
always less than the yielding deformation. Thus the ductility less than 1
conveys the infomation that the overall deformation is greater than cracking
but less than yielding. 2) When the damage index is greater than 1 which
indicates total damage or collaspose. 3) The ductilities are significantly
affected by the seismic input components, types of earthquakes, and building
configurations.

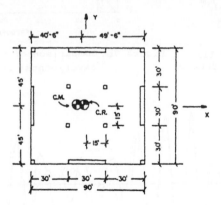

Figure 8a. Two-story Symmetric building

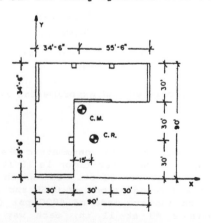

Figure 8b. Two-story Unsymmetric building

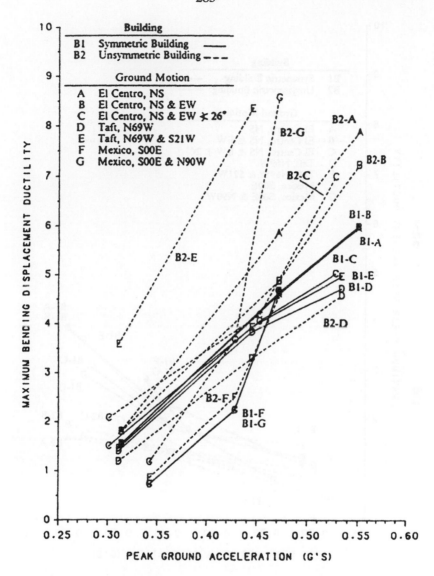

Figure 9. Bending displacement ductilities of shear walls vs. peak ground accelerations

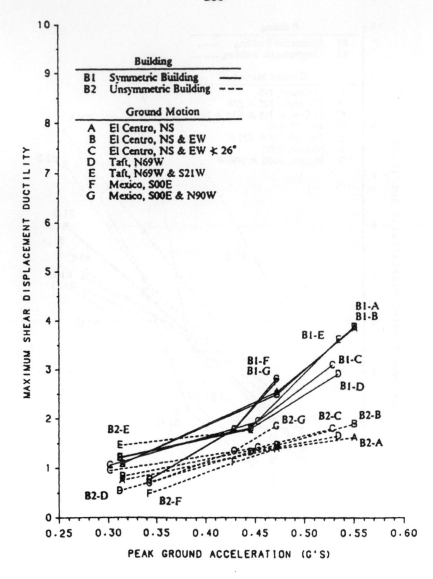

Figure 10. Shear displacement ductilities of shear walls vs. peak ground accelerations

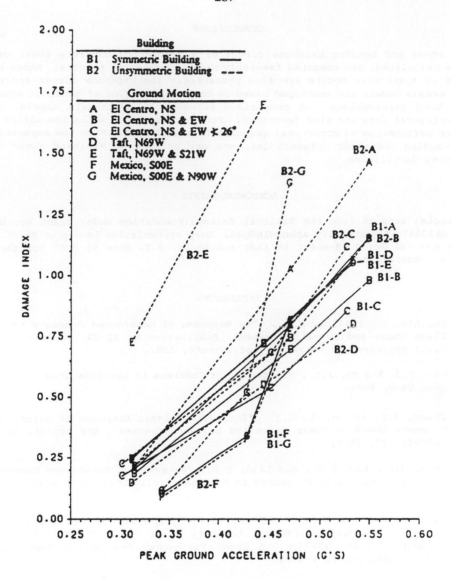

Figure 11. Damage index vs. peak ground accelerations

CONCLUSIONS

The shear and bending backbone curves of a reinforced concrete shear wall were calculated, and compared favorably with experimental results. Shear and bending hysteresis models are also presented. The backbone curves and the hysteresis models are developed based on coupling behavior of bending, shear, and bond deformations. A comparison between the hysteresis models, and experimental data are also favorable. For low-rise shear walls the effect of shear deformation on structural response is significant. From the separation of bending and shear deformations, one may be able to evaluate shear and bending ductilities.

ACKNOWLEDGMENTS

Financial support from the National Science Foundation under Grant No. NSF ECE 8513852 is gratefully acknowledged. Deep appreciation is due to Dr. C.A. Anderson and Dr. J. Bennett at LANL and to Dr. M.S. Sheu at NCKU for their close cooperation.

REFERENCES

1. Vecchio, F., Collins, M.P., "The Response of Reinforced Concrete to In-Plane Shear and Normal Stresses", Publication No 82-03, Department of Civil Engineering, University of Toronto, 1982.

2. Hsu, T.C. and Mo, Y.L., "Softening of Concrete in Low-Rise Shear Walls", UHCE 84-8, 1984.

3. Cheng, F.Y. and Mertz, G.E., "Inelastic Seismic Response of Reinforced Concrete Low-Rise Shear Walls and Building Systems", NSF Report, NTIS, PB90-123217, 1989.

4. Sheu, M.S., Kao, Y.T., and Liao, H.M., "Behavior of Reinforced Concrete Low-Rise Shear-Walls Subjected to Reversed Cyclic Loadings", NCKU Report (in Chinese), 1986.

5. Huang, C.C., Sheu, M.S., and Guo, S.M., "Experimental and Theoretical Study of Low-Rise Reinforced Concrete Shear Walls Without Boundary Elements", Proc. of the 7th Japan Earthquake Engineering Symposium, Tokyo, 1986, pp. 1171-1176.

6. Dove, R.C., Endenbrock, E.G., Dunwoody, W.E., and Bennett, J.G., "Seismic Test on Models of Reinforced Concrete Category-I Buildings", Eighth International Conference on Structural Mechanics in Reactor Technology, 1985.

7. Endebrock, E., Dove, R., and Anderson, C.A., "Margins to Failure-Category-I Structures Program: Background and Experimental Plan", NUREG/CR-2347, LA-9030-MS, 1981.

ANALYTICAL MODELS FOR REINFORCED CONCRETE SHEAR WALLS UNDER REVERSED CYCLIC LOADING

HYUNMOCK SHIN
Assistant Professor
Dept. of Civil Eng.
Sung Kyun Kwan Univ.
Korea

KOICHI MAEKAWA
Associate Professor
Dept. of Civil Eng.
Univ. of Tokyo
Japan

HAJIME OKAMURA
Professor
Dept. of Civil Eng.
Univ. of Tokyo
Japan

ABSTRACT

A computer program called WCOMR was developed to analyze the characteristics of reinforced concrete shear walls under reversed cyclic loading. The WCOMR comprises reinforced concrete element model described by smeared crack model and joint element model described by discrete crack model. The applicability of the WCOMR has been experimentally verified by the element and structural levels.

INTRODUCTION

In making analytical predictions for inelastic behaviors of reinforced concrete walls under reversed cyclic loading, modeling of behaviors of reinforced concrete element, including the cracking habits, from yielding of the reinforcing bars and on through subsequent repetition of loading, is indispensable. As a reinforced concrete shear wall contains steel bars dispersedly disposed in two or more directions, causing numerous cracks to be generated dispersedly, the behaviors of cracks after they have all been generated and attained a stable state are rather more important than the generation and development of individual cracks. For this purpose, the smeared crack model, in which a finite region containing several cracks and reinforcing bars is considered to be a continuum, is quite adequate to describe an reinforced concrete element.

On the other hand, however, reality is that local discontinuities, like pull-out of bars and intrusion of junction planes, can take place due to abrupt changes of rigidity induced at the joints of elements in any composite structural parts composed of several different basic structural elements of which the reinforced concrete shear wall is one. To take these effects into account, the smeared crack model alone is inadequate, and introduction of the discrete crack model becomes necessary. By combining these two models, the authors have developed a computer program called WCOMR developed for FEM analysis of reversed cyclic response of reinforced concrete shear walls.

DEVELOPMENT OF ANALYTICAL MODEL

There are a number of models for FEM analysis of reinforced concrete shear walls. For example, solutions for a reinforced concrete shear wall as a whole can be obtained by formulating empirical results of deformation and fracture of its elements under macroscopic stresses, and by using their characteristics

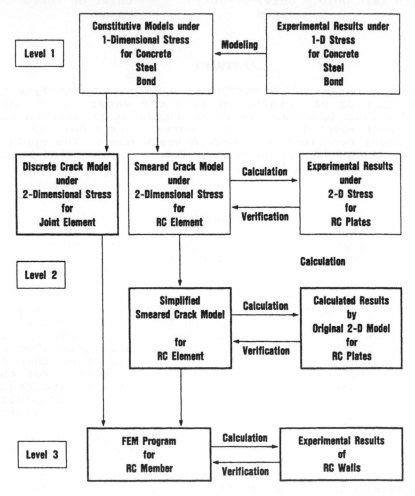

Figure 1. Development of analytical models and verification

as the elements in FEM. It is also possible to solve such a wall by analyzing the behaviors of individual elements, where each element, such as the steel bar, the concrete, cracks, and bond, are modeled independently. Although the method of approach may differ, any model has it in common that everything starts from a set of empirical formulae, the only difference being whence the modeling for analysis has begun [1]. Here, it is only natural that the closer an analytical model has its origin to the problems at hand, the greater is its accuracy. It is also important to note, however, that the closer to the level 1 in Fig.1 the model has been elected to originate, the wider the application it becomes able to claim. It is with these in mind that we have been developing a comprehensive analytical model that starts off at the level 1 of Fig.1 to accurately describe the individual kinetic behaviors, including those under reversed cyclic loading.

Analytical Model for Reinforced Concrete Elements
The analytical model for reinforced concrete elements falls on the level 2 in Fig.1. It may be constructed by combining micro-models in level 1. Here, care should be taken for the fact that, since conditions of experiments for constructing micro-models are often limited and idealized, judgment of its fitness as an reinforced concrete element must be done by comparison with experiments conducted on reinforced concrete element level [1]. This model was verified as capable of simulating both the empirically determined envelopes and internal history curves quite well as discussed in the companion paper [2].
However, one deficiency with this model is that, since its micro-models are empirical formulae, each describing faithfully an individual behavior [3,4,5], the model as a whole becomes too complex, and it is difficult to apply it straight to the FEM analysis of level 3 in a form including the path dependency under cyclic loads. As this called for modifications to make FEM analysis possible, a series of level 2 analyses was conducted to develop a reinforced concrete model applicable to level 3. The outcome has been a comparatively simple model, and this report presents results of analysis conducted for reinforced concrete shear walls based on this model.

Analytical Model for Reinforced Concrete Shear Walls
As mentioned briefly, the kind of reinforced concrete shear walls that are the object of this work are generally composed of such basic structural elements as wall, beam, and column (Fig.2), justifying the FEM analysis to be conducted for reinforced concrete shear walls by combining these reinforced concrete elements. It is to be remembered here that local discontinuities, such as pull-out of steel bars and slipping or intrusion of junction plane, can and do often take place as shown in Fig.2, as a result of abrupt changes in the section stiffness occurring at the joint planes connecting two components of different thicknesses, and that many a FEM analyses have given load-displacement relations that run higher to the observed ones because their influences are overlooked. Based

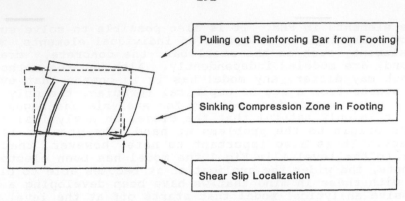

Figure 2. Joint element modeling (discrete crack model)

on the discrete crack model in level 2, we have provided an indispensable joint element that describes the stress versus localized deformation relationship of the junction plane between two reinforced concrete elements of defferent sections.

Modeling for opening: As the opening of discrete crack element represents the pull-out of reinforcing bars from the adjacent members with different thickness, the element behavior in opening mode can be described by the slip versus steel strain relation, which is applicable to both elastic and plastic states [4]. This model is classified into level 1 in Fig.1. In simplifying the original model in the reloading path, we adopted second order polinomials for reversed cyclic model as shown in Fig.3. By combining constitutive model for steel bars under reversed cyclic load, stress versus opening relation for joint element was obtained.

Modeling for closure: Based on two dimensional finite elments, it is impossible to take directly into account the localized strain distribution over the members' thickness as shown in Fig.4. Since ignorance of this three dimensional effect means the underestimated closure between members with different thickness, we took into account this additional displacement in the joint element by introducing a virtual volume equivalent to this three dimensional strain localization. The equivalent height of the virtual volume, which has the less thickness of conecting elements was determined by comparing its effect in two dimensional analysis with three dimensional one. In fact, the analytical contribution of this effect to the displacements of the wall shown in Fig.6(a) were at most 3%.

Modeling for shear slip: The slip along a joint element must be the interface shear displacement plus the shear deformation of the virtual volume mentioned above. Micro-models for

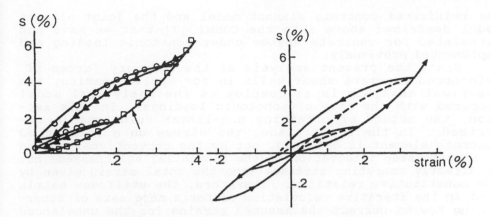

Figure 3. Modeling for opening of joint element
    (a) unloading and reloading   (b) reversed loading

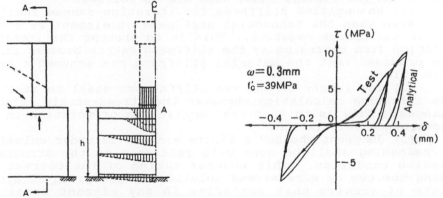

Figure 4. Stress distribution   Figure 5. Model for shear slip
          over the thickness   of the joint element

shear and normal compression transfer along a crack [5] and
concrete under biaxial stress states [6] in level 1 were
incorporated. The slip cyclic response of the joint element
under a fixed crack opening is shown in Fig.5.

As will be discussed later, one significance of clarifying the influences of local displacements at the junction
planes in reversed cyclic FEM analysis of reinforced concrete
shear walls lies in the fact that analytical predictions
become possible for real size structures, whose properties are
difficult to be examined with reduced size test specimens
because of the size effect.

DEVELOPMENT OF THE ANALYSIS PROGRAM

The new FEM analysis program was developed by incorporating

the reinforced concrete element model and the joint element model described above into the COMM2 [7] that we have had formulated for concrete member under monotonic loading and implemented previously.

Since the present analysis of the recovery forces of reinforced concrete shear walls is for cyclic loading, the numerical analysis is as complex as the analytical model compared with the case of monotonic loadings. In this section, the method of obtaining non-linear solutions is described. In the first place, the stress on a reinforced concrete element is obtained, not by the conventional method of step-by-step integration of the tangential stiffnesses, but by directly computing stresses from the total strain given by the constitutive relations. Therefore, the stiffness matrix used in the iterative calculation is for a mere sake of determining how to correct the assumed strains for the unbalanced forces present. In order words, the matrix is determined so as to make the calculation converge most effectively. The matrix and the element model used are as follows:

(1) No negative stiffness for iteration scheme shall be used even when the tangential stiffness of element as a continuum has become negative. This is to prevent the iterative solution from diverging as the stiffness matrix becomes singular provided that the material stiffness has assumed a negative value.

(2) The larger of the two stiffnesses shall be taken in the iteration calculation whenever the stress-strain relation changes. This is to suppress oscillation of solution in the convergence calculation.

(3) Judgment whether a finite element is under unloading or reloading shall be done with reference to the structural loading condition. This ensures taking of the correct one among the two or more stress solutions for a given strain, a state of affairs that can arise in any element model for cyclic loading.

(4) The stress-strain relation of the element model shall conserve its continuity under any loading histories. That the continuity be conserved even at the inflection points in the loading history is one of the most fundamental and the most important among computational conditions for any element model for cycling loading to satisfy.

(5) In calculating stresses from strain, no implicit models, i.e., those that necessarily call for iterative operation to obtain stresses, shall be used. This is to prevent the computation time from prolonging unduly because of the iterative convergence calculations that are conducted for each of the elements, and because of the difficulty of efficiently obtaining stiffness that is inherent in such methods. Another practice adopted was that, even though the stiffness matrix is revised for each round of the iteration routine up to the fifth by which time such large changes in the stiffness as occurrence of cracks or of yielding of the steel bar should have occurred, no revision is to be made for further rounds like in the modified Newton-Raphson method.

FEM ANALYSIS OF REINFORCED CONCRETE SHEAR WALLS

For an object of applying the present method of analysis, a reinforced concrete shear wall of a design shown in Fig.6(a) was chosen. This part of the work was conducted for verification of a non-linear FEM analysis program in collaboration with Dr. T. Shioya and his associates, Shimizu Construction Co.. The alternate horizontal compressive force is applied at both ends of the upper slab in a reversed cyclic mode. Special attention is paid to such defects as pull-out of bars and shear slip of plane occurring at the junctions of columns and walls to the base slab, namely, those defects whose importance remained unrecognized in past studies by conducting precise measurements on them and by clarifying the effects they exert on the entire structural part.

The element discretization and boundary conditions employed in the present analysis are shown in Fig.6(b), where the presence, beside the reinforced concrete elements, of the joint elements that have been provided to account for deformations occurring in columns, beams, walls, and at the junction to the base slab should be noted. In the joint element for the base slab to wall boundary, in particular, the effects of bond-slip of steel bars out from the base slab and those of local compression have been incorporated. Because the effects of apparent pulling out of bars out of walls have been dealt with in the calculation of mean strains of the reinforced concrete plate element, however, they are not included in the joint element. Since the model for bond-slip of steel bars [4] adopted in the present analysis is the one that was obtained under conditions of preventing the occurrence of vertical cracks and eliminating the effects of reduction in bond

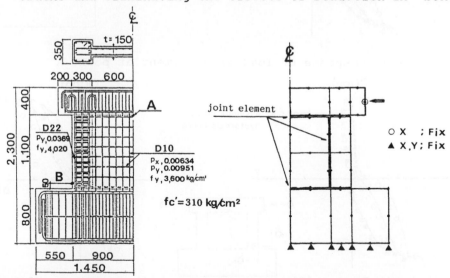

Figure 6. Reinforced concrete shear wall
    (a) specimen   (b) finite element mesh

in the vicinity of loading ends, furthermore, these effects
need be incorporated when reduction in bond is expected to
occur.   In the present analysis, this was taken care of, for
simplicity's sake, by doubling the value of slip given in the
reference.

In the current form, the present analysis program comprises 113 nodes, twenty reinforced concrete plate elements,
and twelve joint elements.   Since it takes about 3 to 8 min to
solve one load step with 16 bit personal computer, tracing of
a whole lifetime expended under monotonic loading, i.e., that
from generation of cracks to the ultimate load, can be performed in about 60 minutes of time.

Figure 7(a) presents the results, both analytical and
experimental, in the form of the relation of the load to the
horizontal displacement at the top (point A) for the aseismic
wall illustrated in Fig.6(a).   Agreement between the two is
excellent both for the envelopes and for the internal history

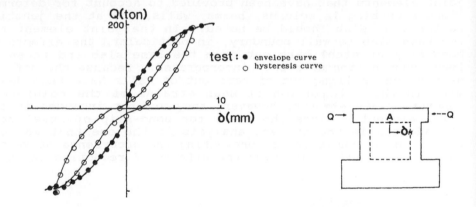

Figure 7(a) Response of load-displacement at point A

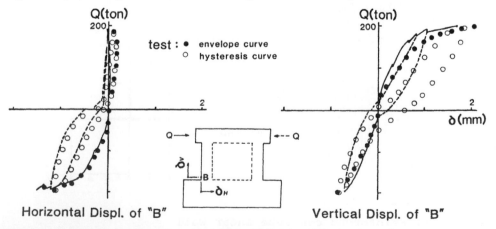

Figure 7(b)(c) Response of load-displacement at point B

curves. The ultimate load the analysis gave was 202ton against 201ton of experiment, while the ultimate displacement was 8.5 mm by analysis versus 7.2 mm by experiment.

The load to displacement relations at the leg (point B, 5 cm above the top surface of the base slab) are shown in Fig.7(b) for horizontal displacements and in Fig.7(c) for vertical displacements. The proportions of the slip at the junction of column and base slab in the overall horizontal displacement measured at point A were calculated to be 5% and 8% respectively for that at one half the ultimate load and that at full ultimate load. The proportion of the vertical displacement at the same place (i.e., pull-out displacement of steel bars in the leg of force application side and intrusion at the leg of the other side) in the point A to the overall horizontal displacement were about 12% both at the one half the ultimate load and the full ultimate load.

These observations mean that, since the proportion of the local discontinuous displacements at junctions of various components, such as walls, beams, columns, and base slab, in the overall displacement can amount to as much as about 20 %, their effects on the apparent recovery or toughness of the wall as a whole should never be ignored. This conclusion clearly indicates the necessity of considering these effects adequately in the analysis.

As an extreme case of experimental verification, another shear wall, which has the same dimension as the steel rein-forced concrete wall as shown in Fig.6(a) but reinforced only by the glass-fiber mesh, was tested. This new fiber mesh has as low elasticity as concrete, no plasticity but approximately two times higher tension-stiffening than deformed steel bars. The effect of these material properties in level 1 to the cyclic structural response was successfully predicted by the WCOMR as shown in Fig.8.

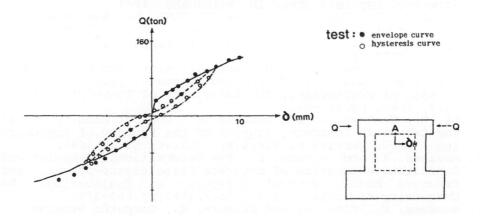

Figure 8. Response of load-horizontal displacement at point A

Furthermore, the authors examined the applicability of the program WCOMR with 5 specimens of reinforced concrete shear walls, which were collected by JCI as its recommendation [8] for check of analytical models. The average ratio of the experimental ultimate load to the analytical one was 1.03 with the coefficient of variation of 0.025. And the average ratio of the area of hysteresis loops to analytical one was 1.02 with the coefficient of variation of 0.147.

CONCLUSIONS

The computational model and numerical procedure for reinforced concrete shear walls under reversed cyclic loads was developed and experimetally verified. In the future, we intend to perform numerical calculations with this analysis program and develop macromodels of reversed cyclic response of reinforced concrete shear walls and new design formulae for them.

ACKNOWLEDGMENT

This work has been partially supported by Grant-in-Aid for scientific research (Research A 61420035) from the Ministry of Education, Japan.

REFERENCES

1. Okamura, H., Maekawa, K. and Sivasubramaniyam,S., Verification of Modeling for Reinforced Concrete Finite Element, Finite Element Analysis of Reinforced Concrete Structures, ASCE, 1986, pp. 528-43.
2. Izumo, J., Shin, H., Maekawa, K. and Okamuara, H., An Analytical Model for RC Panels Subjected to In-Plane Stresses, Concrete Shear in Earthquake, 1991.
3. Miyahara, T., Kawakami, T., and MAEKAWA, K., Nonlinear Behavior of Cracked Reinforced Concrete Plate Element Under Uniaxial Compression, Concrete Library of JSCE, No.11, 1988, pp.131-43.
4. Shima, H., Chou, L. and Okamura, H., Micro and Macro Models for Bond Behavior in Reinforced Concrete, Journal of the Faculty of Engineering, The University of Tokyo(B), Vol.39, No.2, 1987, pp.133-94.
5. Li, B., Maekawa, K. and Okamura, H., Contact Density Model for Cracks in Concrete, Journal of the Faculty of Engineering, The University of Tokyo(B), Vol.40, No.1, 1989.
6. Maekawa, K. and Okamura, H., The Deformational Behavior and Constitutive Equation of Concrete Using Elasto-plastic and Fracture Model, Journal of Faculty of Engineering, The University of Tokyo, Vol.37, No.2,1983,pp.253-328.
7. Maekawa, K., Niwa, J. and Okamura, H., Computer Program "COMM2" for Analyzing Reinforced Concrete, Proc. of JCI 2nd Colloquium on Shear Analysis of RC Structures, 1983.
8. JCI Committee on Shear Analysis of RC Structures, Collected Experimental Data of Specimens for Verification of Analytical Models, JCI, Vol.JCI-C6, 1983.

MIXED MODE FORMULATION FOR MEMBRANE DISPLACEMENT ANALYSIS OF RC PLATE AND SHELL ELEMENTS

T. TANABE, Z. S. WU, and H. NAKAMURA
Department of Civil Engineering
Nagoya University, Japan

ABSTRACT

The unified constitutive equation which considers the coupled effects of tension stiffening and concrete-to-concrete friction of a reinforced concrete element has been derived. Numerical analysis was compared with the experimental results. Based on these results, the finite displacement analysis has been performed for the shear failing beams and walls with stability consideration. On these results, the shear failure mechanism was discussed.

INTRODUCTION

Application of the finite element method to the analysis of a reinforced concrete member or a structure has advanced substantially to the present moment. However, it soon encounters difficulty when the structure meets the occurrence of multiple crackings. In other words, if the crack occurs to three different orientations, the usual FEM code cannot cope with it. The present formulation is specially aimed for those situations and some applications will be treated in the paper. In the second phase of the study, the same formulation, though much simplified, will be applied to the study of shear failure mechanism.

Shear failure is known to be affected by tension stiffness, shear transfer, cracking pattern, strength of intact concrete, bond-slip relation of tensile reinforcement, and so on. Those factors, combined altogether with the configuration of the member, give rise to the total shear failure. The mechanism looks complicated and leads to diversity of different failure modes.

One way of enabling theoretical insight into this diversity of shear failure phenomena will be the adoption of the eigen mode analysis at each critical point with thorough covering of the secondary effect of the displacement.

CONSTITUTIVE EQUATIONS OF CRACKED RC ELEMENTS FOR THE MIXED MODE

In stress conditions when the crack spacing is comparatively wide and frictional displacement at cracks occurs, the concrete close to the cracks is stressed in compression to the normal direction to a crack surface while the region away from the crack is still stressed in tension.

For the cracked reinforced concrete element, the shear forces and normal forces are supposed to be applied. Taking out a representative portion between two cracks as shown in Fig.1a, it is possible to separate the concrete into two regions, one in which compressive force is working to the normal direction for crack surfaces and the other region where tensile stress is working to the normal direction for cracked surfaces as shown in Fig.1b for unidirectional crack system.

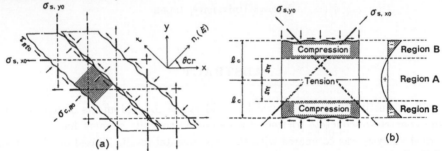

Figure 1. Stress Condition at A Crack in The Mixed Mode Displacement Condition

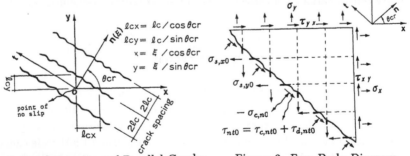

Figure 2. Crack Spacings of Parallel Cracks   Figure 3. Free-Body Diagram for Cracked Element

The transition point of the normal stress from the minus sign (compression) to the plus sign (tension) can be derived as follows. In the previous work[2], we have derived the distribution of the concrete stresses between two adjacent cracks, which is expressed as

$$\sigma_{c,x} = \frac{\sigma_x}{(1+np_x)} + \left(\frac{np_x\sigma_x}{1+np_x} - \sigma_{s,x0} \right) \frac{cosh(x/b_{cx})}{cosh(\ell_{cx}/b_{cx})} \tag{1}$$

$$\sigma_{c,y} = \frac{\sigma_y}{(1+np_y)} + \left(\frac{np_y\sigma_y}{1+np_y} - \sigma_{s,y0} \right) \frac{cosh(y/b_{cy})}{cosh(\ell_{cy}/b_{cy})} \tag{2}$$

$$\sigma_{c,xy} = \tau_{x,y} \tag{3}$$

where the notations ℓ_c and b_c with suffix x, y denote the crack spacings and the bond characteristics coefficients in the x and y directions (shown in Fig.2), respectively, n is a ratio of Young's modulus of steel to that of concrete, and $\sigma_{s,x0}$ and $\sigma_{s,y0}$ are the stresses in the orthogonal reinforcement at the crack, respectively.

By considering the equilibrium equations of the free body shown in Fig.3, we have

$$\sigma_{s,x0} = \sigma_x + \sigma_{c,n0} + (\tau_{xy} + \tau_{nt0})\tan\theta_{cr} \tag{4}$$

$$\sigma_{s,y0} = \sigma_y + \sigma_{c,n0} + (\tau_{xy} - \tau_{nt0})\cot\theta_{cr} \tag{5}$$

where $\sigma_{c,n0}$ is the normal stress of concrete and τ_{nt0} is the shear stress which is the sum of the shear resistance $\tau_{c,nt0}$ due to aggregate interlocking and the shear resistance $\tau_{d,nt0}$ due to the dowel action of reinforcement on the crack plane.

Expressing the state of concrete stress in the element in (n, t) coordinate system, the concrete stress component normal to the crack plane can be obtained as

$$\sigma_{c,n} = c^2 \left\{ \frac{\sigma_x}{1 + np_x} + \left(\frac{np_x\sigma_x}{1 + np_x} - \sigma_{s,x0} \right) \frac{cosh(x/b_{cx})}{cosh(\ell_{cx}/b_{cx})} \right\} +$$
$$s^2 \left\{ \frac{\sigma_y}{1 + np_y} + \left(\frac{np_y\sigma_y}{1 + np_y} - \sigma_{s,y0} \right) \frac{cosh(y/b_{cy})}{cosh(\ell_{cy}/b_{cy})} \right\} + 2sc\tau_{xy} \tag{6}$$

where c and s denote $cos\theta_{cr}$ and $sin\theta_{cr}$, respectively. Hence, the location where concrete stress changes from compression to tension should satisfy the following equation.

$$c^2 \left(\sigma_{s,x0} - \frac{np_x\sigma_x}{1 + np_x} \right) \frac{cosh(\xi/(b_{cx}c))}{cosh(\ell_{cx}/b_{cx})} + s^2 \left(\sigma_{s,y0} - \frac{np_y\sigma_y}{1 + np_y} \right) \frac{cosh(\xi/(b_{cy}s))}{cosh(\ell_{cy}/b_{cy})}$$
$$= \frac{c^2\sigma_x}{1 + np_x} + \frac{s^2\sigma_y}{1 + np_y} + 2sc\tau_{xy} \tag{7}$$

where the substitution of $x = \xi/cos\theta_{cr}$ and $y = \xi/sin\theta_{cr}$ is made by referring to Fig.2. For unidirectional reinforcement which is normal to the crack plane, ξ can be analytically derived as

$$\xi = cosh^{-1} \left[\frac{\sigma_{s,n0} - np_n\sigma_{c,n0}}{\sigma_{s,n0} - \sigma_{c,n0}} cosh(\ell_c/b_c) \right] \tag{8}$$

where $\sigma_{s,n0}$ is the stress in the reinforcement at the cracks. If $\xi \leq 0$, there exists no tension zone. However, if $0 < \xi < \ell_c$ the stress state of an element is in both compression and tension . Moreover, if $\xi \geq \ell_c$ the stress state is in tension only, and there exists no shear slip on the crack plane.

The portion where concrete is stressed in tension to the ξ direction, has to be treated as having slip between steel and concrete and its situation is exactly the same as the case of the S-mode(Separation mode)[1],[2],[3]. We separate the concrete portion along the ξ direction to the region B where $\sigma_{c,n}$ is in compression and, to the regions A where $\sigma_{c,n}$ is in tension. The concrete stress is zero at the boundary between A and B. Hence it can be considered that the constitutive equation which predicts the tension stiffness is applicable in the region A. For the region B, the stress and displacement field is mainly influenced by aggregate interlocking and the dowel action on the crack plane and the tension stiffening effect never occurs because the plane stress field of concrete is stressed in compression so that the constitutive equations developed for the F-mode(Frictional contact mode)[3] are considered to be applicable. However, the slipping out of a bar at the region A contributes to the crack opening of the region B, the total crack opening δ_n should be the sum of the contributions from the region A and from the region B. The total crack slip δ_t is also contributed by the two regions. Based on these considerations, we develop the constitutive equations for the mixed mode. Obviously, at the boundary of two regions, the stress equilibrium should be satisfied and also the stresses at the boundary of two regions should correspond to the applied stresses. Hence $\{\sigma\}_A = \{\sigma\}_B = \{\sigma\}$, and the total elongation of the portion between two cracks is the sum of the elongation of each region of A and B, and the average strain of the total portion $\{\varepsilon\}$ is written as

$$
\left\{ \begin{array}{c} \varepsilon_n \\ \varepsilon_t \\ \gamma_{nt} \end{array} \right\} = \left[\begin{array}{ccc} \eta & 0 & 0 \\ 0 & \eta & 0 \\ 0 & 0 & \eta \end{array} \right] \left\{ \begin{array}{c} \varepsilon_n \\ \varepsilon_t \\ \gamma_{nt} \end{array} \right\}_A + \left[\begin{array}{ccc} 1-\eta & 0 & 0 \\ 0 & 1-\eta & 0 \\ 0 & 0 & 1-\eta \end{array} \right] \left\{ \begin{array}{c} \varepsilon_n \\ \varepsilon_t \\ \gamma_{nt} \end{array} \right\}_B \tag{9}
$$

$$
= [\eta]\{\varepsilon\}_A + [\zeta]\{\varepsilon\}_B
$$

where η is the fraction to the crack spacing of the length of the area where the concrete is in compression along the ξ direction and frictionless mode is predominant. This is written in the linear case as $\eta = \xi/\ell_c$. For multiple crack systems, the average values of ξ and ℓ_c to each crack of different orientation are adopted. At this stage, it should be recalled that the constitutive equation for the S-mode is written as

$$
\{\sigma\} = ([I] - [\Omega]_S)\,[D]_c\{\varepsilon\} \tag{10}
$$

and for the F-mode, the similar equation is written as

$$
\{\sigma\} = ([I] - [\Omega]_F)\,[D]_c\{\varepsilon\} \tag{11}
$$

where $[\Omega]_S$ and $[\Omega]_F$ denote the stress reduction tensors in the S-mode and in the F-mode respectively. The detailed discussion on these tensors will be found in the ref.[1],[2]. As we already have the constitutive equations (10) and (11) for the regions A and B, Eq.(9) is rewritten as

$$
\{\varepsilon\} = [\eta][D]_c^{-1}\,([I] - [\Omega]_S)^{-1}\{\sigma\} + [\zeta][D]_c^{-1}\,([I] - [\Omega]_F)^{-1}\{\sigma\} \tag{12}
$$

Hence,

$$
\{\sigma\} = ([I] - [\Omega])\,[D]_c\{\varepsilon\} \tag{13}
$$

where

$$
\{\Omega\} = [I] - \left([\eta]([I] - [\Omega]_S)^{-1} + [\zeta]([I] - [\Omega]_F)^{-1} \right)^{-1} \tag{14}
$$

The relation ship of the F-mode and the S-mode and the Mixed mode can be expressed as Table 1.

It should be noted that we cannot have the frictional mode from the beginning since crack initiation is always to the principal tensile direction and the first mode should be

TABLE 1
Relationship of The S-Mode, The F-Mode and The Mixed Mode

Region	Idealized Mode	Applicable Equation
A	S-mode	Eq.(10)
B	F-mode	Eq.(11)
Whole Element	Mixed Mode	Eq.(12)

the frictionless mode. After a small crack width is formed, then the frictional mode or the mixed mode can exist. At the initiation of the first step of the friction mode, the stress equilibrium requires that $\{\sigma\}_A = \{\sigma\}_B = \{\sigma\}$ and the constitutive Eq.(12) of the first step must satisfy this condition.

FINITE DISPLACEMENT FORMULATION OF PLATE ELEMENT WITH OUT OF PLANE DISPLACEMENT

Formulation of Strain

The strain and displacement relation with second order terms is written in the following form in view of the notations shown in Fig.4.

$$\varepsilon_x = \left(\partial u/\partial x - z\partial^2 w/\partial x^2\right) + 1/2\left[(\partial u/\partial x)^2 + (\partial v/\partial x)^2 + (\partial w/\partial x)^2\right]$$

$$\varepsilon_y = \left(\partial v/\partial y - z\partial^2 w/\partial y^2\right) + 1/2\left[(\partial u/\partial y)^2 + (\partial v/\partial y)^2 + (\partial w/\partial y)^2\right]$$

$$\gamma_{xy} = \left(\partial v/\partial x - z\partial^2 w/\partial x\partial y\right) + \left(\partial u/\partial y - z\partial^2 w/\partial x\partial y\right)$$

$$+ [(\partial u/\partial x)(\partial u/\partial y) + (\partial v/\partial x)(\partial v/\partial y) + (\partial w/\partial x)(\partial w/\partial y)] \tag{15}$$

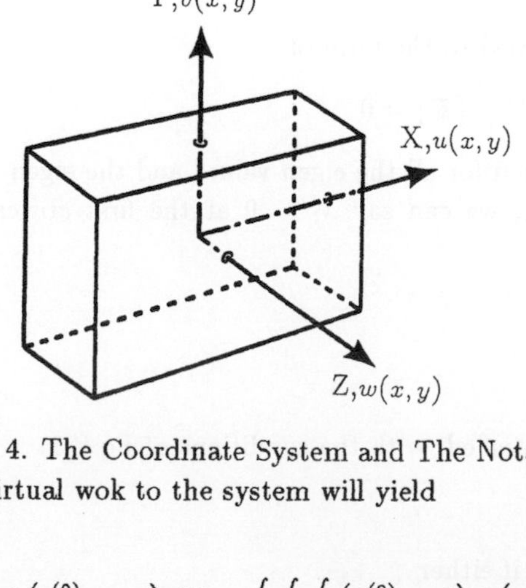

Figure 4. The Coordinate System and The Notations

The application of virtual wok to the system will yield

$$\int\int\int \left(\sigma_{ij}^{(0)} + \sigma_{ij}\right)\delta\left(e_{ij}^{(0)} + e_{ij}\right) dV - \int\int\int \left(f_i^{(0)} + f_i\right)\delta\left(u_i^{(0)} + u_i\right) dS = 0 \tag{16}$$

where, the summation convention rule is used and body force is neglected. The super subscript (0) denote the condition functions at ith step and the quantities without super script denote the incremental quantities in the $(i+1)$th step.

With proper use of displacement function and rewriting in the Cartesian Coordinate System, Eq.(16) will be reduced to the incremental form of

$$[K + K_0 + K_g]^{(n+1)}\{\Delta d\} = \{\Delta F\}^{(n+1)} + \{F_r\}^n \tag{17}$$

where $[K]$ denotes the incremental stiffness matrix, $[K_0]$ denotes the incremental initial strain matrix and $[K_g]$ denotes the incremental geometrical stiffness matrix. The vector $\{F_r\}^n$ denote the unbalanced force vectors at the nth step calculation and should be added at the $(n+1)$th step. The incremental stress and strain relation will be the one discussed in the previous section.

Stability Analysis
At each critical point, stability analysis has been performed[4]. Eq.(17) is written in the form of

$$[K^*]\{\Delta d\} = \mu\{F\} \tag{18}$$

$$K^* = K + K_0 + K_G \tag{19}$$

where, K^*, F, d, and μ denote tangential stiffness matrix, load vector, incremental displacement vector and load parameter respectively. Diagonalizing the K^* matrix using the eigen vectors, Eq.(18) is reduced to

$$[V_i]^T[K^*][V_i][V_i]^T\{\Delta d\} = \mu[V_i]^T\{F\} \tag{20}$$

where $V_i = [v_1, v_2, \cdots, v_n]_i$.
In view of the eigen equation of

$$[K]\{V_i\} = \lambda\{V_i\}, \quad for \ i = 1, \cdots, n \tag{21}$$

Eq.(20) will be transformed to the form of

$$\lambda_i[V_i]^T\{\Delta d\} - \mu[V_i]^T\{F\} = 0 \tag{22}$$

Eq.(22) should be satisfied for all the eigen values and the eigen vectors. Supposing that $\lambda_1 < \lambda_2 < \lambda_3 \cdots < \lambda_n$, we can say $\lambda_1 = 0$ at the first critical point without loss of generality,i.e.,

$$\lambda_1 = 0 \tag{23}$$

$$\lambda_2\lambda_2 \cdots \lambda_n \neq 0 \tag{24}$$

For the Eq.(22) to be satisfied, with the conditions of Eq.(23)

$$\mu[V_1]\{F\} = 0 \tag{25}$$

Eq.(25) will be satisfied if either

$$\mu = 0 \tag{26}$$

or

$$[V_1]^T \{F\} = 0 \tag{27}$$

Eq.(26) denotes the limit point and Eq.(27) denotes the bifurcation point. The displacement in these critical point is given in the following way. At the limit point it is self evident that

$$\Delta d = A * V_1 \tag{28}$$

At the bifurcation point, reducing the rank of K^* matrix, we will be able to obtain the particular solution which will satisfy the following equation

$$[K^*]\{\Delta d^*\} = \mu\{F\} \tag{29}$$

Substituting Eq.(29) to Eq.(22) yields

$$\lambda_i[V_i]^T\{\Delta d\} - [V_i]^T[K^*]\{\Delta d^*\} = 0 \tag{30}$$

Rewriting Eq.(30) with eigen vectors,

$$\lambda_i\{Vi\}^T\left[\{\Delta d\} - \{\Delta d\}^*\right] = 0 \tag{31}$$

Since we can assume that $\lambda_1 = 0$, λ_2, λ_3, $\cdots$, $\lambda_n > 0$ without loss of generality, for Eq.(31) to be consisting for all the i,

$$\{\Delta d\} = \{\Delta d\}^* + \beta\{V_1\} \tag{32}$$

should be satisfied.

With these consideration, incremental displacement is written as

$$\Delta d = \alpha[\{\Delta d\}^* + \beta\{V_1\}] \tag{33}$$

where α and β are the arbitrary constants as far as K^* matrix remain constant. Since any small increase of displacement will change the K^* matrix, α and β converges to certain values. In these ways, it is possible to find out the bifurcation point, as well as their mode of displacement. Especially, at the limit point, the displacement vector agrees with the eigen vector and it is considered that the first eigen vector strongly suggests the failure mode. At the bifurcation point, the linear combination of the fundamental mode and the first eigen vector suggest the failure mode.

THE CALCULATION ALGORITHM

In performing the above-mentioned calculation, the numerical algorithm to obtain the good convergence is far more important than the other factors. In this calculation, cylindrical arch length method is adopted in which the scaling factor was taken as zero. The convergence criteria was

$$\frac{\| \boldsymbol{\Psi}_i \|}{\| \boldsymbol{F}_i \|} < 10^{-3} \tag{34}$$

where $\boldsymbol{\Psi}$ is the norm of unbalanced forces of ith step and $\boldsymbol{F}_i$ is the norm of applied laod increment of ith step.

SOME NUMERICAL CALCULATION FOR THE MIXED MODE EXPERIMENT [5]

Experiments corresponding to the mixed mode are very scarce, due to its complexity[6],[7]. However, Millard and Johnson [6] carried out this type of experiment using the specimen shown in Fig.5(a). They introduced a crack at the center of a specimen by applying tensile forces at both ends.

Maintaining the tensile stress, the shear force was applied at the center. The stress condition of concrete is such that the compressive stress is working at the crack to the normal direction to a crack surfaces while the tensile force is working at the ends to the same direction. The experimental relations between the shear stress and the shear displacement are shown in comparison with the calculated relations in Figs.5 (specimen mark: 2-5-8). In the calculated procedure, it is found that the shear rigidity is very dependent on the extent of fraction of the region A. In Fig.6, the differences in the shear rigidity due to the extent of the fraction of the region A of the total area, which is assumed as given constant, were shown.

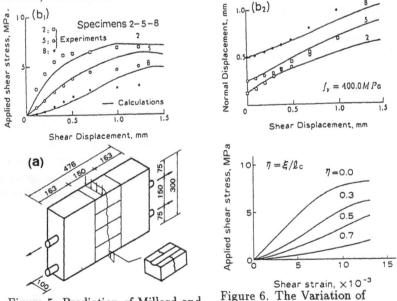

Figure 5. Prediction of Millard and Johnson's Experiment(1985)

Figure 6. The Variation of The Shear Rigidity

The numerical calculations show that the greater the fraction of the region A, the softer is the shear rigidity.

SOME MECHANICAL CONSIDERATION FOR SHEAR FAILURE

To obtain the better insight to the shear failure mechanism, two examples are analyzed in detail. One example is the beam the dimension of which is shown in Fig.7(a). The load and the displacement at the bottom relation is shown in Fig.7(c). The calculation algorithm enabled us to reach even the points on snapping back hysterisis. The crack pattern in the ultimate strength stage was shown in Fig.7(b). The crack pattern together with the information of shear span and beam depth ratio of 2.3 suggested that the specimen failed in the shear compression failure mode.

The analysis, however, showed that there exists a good possibility that the maximum loading point was due to the bifurcation and not the limit point in the stability sense.

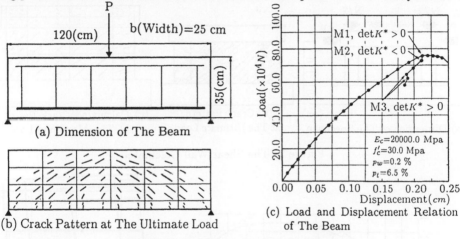

(a) Dimension of The Beam

(b) Crack Pattern at The Ultimate Load

(c) Load and Displacement Relation of The Beam

Figure 7. Analysis of A Simple Beam

In other words, the failure was possibly caused by the buckling at the top cord of compression in the plane. The eigen mode at the maximum loading point M1 was shown in Fig.8(a) and the eigenmode at the next loading point M2 was shown in Fig.8(b) which is very simillar to the one numerically obtained(Fig.8(c)), however, quite different in its

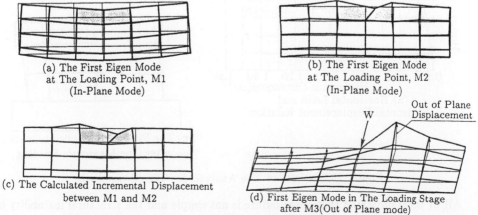

(a) The First Eigen Mode at The Loading Point, M1 (In-Plane Mode)

(b) The First Eigen Mode at The Loading Point, M2 (In-Plane Mode)

(c) The Calculated Incremental Displacement between M1 and M2

(d) First Eigen Mode in The Loading Stage after M3(Out of Plane mode)

Figure 8. Eigen Modes of The Beam at The Maximum Point

shape from the one of the previous loading step. These indicate that the maximum point load be the bifurcation point since the fundamental path there would be symmetrical in terms of center section of the specimen. From the figure, it is also understood that the two compressed elements at the central upper portion have undergone the unsymmetrical deformation, i.e., one in volume expansion and the other in volume contraction. This also suggests that the instability has been restored by the unsymmetrical deformation. If this is the case, the true maximum point cannot be obtained by imposing the symmetrical condition at the center section since the unsymmetrical eigen mode is impossible in the ultimate condition of the structure. Another interesting fact is that the out-of-plane bifurcation appears as shown in Fig.8(d) at the loading point M3 which is rather close to the maximum point.

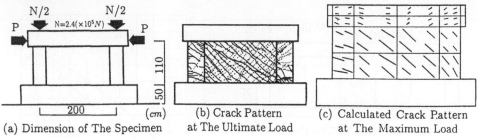

(a) Dimension of The Specimen

(b) Crack Pattern
at The Ultimate Load

(c) Calculated Crack Pattern
at The Maximum Load

Figure 9. The Shear Wall Calculated

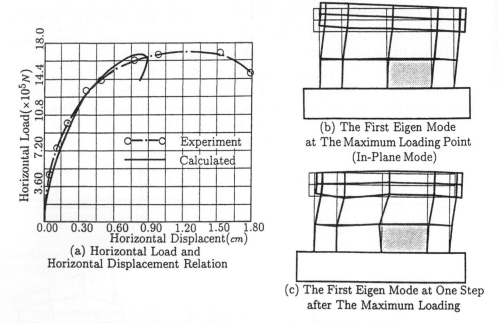

(a) Horizontal Load and
Horizontal Displacement Relation

(b) The First Eigen Mode
at The Maximum Loading Point
(In-Plane Mode)

(c) The First Eigen Mode at One Step
after The Maximum Loading

Figure 10. Results of The Analysis for The Shear Wall

All of these indicated that shear failure is not simple and the structural instability be related in this case.

The other example treated is the shear wall dimension of which is shown in Fig.9(a). The relations between the horizontal load and the horizontal displacement of the loading point was shown in Fig.10(a). The estimated displacement was somewhat smaller than the experimental value. However, the numerical results again traces the hysterisis to sufficient amount after the softening of structure started. The stability analysis in this case showed that the eigen mode had changed continuously from one step to the next around the maximum point and there is a strong indication that the maximum point be the limit point. The eigen mode just before and right after the the limit point was shown in Figs.10(b) and 10(c). The heavy damage seems to be existing at around the loading point. However, the bifurcation seems not existing and rather smooth transition is observed from the initial loading to the maximum point and to softening region. It is, however, noted that the snap back is also existing in this case.

CONCLUSION

The detailed constitutive equation to treat the cracked reinforced concrete element was derived basing on the mixed mode deformation analysis. It is found that the model can predict the behavior of the element cracked to multiple directions. To study the shear failure mechanism, then, the finite displacement formulation which includes stability analysis were developed. From the analysis of two simple examples, the shear failure is considered to be due to the instability of the system in some cases. It was indicated that the shear problem still has unexplored expansion in which shear failure is occurring.

REFERENCES

[1] Tanabe, T. and Yoshikawa, H., Constitutive equation of a cracked reinforced concrete panel, IABSE Colloquium on Computational Mechanics of concrete structures, DELFT, pp.17-34, 1987.

[2] Z.S. Wu, Yoshikawa,H. and Tanabe,T., Tension stiffness model for cracked reinforced concrete, J. Struct. Div., ASCE, to appear.

[3] Yoshikawa,H., Z.S. Wu and Tanabe,T., An analytical model for shear slip of cracked concrete, J. Structure Division, ASCE, Vol.115, No.ST4, pp.771-788, 1989.

[4] H. Nakamura, J.Niwa, and T.Tanabe, Analytical Study to Predict the Ultimate Failure Point of RC structures, Proc. of JCI Symposium on Design of Reinforcement and Ductility of Concrete Structures, May, 1990, pp.181-187.

[5] Z.S. Wu, Development of Computational Models for Reinforced Concrete Plate and Shell Elements, Doctoral Dissertation presented to Nagoya University, March, 1990

[6] Millard, S.G. and Johnson, R.P., Shear transfer in cracked reinforced concrete, Magazine of Concrete Research, Vol.37, No.130, 3-15, 1985.

[7] Oesterie, H.G. and Russel, H.G., Shear transfer in large scale reinforced concrete containment elements, Construction Technology Laboratories, NUREG/CR-1374, 1980.

THE SHEAR DEFORMATION ANALYSIS OF RC MEMBRANE ELEMENTS USING THE CRACK STRAIN DISTRIBUTION FUNCTION

JUNICHIRO NIWA   TADA-AKI TANABE
Associate Professor   Professor
Department of Civil Engineering
Nagoya University
Furo-cho, Chikusa-ku, Nagoya, 464-01, Japan

ABSTRACT

There are several approaches that can be used to predict shear deformation of reinforced concrete plates, such as the smeared crack model. These approaches treat the phenomenon of tension stiffening by assuming a simple and empirical equation without mechanical considerations. Estimation of the tension-stiffening effect of reinforced concrete is very important, however, and thus a method which uses the crack strain distribution function has been derived theoretically, to predict it based on the differential equations concerning the bond-slip relationship. This paper deals with the analysis of reinforced concrete membrane elements. The analytical method has been extended to two-dimensional problems. Obtained analytical results can predict the experimental data fairly well.

INTRODUCTION

Many important structures, such as containment vessels of electric power plants or liquefied natural gas tanks, are constructed using reinforced concrete structures. For these structures, a reasonable design procedure is required to consider not only strength but also deformation due to design loads. However, deformation of reinforced concrete membrane elements subjected to in-plane forces shows complicated nonlinear behavior. It is well known that the influence of cracks, and following slips, between reinforcing bars and concrete, on the deformation is especially large.

A substantial amount of research has been carried out attempting to predict this nonlinear phenomenon. The smeared crack model is very useful because it is quite easy to consider the tension-stiffening effect in the analysis and, because it is easy to apply in finite element analyses. In most smeared crack models, however, the tension-stiffening effect is determined regardless of the bond-slip relationship. These kinds of tension-stiffening models cannot represent the actual mechanical behaviors within reinforced concrete members.

Attempts to evaluate the bond-slip relationship precisely have also been made. In these methods, however, many calculations are required because many nonlinear parameters are included in these rigorous methods.

The crack strain method is also proposed [1]. This method uses the conceptional nonlinear strain distribution function for which we have defined the crack strain distribution function. This method can express nonlinear behaviors, such as strength or deformation, of cracked reinforced concrete members reasonably based on the bond mechanism. In this paper, the analytical model for two-dimensional problems is described.

CRACK STRAIN AND CONCRETE STRAIN

In cracked reinforced concrete elements, the total strain $\{\varepsilon\}$ of Eq. (1) does not coincide with the local concrete strain $\{\varepsilon_c\}$ because of the existence of cracks and the slip between reinforcing bars and concrete.

$$\{\varepsilon\} = [B]\{\delta\} \tag{1}$$

where, [B] is the nodal displacement vector - strain matrix and $\{\delta\}$ is the nodal displacement vector.

The crack strain $\{\varepsilon_{cr}\}$ defined by the difference between total strain $\{\varepsilon\}$ and local concrete strain $\{\varepsilon_c\}$ is introduced, as shown in Eq. (2). Assuming the crack strain, the nonlinear behavior of cracked concrete can be treated as continuous material.

$$\{\varepsilon\} = \{\varepsilon_c\} + \{\varepsilon_{cr}\} \tag{2}$$

This concept of crack strain applies to the orthogonally cracked reinforced concrete membrane element in Fig. 1. In the research, we determined the crack directions to be perpendicular to the x-y coordinates for convenience. This assumption is equivalent for the element to have only principal stresses on the boundaries. On the other hand, directions of orthogonal reinforcing bars are arbitrary and they are defined as the r-s directions, respectively. For a further simplification, we ignored the influence of shear crack strain.

Concrete strain does not exist at the location of cracks because concrete becomes discrete. As a result, the total strain is coincident to the crack strain. The crack strain has the maximum value at the location of cracks and it can be assumed to distribute within a certain range $\pm l_x$, $\pm l_y$. Moreover, it is assumed that the integrated value of crack strains between two parallel cracks is equal to the crack width.

By the assumptions described above, the crack strain vector $\{\varepsilon_{cr}\}$ for the idealized element of Fig. 1, in which the location of cracks is designated as x=0 and y=0, can be defined. According to Eq. (2),

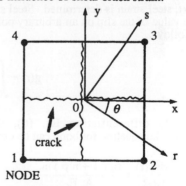

Figure 1. Reinforced concrete membrane element and the definition of coordinates

$$\{\varepsilon_c\} = \{\varepsilon\} - \{\varepsilon_{cr}\} = [[B] - [B_{cr}]]\{\delta\} = [B_c]\{\delta\} \tag{3}$$

At the location of cracks, concrete strain does not exist and total strain is equal to the crack strain.

$$\{\varepsilon_{cr} (x=0, y=0)\} = \{\varepsilon (x=0, y=0)\} = [B (x=0, y=0)]\{\delta\} \tag{4}$$

It may be considered that crack strain will occur owing to slip between reinforcing bars and concrete. Therefore, the shape of the distribution of crack strain was estimated from the differential coefficient of the slip and the crack strain assumed was required to satisfy the boundary conditions of Eq. (4). The influence of crack strains is generally changed according to specified members. Considering this influence, one parameter "a" was introduced into the crack strain distribution function. The influence of the crack strain increases with the decrease of "a". This also means decrease in tension stiffening. Finally, the crack strain can be obtained as follows.

$$\{\varepsilon_{cr}\} = [B_{cr}]\{\delta\} \tag{5}$$

where,

$$[B_{cr}] = \begin{bmatrix} L_1 & 0 & 0 \\ & L_2 & 0 \\ \text{symm.} & & 0 \end{bmatrix} [B\ (x=0,\ y=0)]$$

$$L_1 = \frac{\cosh(a(x-l_x)/b_{c,x})}{\cosh(al_x/b_{c,x})},\ \left(0 \le x \le L_x\right) \quad L_2 = \frac{\cosh(a(y-l_y)/b_{c,y})}{\cosh(al_y/b_{c,y})},\ \left(0 \le y \le L_y\right)$$

$$= \frac{\cosh(a(x+l_x)/b_{c,x})}{\cosh(al_x/b_{c,x})},\ \left(-L_x \le x \le 0\right) \quad = \frac{\cosh(a(y+l_y)/b_{c,y})}{\cosh(al_y/b_{c,y})},\ \left(-L_y \le y \le 0\right)$$

Note that $2L_x$ and $2L_y$ are the crack spacing in the x-y directions, respectively, and $b_{c,x}$ and $b_{c,y}$ are parameters representing the bond characteristics. With decrease in these values, the bond characteristics will be improved. The precise explanation for this value is given in the next chapter.

BASIC DIFFERENTIAL EQUATION FOR BOND AND STEEL STRAIN

Next, steel strain is determined. Steel strain should be derived from the bond-slip relationship. The value of the slip on an arbitrary point of a reinforcing bar in the r-direction can be defined as follows.

$$g(r) = \int_0^r \varepsilon_{s,r}\ dr - \int_0^r \varepsilon_{c,r}\ dr \tag{6}$$

By differentiating Eq. (6), the basic differential equation for bond can be obtained.

$$\frac{d^2g(r)}{dr^2} = \frac{\phi_s\ (1 + n\,p\,)\ u_b}{A_s\ E_s} \tag{7}$$

where, $g(r)$ is the slip; ϕ_s is the perimeter of a reinforcing bar; u_b is the local bond stress; n is the ratio of Young's modulus; p is the reinforcement ratio.

Eq. (7) can be solved, provided the relationship of the local bond stress and the slip, $u_b = u_b(g)$, and the boundary condition, are given. The relationship of local bond stress and slip generally shows a nonlinear behavior, as shown in Fig. 2 [2]. In this paper, however, the simplified linear relationship was adopted for convenience.

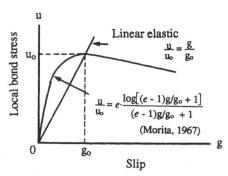

Figure 2. Relationship of local bond stress - slip

$$u_b = \frac{u_o}{g_o} g(r) = k_1 g(r) \qquad (8)$$

where, u_o and g_o are the bond strength and the corresponding slip, respectively. By substituting Eq. (8) into Eq. (7), Eq. (9) can be obtained.

$$\frac{d^2 g(r)}{dr^2} = \frac{\phi_s (1 + n p) k_1 g(r)}{A_s E_s} = \frac{g(r)}{b_{c,r}^2} \qquad (9)$$

where, $b_{c,r} = \sqrt{\dfrac{A_s E_s}{\phi_s (1 + n p) k_1}}$

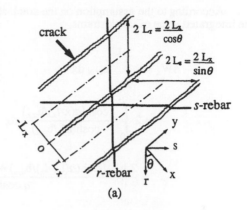

(a)

This is the value for the bond characteristics used in $[B_{cr}]$. We consider the relationship of cracks perpendicular to the x-axis and the intersectional bars as shown in Fig. 3. Steel strain can be represented by differentiating Eq. (6).

$$\varepsilon_{s,r} = \varepsilon_{c,r} + \frac{dg(r)}{dr}, \quad \varepsilon_{s,s} = \varepsilon_{c,s} + \frac{dg(s)}{ds} \qquad (10)$$

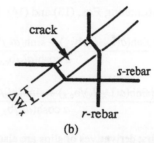

(b)

Figure 3. Assumption for the deformation of reinforcing bars at the location of cracks

To obtain the second term of right-hand side in the above equations, Eq. (9) should be solved. The general solutions of Eq. (9) for the location of cracks to be x=0 and y=0 are as follows.

$$g(r) = C_{r1} \exp((r-L_r)/b_{c,r}) - C_{r2} \exp(-(r-L_r)/b_{c,r})$$
$$g(s) = C_{s1} \exp((s-L_s)/b_{c,s}) - C_{s2} \exp(-(s-L_s)/b_{c,s}) \qquad (11)$$

To determine the unknown constants in Eq. (11), it is necessary to assume the deformation of reinforcing bars at the location of cracks. Therefore, it was assumed that reinforcing bars could deform to be perpendicular to cracks. According to this assumption, the absolute value of slip of each reinforcing bar at the location of cracks becomes constant and equal to the half of the crack width, regardless of the intersectional angles. In the intermediate region between adjacent cracks, it was also assumed that the point in which the slip does not occur should exist. From these two assumptions, the boundary conditions for g(r) and g(s) can be determined as follows.

$$g(r = -L_r, L_r, s = -L_s, L_s) = 0$$
$$g(r = 0, s = 0) = -\frac{\Delta W_x}{2} \qquad (12)$$

where, ΔW_x is the crack width. The slip can also be obtained.

$$g(r) = \frac{\Delta W_x}{2} \frac{\sinh((r-L_r)/b_{c,r})}{\sinh(L_r/b_{c,r})}$$
$$g(s) = \frac{\Delta W_x}{2} \frac{\sinh((s-L_s)/b_{c,s})}{\sinh(L_s/b_{c,s})} \qquad (13)$$

According to the assumption on the crack strain, the crack width ΔW_x can be obtained as the integrated value of crack strains.

$$\Delta W_x = 2 \int_0^{L_x} \varepsilon_{cr,x}\, dx$$

$$= 2 \int_0^{L_x} \frac{\cosh(a(x-l_x)/b_{c,x})}{\cosh(al_x/b_{c,x})}\, dx$$

$$= \frac{2\, b_{c,x}\,(\sinh(a(L_x-l_x)/b_{c,x}) + \sinh(al_x/b_{c,x}))\,[B_x\,(x=0)]\,\{\delta\}}{a\,\cosh(al_x/b_{c,x})} \tag{14}$$

By substituting Eqs. (13) and (14) into Eq. (11), the slips can be obtained as follows.

$$g(r) = \frac{b_{c,x}\,(\sinh(a(L_x-l_x)/b_{c,x}) + \sinh(al_x/b_{c,x}))\,\sinh((r-L_r)/b_{c,r})\,[B(x=0)]\,\{\delta\}}{a\,\cosh(al_x/b_{c,x})\,\sinh(L_r/b_{c,r})} = [B_{bs,r}]\,\{\delta\}$$

$$g(s) = \frac{b_{c,x}\,(\sinh(a(L_x-l_x)/b_{c,x}) + \sinh(al_x/b_{c,x}))\,\sinh((s-L_s)/b_{c,s})\,[B(x=0)]\,\{\delta\}}{a\,\cosh(al_x/b_{c,x})\,\sinh(L_s/b_{c,s})} = [B_{bs,s}]\,\{\delta\} \tag{15}$$

The first derivatives of slips are also obtained.

$$\frac{dg(r)}{dr} = \frac{b_{c,x}\,(\sinh(a(L_x-l_x)/b_{c,x}) + \sinh(al_x/b_{c,x}))\,\cosh((r-L_r)/b_{c,r})\,[B(x=0)]\,\{\delta\}}{a\cdot b_{c,r}\,\cosh(al_x/b_{c,x})\cdot\sinh(L_r/b_{c,r})} = [B_{ss,r}]\,\{\delta\}$$

$$\frac{dg(s)}{ds} = \frac{b_{c,x}\,(\sinh(a(L_x-l_x)/b_{c,x}) + \sinh(al_x/b_{c,x}))\,\cosh((s-L_s)/b_{c,s})\,[B(x=0)]\,\{\delta\}}{a\cdot b_{c,s}\,\cosh(al_x/b_{c,x})\cdot\sinh(L_s/b_{c,s})} = [B_{ss,s}]\,\{\delta\} \tag{16}$$

Finally, the steel strain can be expressed as follows.

$$\{\varepsilon_{s,r}\} = [\,[B_{c,r}] + [B_{ss,r}]\,]\,\{\delta\} = [B_{s,r}]\,\{\delta\}$$

$$\{\varepsilon_{s,s}\} = [\,[B_{c,s}] + [B_{ss,s}]\,]\,\{\delta\} = [B_{s,s}]\,\{\delta\} \tag{17}$$

STIFFNESS MATRIX FOR CRACKED REINFORCED CONCRETE MEMBRANE ELEMENT

Potential energy can be generally expressed as follows.

$$\Pi = U - W \tag{18}$$

U and W designate the strain energy and the work of external loads, respectively. For a reinforced concrete membrane element concerned, strain energy is shown in Eq. (19).

$$U = U_c + U_s + U_{bs} \tag{19}$$

where, U_c is the strain energy of concrete; U_s is the strain energy of reinforcing bars; U_{bs} is the strain energy of the bond between concrete and reinforcing bars existing in the surface area on reinforcing bars (b_s).

Constitutive equations required for the determination of strain energy are as follows.

$$\{\sigma_c\} = [D_c]\{\varepsilon_c\}, \quad \{\sigma_s\} = [D_s]\{\varepsilon_s\}, \quad \{u_b\} = [D_{bs}]\{g\} \tag{20}$$

where,

$$[D_c] = \frac{E_c}{1 - v^2}
\begin{bmatrix}
1 & v & 0 \\
 & 1 & 0 \\
\text{symm.} & & \dfrac{1 - v}{2}
\end{bmatrix}$$

$$[D_s] = \begin{bmatrix}
E_s & 0 & 0 \\
 & E_s & 0 \\
\text{symm.} & & 0
\end{bmatrix}, \quad
[D_{bs}] = \begin{bmatrix}
k_1 & 0 & 0 \\
 & k_1 & 0 \\
\text{symm.} & & 0
\end{bmatrix}$$

Using the nodal force vector $\{P\}$, the work of external loads can be expressed as follows.

$$W = \{\delta\}^T \{P\} \tag{21}$$

Moreover, potential energy can be obtained by Eq. (22).

$$\Pi = \frac{1}{2} \int_{V_c} \{\varepsilon_c\}^T \{\sigma_c\} \, dV_c + \frac{1}{2} \int_{V_s} \{\varepsilon_s\}^T \{\sigma_s\} \, dV_s + \frac{1}{2} \int_{b_s} \{g\}^T \{u_b\} \, db_s - \{\delta\}^T \{P\} \tag{22}$$

Eq. (22) can be rewritten by using nodal displacement - strain matrices.

$$\Pi = \frac{1}{2} \int_{V_c} \{\delta\}^T [B_c]^T [D_c] [B_c] \{\delta\} \, dV_c + \frac{1}{2} \int_{V_s} \{\delta\}^T [B_s]^T [D_s] [B_s] \{\delta\} \, dV_s$$

$$+ \frac{1}{2} \int_{b_s} \{\delta\}^T [B_{bs}]^T [D_{bs}] [B_{bs}] \{\delta\} \, db_s - \{\delta\}^T \{P\} \tag{23}$$

Using the minimum principle of potential energy, the stiffness equation is obtained as follows.

$$\{P\} = \left[\int_{V_c} [B_c]^T [D_c] [B_c] \, dV_c + \int_{V_s} [B_s]^T [D_s] [B_s] \, dV_s + \int_{b_s} [B_{bs}]^T [D_{bs}] [B_{bs}] \, db_s \right] \{\delta\}$$

$$= [K]\{\delta\} \tag{24}$$

RESULTS OF NUMERICAL ANALYSIS AND CONSIDERATIONS

At first, the results obtained from numerical analyses were considered qualitatively. The element in Fig. 4 was chosen as the subject of this numerical analysis. Fig. 5 (a)~(c) show the applied stress - total strain relationship on this element.

The property which is common to all of calculated curves is the existence of flat parts, like a staircase. The stresses of these flat parts correspond to the cracking stresses of reinforced concrete elements. The criterion of cracking used was to assume the maximum value of tensile stresses in the element to reach the tensile strength of concrete.

Flat parts of calculated curves corresponding to the occurrence of cracks show appropriately the typical mechanical behavior of reinforced concrete, that is the sudden slip of reinforcing bars because of cracking within the element. With the occurrence of cracks, the influence of crack strains will increase while, on the contrary, the effect of tension stiffening will decrease. The stiffness of the element will decrease gradually compared with the initial stiffness due to the occurrence of cracks and if reinforcing bars will not yield, the stiffness of the element approaches to the stiffness of reinforcing bars themselves.

Next, the influences of the ratio of principal stresses ($K = \sigma_2/\sigma_1$), the deviation angle between the directions of reinforcement and principal stresses (θ), and the ratio of reinforcement (p) on the analytical results are considered. The influence of K is shown in Fig.5 (a) in which the value of K changes from -0.5 to 0.5. In the case of $K = -0.5$, the level of cracking stresses becomes larger.

The influence of variation of the deviation angle, θ, is shown in Fig. 5 (b). From this figure, the decrease in the stiffness and also the increase in the deformation according to the

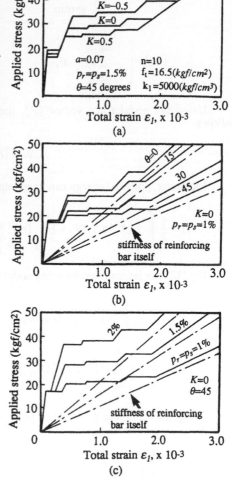

Figure 5. Relationship of applied stress - total strain

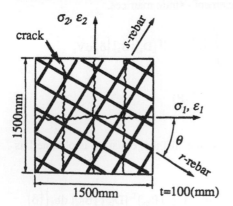

Figure 4. Subject for numerical analysis

increase in the deviation angle can be observed. The stiffness in the case of $\theta = 45$ degrees is approximately one-half of that of $\theta = 0$. Accordingly, the influence of θ on deformation is extremely great.

Fig. 5 (c) shows the influence of the change of the reinforcement ratio. In these calculations, the value of "a" was fixed to 0.07. As a result, for the small reinforcement ratio ($p = 1\%$), cracks occur successively from the occurrence of the initial crack and the stiffness rapidly approaches the stiffness of the reinforcement itself, under the small level of applied stress. On the contrary, in the case of $p = 2\%$, the effect of tension stiffening remains, even though the stress level reaches twice the initial cracking stress.

Fig. 6 shows the variation of local stresses of concrete, reinforcing bars and bond for the reinforced concrete membrane element subjected to uniaxial tension. It can be easily seen from Fig. 6 that the distribution of local stresses changes dramatically at the location of cracks when cracks occur.

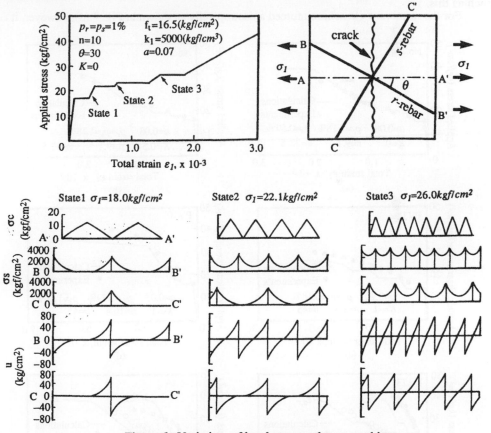

Figure 6. Variation of local stresses due to cracking

COMPARISON OF CALCULATED RESULTS WITH PREVIOUS EXPERIMENTAL DATA

To investigate the validity of this analytical method, the calculated results were compared with previous experimental data. Experimental data reported by Aoyagi & Yamada [3] and Vecchio

& Collins [4] were selected for verification.

From Fig. 7 (a) and (b), it is shown that the applied stress - total strain relationship calculated can predict the experimental results reported by Aoyagi & Yamada fairly well. In these experimental data, the value of K and the reinforcement ratio in two directions were varied. As for the applied stress - total strain relationship, the crack strain method can cope with the variation of these parameters.

Fig. 7 (c) and (d) show the comparison concerning the applied shear stress - total shear strain relationship. The experimental data were obtained from pure shear tests carried out by Vecchio & Collins. In the case of pure shear, principal compressive stress should be generated in the element. In this analysis, we do not incorporate the softening effect of concrete, and treat concrete subjected to compression as being elastic. According to Fig. 7 (c) and (d), the calculated results can follow the experimental data approximately. Therefore, it is considered that the effect of the softening of concrete is not serious for a low compressive stress level, such as this.

For the membrane element reinforced anisotropically as shown in Fig. 7 (d), however, it

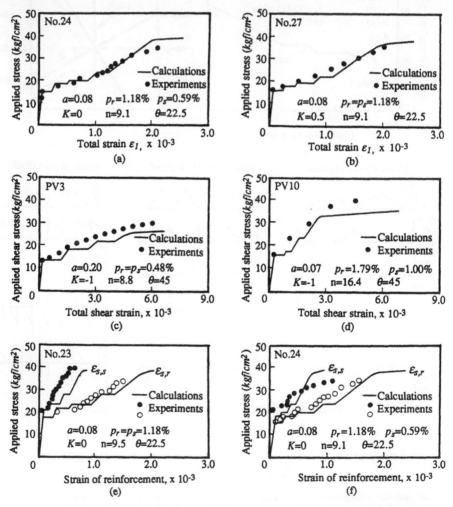

Figure 7. Comparison of calculated results with experimental data

319

can be observed that the accuracy of estimation decreases to some extent. For the simplification, the influence of shear crack strain is not considered in this paper. For anisotropic reinforcement, the estimation for the change of shear stiffness due to cracks will be required.

Fig. 7 (e) and (f) show the comparison concerning the average strain of reinforcing bars. For the r-direction reinforcement that intersects the direction of the maximum principal stress with the smaller angle, the average strain can be predicted almost accurately. However, for the same reason as for the prediction of total shear strain, the problem in calculation of the average strain of the s-direction reinforcement remains.

CONCLUSIONS

In this paper, the analytical method for the reinforced concrete membrane element subjected to in-plane stresses has been formulated by using the crack strain distribution function. Comparing calculated results with the experimental results, the following conclusions can be made.

(1) Due to the formulation of tension stiffness for a cracked reinforced concrete membrane element by using the crack strain concept, it is shown that the deformation behavior of the element subjected to normal stresses can be predicted fairly well.

(2) As for pure shear, the relationship of applied shear stress - total shear strain can be estimated approximately. To improve the accuracy of prediction, it is required to introduce the shear crack strain.

REFERENCES

[1] Yoshikawa, H., and Tanabe, T., "Analytical Study on the Tension Stiffness of Reinforced Concrete," Proc. of JSCE, No. 366/V-4, pp.93~102, Feb. 1986

[2] Muguruma, H., Morita, S., and Tomita, K., "Basic Studies on the Bond Mechanism between Steel and Concrete," Proc. of AIJ, No.131, 132, 134 and 139, Jan., Feb., Apr. and Sept. 1967

[3] Aoyagi, Y., and Yamada, K., "Strength and Deformation of Reinforced Concrete Shell Elements Subjected to In-Plane Forces," Proc. of JSCE, No. 331, pp.167~180, 1983

[4] Vecchio, F. J., and Collins, M. P., "The Response of Reinforced Concrete to In-Plane Shear and Normal Stresses," Publication No. 82-03, Department of Civil Engineering, University of Toronto, Mar. 1982, 332pp.

COMPUTATION OF RC WALL RESPONSE

W. C. SCHNOBRICH, CHUNJIAN XU, AND C. CHESI
Department of Civil Engineering
University of Illinois
Urbana, Illinois USA 61801

ABSTRACT

A number of different analysis models have been proposed for the computation of the cyclic response of reinforced concrete walls. These models span a broad spectrum of viewpoints regarding how the wall response should be determined. Both micro and macro versions of these models are reviewed and their appropriateness regarding the wall response problem is discussed in so far as accuracy and efficiency are concerned. Material models used in the development of the various computational models are also reviewed in the process of the global review. In particular the material model for use in the finite element micro models is singled out for a more detailed study because there have been major strides made in the experimental basis and refinement of this aspect of the computational model. In the course of the discussion some recommendations are made relative to the model (micro or macro, material etc.) to select for application to particular wall problems.

Structural wall systems occur in a variety of forms in building structures. Shear walls (isolated or coupled) form the lateral load resisting mechanism for many multi–story reinforced concrete structures. The computation of the response of these wall structures to base–induced motion and/or cyclic motion can be addressed at several levels ranging from a global to a macroscopic, even a microscopic level.

With the development and refinement of the various concrete models for use in the finite element method, this technique seems to provide the analyst with all the capability that he needs to evaluate the response of these wall structures.

INTRODUCTION

The computation of the response of reinforced concrete walls has been approached by a number of different avenues. Line elements with various forms of inclusion of inelastic behavior have been successful. Keshavarzian [7], and Emori, for example, have achieved good results for frame–wall structures, while Takayangi [2] solved coupled walls including the effect of changing axial forces. For cases where more detail is reasonable, two– and three–dimensional finite element models are possible. These can include bending and twisting effects but must be approached with realism.

Finite Element Model

Concrete material model: A wide spectrum of material models have been developed for use with the finite element method in an effort to model the behavior of concrete structures. On the tension side, the possibilities include the smeared crack or the discrete crack alternatives and these can be driven by a simple stress level criterion or one based on fracture mechanics principles. Compression behavior can be based on a plasticity theory with or without hardening and governed by an associated or a nonassociated flow rule, a hyperelastic, hypoelastic, or brittle fracture criterion, an endochronic theory, etc. Of the behavior characteristics, cracking probably is the most important, having the most significant impact upon what response is computed by an element analysis. The ASCE state–of–the–art report [1] gives a comprehensive overview of this method of computing or estimating behavior. This is however somewhat dated and lean in its coverage of cyclic characteristics. An update is anticipated.

Plane stress elements: Two dimensional models of the plane stress finite element type are capable of analyzing any planer wall. They make most sense when applied to a wall that has a low aspect ratio (is squatty). There the distribution of strain may have departed significantly from linear. The material properties to be assigned to the E matrix can range from those of smeared cracking with tension stiffening to complex fracture mechanics concepts in a smeared crack with localizations or discrete with local dislocations. Standard 4–node 1 point integration elements with hour glass mode suppression or 8–node 2x2 integration elements should function adequately. Thickened flanges and/or rod elements can be placed at the edges to account for bundled reinforcing and the presence of a barbell or I–section wall. So long as membrane forces remain the dominant force system, even box, I, U, or channel shaped sections can be handled with plane stress systems. Transfer across the junction lines, intersection lines of the plates lying in different planes, is just the common displacement. This should be sufficient to produce the requisite axial force shear transfer mechanism. Of course if one wishes to view deflected shapes it will be necessary to supplement the planer displacements with any rigid body motions normal to the planes of the various plane stress systems. This to be done through the use of constraint equations setting the rigid body movements equal to the proper displacements found on the common edge. If slab bending also has a significant role, then some form of shell element with its combined direct stress and bending response must be utilized. Even these elements will experience some theoretical problems at the web–flange junction; however, in spite of theoretical errors present in these models, their results should be adequate as some local discontinuities are not important in establishing the behavior of such reinforced concrete members.

The choice between a discrete crack and a smeared crack representation of the effects of the cracking process does depend upon the nature of the cracking. For most cases the cracks that develop do so in number, producing a reasonable array or network of cracks. The specific geometry changes that develop at any one crack are not critical to the overall response. Globally what is important is the degradation of the strength and stiffness of the member as a consequence of the development of the cracks. The steel present in these plane stress models does not contain any flexural characteristics. Steel is normally incorporated through orthotropic layers with directional stiffnesses. Thus even with nonlinear geometric considerations such models cannot anticipate any dowel action contribution or local bar buckling as a consequence of excessive steel compression forces following concrete breakup. Even with a discrete crack system, the inclusion of dowel action across the crack presents serious computational problems. It would be necessary to include steel bars with flexural stiffness with the length of the bars variable being set by anticipated bond breakdown. The question is, Is this level of detail necessary to the global determination of response? If it is, then a very detailed and complex microscopic analysis should be performed many of whose details we are not in a position to quantify.

Cracking along the base of the wall with the wall then going into a rocking mode with the pivot axis shifting from near one outer edge to the other during that rocking process. The opening and closing of such a crack is a significant behavior occurrence that has a major impact upon how the overall response of the structure plays. Whether this initial crack will totally dominate the behavior depends upon the reinforcing present across the junction of the wall with the foundation or base. If the reinforcing is light or nominal, then sufficient rotation can take place that a large deflection response can ensue. This can involve significant vertical displacements of the wall edges at the first floor and above. If there are floor beams framing into that column joint perpendicular to the plane of the wall, these beams will transmit that deflection as a vertical or shear force to the adjacent columns. This brings adjacent columns into the T–C couple moment resisting mechanism which can be a major contributor to the response mechanism. The inclusion of some recognition of such three dimensional characteristics of the framing system in the structural response model does improve the estimated global response. In the test structure that made up the US–Japan model this response mechanism made up some 5–10% of the peak resisting capacity.

Line or stick models: Stick models still represent a computationally efficient model to represent frame–wall and coupled wall structural response, if the walls in those structures have sufficiently high aspect ratios so that a linear variation of strain assumption is viable. These line element models can be spring–elastic beam models, multiple spring models if the nonlinear or plastic hinge zone is considered to have a deep or wide spread depth of penetration, layered models in which one section is replaced by a network of fibers or filaments parallel to the wall's axis or beam–columns with a variable length plastic zone. These later variable length models are the most adaptable of the wall simulations. About the only problem with the stick models is that the incorporation of the rocking motion associated with major opening and closing of the crack along the base is not natural to the geometry of that model. All line or stick models thus do require some form of rigid link arrangement placed at the floor levels and directed from the axis of the stick model to the end faces of the wall to try to provide the lifting mechanism associated with the rocking. However the geometry is not the same as it is difficult to achieve the asymmetric nature of the motion with basically a symmetric model. One solution would be to alternately move the wall member stick model back and forth between the two extreme positions to maintain correct links for the uplift. Such a shifting although confusing presents no major problems.

The incorporation of inelastic behavior into the response of wall stick models is easily accomplished through appropriate modifications of the moment–curvature relations. Using stress–strain and strain–displacement relations moment curvature relations can be established with cracking yielding and ultimate moments defining the transition points on the curves. These can be transformed to secant or tangent stiffnesses for incorporation into a time–history response. Inclusion of shear deformation is through a Timoshenko like theory using an average value. When stick or line models are appropriate beam behavior is dominant so shear behavior should not depart significantly from that of a beam. The incorporation of a simple reduction factor to account for shear deformation can provide sufficient accuracy. Therefore the inelastic value of shear rigidity can be assumed to be reduced in direct proportion to the flexural rigidity. The equation stating this assumption can be expressed as:

$$GA^* = (GA_e) \times (EI^*)/(EI_e)$$

(1)

where EI^*, GA^* : inelastic flexural and shear rigidities;

EI_e, GA_e : elastic flexural and shear rigidities.

Walls are subject to combined axial force and bending. Because of the large cross sectional area of shear walls they normally have axial compression forces well off the balanced failure point on an interaction diagram with the moment capacity usually being increased due to the presence of axial gravity forces in the wall. Ties to provide confinement to the concrete and constraint against any buckling of the reinforcement must be designed into the wall. The beneficial effect of axial force in a shear wall must not be overestimated. Park and Paulay recommend only real gravity load with a reduction of say 20% to account for vertical accelerationbe considered.

A factor to account for rotation due to bond slippage of the embedded steel must included in the model. Due to the significant contribution of the fixed–end rotation resulting from the reinforcement slippage at the joint to thetotal element deformation, a non-linear rotational spring, as an additional flexibility for the element, should be provided as a bond slip mechanism.

EFFECTS OF THREE DIMENSIONAL WALL RESPONSE

Three Dimensional Analysis

A full scale seven–story reinforced concrete frame–wall structure was tested during the US–Japan program. The structure consists of two moment resisting frames surrounding a central wall–frame Fig. 1 . The contribution of the 3–dimensional effects to the

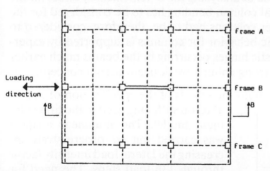

building's ultimate resistance has a simple yet significant contribution to a collapse analysis. Throughout the loading process, the response is controlled mainly by the central shear wall's deformation. As the loading proceeds, bending in the plane of the wall causes extremely large elongations to occur at the tension side of the wall as cracking causes the wall to rock about the compression side column. In this mode the transverse beams that frame into that wall undergo large relative vertical

Figure. 1 Floor Plan for US–Japan Structure

displacement differentials between the ends attached to the inner frame (frame B) and those connected to the outer frames. The shear forces thus generated have a stabilizing effect on the shear wall, at the same time increasing, the overturning moment present in the outer frames. This is the so–called "3 dimensional effect".

A finite element model of this structure was also tested for the 3–dimensional effects by Chesi [5]. The model focused on the first-story behavior so the upper-story effects were applied as constraints and loads at the midheight of the second story.

The following two kinds of elements have been used during the discretization of the different structural components present in the finite element model:

1. For the shear wall: a grid of 9–node Lagrangian shell elements [8]. Due to symmetry the wall is subjected to in–plane forces only, so the element is used mainly to describe direct stress or membrane behavior. The selection was made on the basis of the material model incorporated in the element rather than a need of any flexural degrees of freedom to model the wall response.
2. For the slab: the same kind of shell element is used, but here that element is subjected to both in–plane and out–of–plane forces. The flexural behavior is important.
3. For the T–beams resulting from an effective slab width plus eccentric web: the webs have been represented by an eccentric shell stiffener beam element [8], which is in the form of a 3–node Lagrangian beam element. This last element, if used in conjunction with the 9–node Lagrangian shell element to form the T–beam. This combination is expected to provide an accurate description of the T–beam behavior. This feature was of special interest for the present analysis, as there was experimental evidence of the beam–slab interaction having a much more pronounced effect on the overall response than conventionally assumed.
4. For the columns: the same elements, as used in (3), these elements simply work as beam column elements, if they are not attached or combined with shell elements.
5. For the shear wall's edge columns: again, the same beam elements as in (4), but now with a zero bending stiffness. Their response is then basically that of a rod or flange with the shear wall as the web.

The possibility of non–linear material behavior was specified for all the elements, except the columns of item 4. Only those columns which form the edge of the wall were allowed to undergo nonlinear material behavior. In the Milford and Schnobrich formulation [8], beams are handled as a layered system, so their response is limited to one plane. Thus, these nonlinear beam elements have non–zero bending stiffness in one direction only, ie. are planer beams. For the purpose of analyzing the 3–dimensional effect on the structure this is sufficient, however, biaxial column bending has to be considered for the columns in the outer frames. These outer columns had to be, therefore, considered to remain elastic. This assumption of elastic behavior for columns is supported by experimental results, those results showing plastic hinges occurring in the beams much earlier than in the columns consistent with the strong column weak beam design concept..

This study investigated among other items the effects of the specification of the tension stiffening used in the finite element model. The nonlinear analysis was carried out by progressively increasing the lateral load intensity factor through 136 load steps. The need for small incremental load steps was adhered to during behavioral studies of RC members. Care was required to follow crack opening and propagation throughout the wall. A high accuracy in the solution is mandatory, as the shear wall is by far the stiffest element in the model, so much so that most of the lateral load is carried by the wall, while the lateral resistance contribution from the other elements is relatively small.

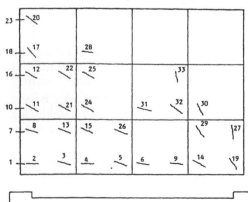

Figure. 2 Crack Pattern at Base of Wall

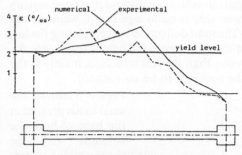

Figure 3 Strain Distribution at Base of Wall

The wall is subjected to in-plane or membrane forces only. The linear elastic range extends up to   a lateral load intensity factor of 0.113 (ratio betweeen the total horizontal force and the total building weight). Nonlinearity begins with the cracking at the base of the wall with the end column in tension (see Fig.2). As a consequence of this, cracking soon extends to the adjacent points along the base of the wall. During the next load steps, other points in the wall end column crack, followed immediately by their adjacent wall points. At the same time, cracking at the base of the wall propagates towards the compressed end causing an axis migration" to take place. Note that, because of this last phenomenon, an increase in moment capacity is produced more by this lever arm alteration than by an increase in tension stresses. At load step n. 84, the wall base and the end column are completely cracked (26 cracks opened or integration points cracked, see Fig. 2); then, the axial load in the left tension column starts increasing again, until reinforcement yielding occurs in the end column at load step n. 119. This step corresponds to a lateral load factor of 0.244. The corresponding crack distribution is given in Fig. 2. The strain profile over a cross section at the base of wall is shown in Fig. 3. Agreement with the experimental profile, as given by Yoshimura and Kurose [13], is satisfactory; note, moreover, the few wall bars had already yielded at previous load steps.

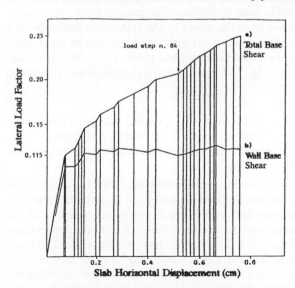

Figure 4 Horizontal Force Displacement US–Japan

A global description of the structural response is given in Fig. 4 , in terms of a lateral load–horizontal deflection curve. Sudden, but temporary, decreases in stiffness can be observed at the cracking of each of the integration points of the wall end column that is in tension. The figure also shows the high degradation of the wall's lateral stiffness (case b shows the base shear contribution stemming from just the wall), resulting in the transfer of the additional shear forces from going into the wall to going into the adjacent columns as the lateral load increases. Case a is the total base shear.

    In order to investigate the 3–dimensional effect, attention has been focused on the elongation of the tension side of the wall, which is the key parameter controlling the bending of the transverse beams. With cracks opening along the base of the wall to a location that is nearly at the compression column, the wall tends to pivot around that compression column. Thus, upward movement on the tension side is quite significant, while on the compression side the movements are nominal. The load deflection curve, relating the lateral load factor to the vertical displacements at the joint where the beams frame into the wall's tension side indicates a little stiffer response than the test structure, mainly at the lower load levels (where the model assumes the concrete to be uncracked).

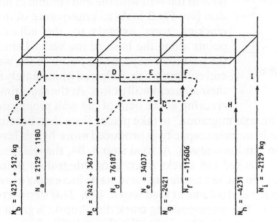

Figure 5 First Story Column Forces

Figure 5 shows the axial loads present at the base of both the columns and the shear wall for a lateral load factor of 0.244. The effect of gravity is not included in the values given in Fig. 5. Columns A and C, connected by beams to the wall's tension side, undergo much higher loads than the corresponding columns connected to the wall's compression side (as expected based on the observed displacements described in the previous paragraph). The effect is most pronounced in column C; the increase in axial load ($\Delta N_c = N_c - N_g$), which comes from the shear in the transverse beam, gives a measure of the 3–dimensional effect. As expected, the axial load increase in the column and the shear in the transverse beam are about the same and grow proportionally to wall elongations. Examining the situation of Fig. 5 shows that vertical equilibrium is satisfied by the following set of axial loads:

$$N_f = N_e + N_d + \Delta N_a + \Delta N_b + \Delta N_c \tag{2}$$

A portion of the wall's boundary column's axial load (N_d), indeed, has been transferred to the surrounding columns as a consequence of the wall's tension side elongation. The wall moment resistance is thus enhanced by the formation of a moment resisting mechanism, including both the wall and the columns.

A measure of the contribution developed by the 3–dimensional effect to the total resistance can be defined by comparing the moment developed by ΔN_c to the total overturning moment. The result of this comparison is shown in Fig. 6, where the beams at all the floors are supposed to develop the same contribution at collapse. Again, the dependence on the end column elongation is evident. After this member has entirely cracked, the percent contribution becomes almost constant. Note that a meaningful value of this effect is reached with the wall main reinforcement still elastic; the deformations activating the 3–dimensional effect come more from the concrete cracking than from column bar yielding.

The 3–D contribution to resist the overturning moment reached a value of 6.12% at load step n. 136, at which point the analysis was terminated. It is not difficult to give an approximate evaluation of the maximum value for this 3–D contribution. A further increase of the lateral load would produce plastic elongation of the reinforcement in the wall's end column and formation of plastic hinges at both ends of the transverse beams (this was observed in the experimental analysis as well). In such a situation, the limit value for shear in the transverse beams has been estimated as T = 6t (metric tons), or 13.2 kips, corresponding to a value of 6.87% for the 3–D contribution. (The average value of the shear stress in the transverse beam is about 5 kg/cm^2 or 70 psi). This result is not far from the one given in (7), where the contribution of the transverse beams to building moment resistance is estimated at a value of 8%.

The slab width which is effective with the beam in carrying moment is significantly larger than that prescribed by design codes (ACI 318–83 included). In the present analysis, eccentric shell stiffener beam elements have

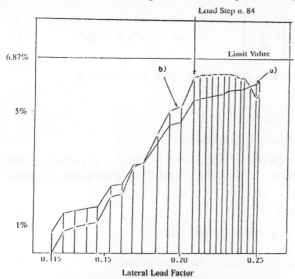

Figure 6 Contribution of 3–D Effects on Structure

been used to model the transverse beams. This modelling allows a good representation of the beam–slab interaction phenomenon, and the same conclusion has been achieved. A reasonable value for the effective slab width (w) is 215cm while a value w = 150cm (4'11") would be obtained following the ACI 318–83 Code (section 10.6.6). When computing the limit value for the shear force, w = 215cm was used. Note, however, that w is a function of the lateral displacement (as also remarked in [2]), in the sense that a higher value should be expected for higher displacement values.

A CONSTITUTIVE MODEL FOR CYCLIC LOADING

Cracking is an important feature of reinforced concrete structures. It can affect overall structural behavior considerably. So, a realistic crack model,which can monitor the opening and closing of cracks in the case of cyclic loadings, is vital to the analytical simulation of the behavior of reinforced concrete. The following constitutive relationship is based on the smeared crack concept and multi–crack model [2,3]. A particular crack law under cyclic loadings will be developed and incorporated in the constitutive equations in an explicit way.

Crack Interface Behavior

Let X and Y refer to global axes, t_i and n_i refer to local coordinate system for crack i with t_i and n_i tangential and normal to the crack respectively. The local crack strains $\epsilon^{cr}_{n_i n_i}$ and $\gamma^{cr}_{n_i t_i}$ can be transformed to global axes as follows:

$$\begin{bmatrix} \epsilon^{cr}_{xx} \\ \epsilon^{cr}_{yy} \\ \gamma^{cr}_{xy} \end{bmatrix}_i = \begin{bmatrix} c^2 & -sc \\ s^2 & sc \\ 2sc & c^2-s^2 \end{bmatrix} \begin{bmatrix} \epsilon^{cr}_{n_i n_i} \\ \gamma^{cr}_{n_i t_i} \end{bmatrix} \quad \text{or} \quad [\epsilon^{cr}]_i = [N]_i [\epsilon^{cr}]_i \quad (3)$$

in which c = $\cos\theta_i$; s = $\sin\theta_i$, θi = angle from the global X to the normal of the ith crack.

Similarly, the crack normal stress $\sigma^{cr}_{n_i n_i}$ and the crack shear stress $\tau^{cr}_{n_i t_i}$ can be related to global concrete stresses by $[N]_i$.

$$\begin{bmatrix} \sigma^{cr}_{n_i n_i} \\ \tau^{cr}_{n_i t_i} \end{bmatrix} = \begin{bmatrix} c^2 & s^2 & 2sc \\ -sc & sc & c^2 - s^2 \end{bmatrix} \begin{bmatrix} \sigma^{cr}_{xx} \\ \sigma^{cr}_{yy} \\ \tau^{cr}_{xy} \end{bmatrix} \qquad \text{or} \qquad [\sigma^{cr}]_i = [N]^T_i[\sigma^{cr}] \tag{4}$$

The relation between crack interface stresses and crack strains reads,

$$[d\sigma^{cr}]_i = [D^{cr}]_i[de^{cr}]_i \tag{5}$$

where $[D^{cr}]_i$ is determined by a particular crack law.

Constitutive Relationship with the Multi–Crack Model

The basic assumption in the multi–crack model is that the total strain ϵ at a cracked point can be decomposed into intact concrete total strain ϵ^{co} and crack total strain ϵ^{cr}, which gives

$$[d\epsilon]_{x, y} = [d\epsilon^{co}]_{x, y} + [d\epsilon^{cr}]_{x, y} \tag{6}$$

$[d\epsilon^{cr}]_{x,y}$ is the sum of transformed local crack strains, that is

$$[d\epsilon^{cr}]_{x, y} = \sum_{i=1}^{m} [N]_i[de^{cr}]_i \qquad \text{or} \qquad [d\epsilon^{cr}]_{x, y} = [N][de^{cr}] \tag{7}$$

Where:

$$[N] = [N_1, \ldots\ldots N_i, \ldots\ldots N_m] \qquad \text{and} \qquad [de^{cr}] = [de^{cr}_1, \ldots\ldots de^{cr}_i, \ldots\ldots, de^{cr}_m] \tag{8}$$

Similarly the local crack interface stress increment vector $[d\sigma^{cr}]$ can be related to the global concrete stress increment vector as follows:

$$[d\sigma^{cr}] = [N]^T[d\sigma^{cr}]_{x, y} \tag{9}$$

Where $[d\sigma^{cr}]$ is

$$[d\sigma^{cr}] = [d\sigma^{cr}_1, \ldots\ldots d\sigma^{cr}_i, \ldots\ldots d\sigma^{cr}_m] \tag{10}$$

The constitutive relationship between $[d\sigma^{cr}]$ and $[de^{cr}]$ reads

$$[d\sigma^{cr}] = [D^{cr}][de^{cr}] \tag{11}$$

while the stress–strain relationship for intact concrete is

$$[d\sigma^{c}]_{x, y} = [D^{co}][d\epsilon^{co}]_{x, y} \tag{12}$$

Where $[D^{cr}]$ is a diagonal matrix with submatrices $[D_i]_{2 \times 2}$ at its diagonal.

By proper mathematical manipulation, the constitutive relation between global stresses and global strains can be derived to be

$$[d\sigma^{c}]_{x, y} = ([D^{co}] - [D^{co}][N]([D^{cr}] + [N]^T[D^{co}][N])^{-1}[N]^T[D^{co}])[d\epsilon]_{x, y} \tag{13}$$

Equation (11) shows that cracking causes softening of the material which is reflected by the reduction in the stiffness of the intact concrete stiffness $[D^{co}]$.

Concrete Material Model

In equation (13), the incrementally orthotropic model proposed by Darwin and Pecknold [6] is used for $[D^{co}]$. This model is based on the equivalent uniaxial strain theory. The principal stress increments are a function of the equivalent uniaxial strain which is not a true strain but a fictitious quantity used as a measure on which to base the variation of material properties.

The orthotropic constitutive equation in material coordinates reads

$$
\begin{bmatrix} d\sigma_1 \\ d\sigma_2 \\ d\tau_{12} \end{bmatrix} = \frac{1}{1-\nu^2} \left\{ \begin{bmatrix} E_1 & \nu\sqrt{E_1E_2} & 0 \\ & E_2 & 0 \\ & & \frac{1}{4}(E_1 + E_2 - 2\nu \sqrt{E_1E_2}) \end{bmatrix} \begin{bmatrix} d\epsilon_1 \\ d\epsilon_2 \\ d\gamma_{12} \end{bmatrix} \right. \tag{14}
$$

The equation used to calculate equivalent uniaxial strain increment in reference [6] is only applicable for proportional loadings. In the case of cyclic loadings, the error in the equivalent uniaxial strain gets bigger with loading cycles. In order to overcome this problem, stresses corresponding to the common point and turning point are used to monitor the stiffness changes during unloading from compression. This approach seems to work well for both cyclic and monotonic,proportional and nonproportional loadings.

Steel Material Model

It is assumed that average strains of the reinforcing steel bars are the same as those of cracked concrete as whole. This implies that there is local bond slip near the cracks because the strain of the bars is assumed not to be equal to that of solid concrete.

For cyclic loadings, the basic characteristics to be modeled to accurately represent the stress–strain behavior of the steel bars are yielding and strain hardening for the first quarter cycle, and the nonlinear Bauschinger effect for subsequent cycles. The Bauschinger effect reduces the apparent yield stress for cyclic loadings after first yielding is reached. Because the behavior of reinforced concrete subject to cyclic loadings in the inelastic range is strongly influenced by the characteristics of the steel stress–strain relationship, the Bauschinger effect must be included. An expression based on Richard and Abbott type formulation [11], which gives stresses in terms of strains, is adopted in this study because of its simplicity and accuracy.

A Crack Law For Cyclic Loadings

So far, some sophisticated crack laws have been proposed for monotonic loadings [3]. Most of them express crack interface stresses in terms of local crack displacements. In equation (13), the crack stiffness matrix $[D^{cr}]$ which relates crack interface stresses to local crack strains is needed. The crack law proposed in [3] is modified and extended to cyclic loading range. $[D^{cr}]$ assumes the following incremental form:

$$
[D^{cr}] = \begin{bmatrix} D_{nn} & D_{nt} \\ D_{tn} & D_{tt} \end{bmatrix} \tag{15}
$$

In order to obtain the stiffness coefficients in $[D^{cr}]$, it is assumed that σ_{nn}^{cr} and τ_{nt}^{cr} can take on the following functional form.

$$
\sigma_{nn}^{cr} = f_n(d\epsilon_{nn}^{cr}, d\gamma_{tt}^{cr}) \quad , \quad \tau_{tt}^{cr} = f_t(d\epsilon_{nn}^{cr}, d\gamma_{tt}^{cr}) \tag{16}
$$

Functions (16) are shown on Figure 7.

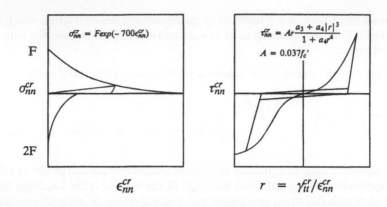

Figure 7 Stress vs. Crack Strain Functions

Analytical Examples

Three tests on reinforced concrete panels under cyclic loadings have been conducted at the University of Toronto [10].The test results showed that the modelling of tension stiffening and strength degradation due to both cycling and coexisting tensile strain normal to the compressive direction is vital to the simulation of cyclic behavior of reinforced concrete. The proposed constitutive model was applied to analyze the three panels. The analytical results are compared with experimental ones on Figures 8–10. It is seen that the model is capable of reproducing the experimental results,especially at low levels of shear stresses. Because small errors in material strength and predicted degraded strength can have large effect on the hysteresis loops at high load levels,it is unrealistic to expect the exact match of analytical hysteresis loops with the observed ones.

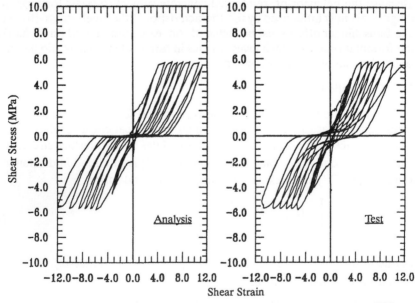

Figure 8   Shear Stress vs. Shear Strain Response for SE8

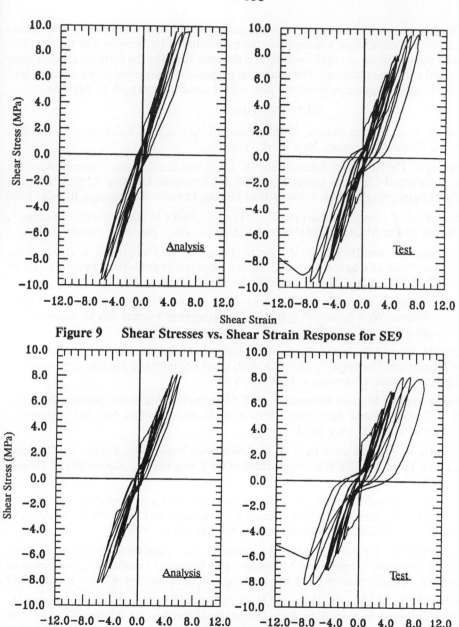

Figure 9 Shear Stresses vs. Shear Strain Response for SE9

Figure 10 Shear Stresses vs. Shear Strain Response for SE10

The deterioration of the concrete compressive strength as a function of the coexisting tensile strain for monotonic loadings has been formulated by Vecchio and Collins. The three tests showed more strength degradation than predicted by the formula. It is not clear what caused additional damage. For analytical purpose, it is assumed that maximum tensile strain in the compressive direction also causes concrete strength to degrade.

REFERENCES

1. ASCE, Task Comm Report, "Finite Element Analysis of Reinforced Concrete", A. H. Nilson, Chairman., New York, 1982.

2. Barzegar, Fariborz and Schnobrich, W. C., " Nonlinear Finite Element Analysis of Reinforced Concrete under Short Term Monotonic Loading ", SRS No. 530, Civil Engineering Studies, University of Illinois, Urbana–Champaign, Illinois, 1986

3. Bazant, Z. P., and Gambarova, P., " Rough Cracks in Reinforced Concrete ", Journal of the Structural Division, ASCE, Vol. 106, No. ST4, April, 1980

4. Charney, F.A. and Bertero, V. V, "An Evaluation of the Design and Analytical Seismic Response of a Seven Story Reinforced Concrete Frame–Wall Structure" UCB/EERC 82/08, Univ. Calif @ Berkeley Berkeley, Calif., 1982

5. Chesi, C. and Schnobrich, W. C. "Three–Dimensional Effects in the Lateral Behavior of Frame–Wall Systems" Structural Engineering Journal, ASCE, accepted for publication

6. Darwin, D. and Pecknold, D. A. W., " Inelastic Model for Cyclic Biaxial Loading of Reinforced Concrete ", SRS NO. 409, Civil Engineering Studies, University of Illinois, Urbana–Champaign, Illinois, 1974

7. Keshavarzian, M. and Schnobrich, W.C. "Computed Nonlinear Seismic Response of R/C Wall–Frame Structures" Structural Research Series No. 515, University of Illinois Urbana, May 1984

8. Milford, R. V. and Schnobrich, W. C., "Nonlinear Behavior of Reinforced Concrete Cooling Towers", SRS 514, Department of Civil Engineering, University of Illinois, Urbana, Ill., 1984.

9. Okamura, H. and Maekawa, K. " Non–Linear Analysis and Constitutive Models of Reinforced Concrete ", 2nd International Conference on Computer Aided Analysis and Design of Concrete Structures, Zel Am See, April, 1990

10. Stevens, N. J., Uzumeri, S. M. and Collins, M. P., " Analytical Modelling of Reinforced Concrete Subjected to Monotonic and Reversed Loadings ", University of Toronto, Department of Civil Engineering, Publication No. 87–1, January 1987

11. Suharwardy, M.I.H. and Pecknold, D.A., " Inelastic Response of Reinforced concrete Columns Subjected to Two–Dimensional Earthquake Motions ", SRS No. 455, Civil Engineering Studies, University of Illinois, Urbana–Champaign, Illinois, 1978

12. Takayangi, T. and Schnobrich, W.C. "Nonlinear Analysis of Coupled Wall Systems" Earthquake Engineering and Structural Dynamics, Vol. 7, No. 1 Jan–Feb 1979

13. Yoshimura, M., and Kurose, Y., "Inelastic Behavior of the Building", in Earthquake Effects on Reinforced Concrete Structures–US–Japan Research", Wight, J. K. Ed. ACI , SP–84, Detroit, Mich. 1985

MICROSCOPIC AND MACROSCOPIC ANALYSES OF REINFORCED CONCRETE FRAMED SHEAR WALLS

NORIO INOUE
Senior Research Engineer/Building Engineering Department
Kajima Institute of Construction Technology/Kajima Corporation
19-1 Tobitakyu 2-Chome Chofu-shi, Tokyo 182, Japan

NORIO SUZUKI
Research Engineer/Research Department
Kobori Reseach Complex/Kajima Corporation
5-30 Akasaka 6-Chome Minato-ku, Tokyo 107, Japan

ABSTRACT

Macroscopic models are proposed to obtain not only the ultimate strength but also the load-deflection relationships of reinforced concrete framed shear walls. They are modeled by referring to the results obtained from the microscopic model using Finite Element Method (FEM) by which the distributions of stress and strain are clarified and the resisting mechanisms of shear walls are investigated. These macroscopic models are useful to define the skeleton curve for dynamic response analyses and to evaluate the deformability.

INTRODUCTION

When dynamic response analyses are performed and deformability is evaluated for reinforced concrete structures with shear walls, it is important to know not only the ultimate strength but also the load-deflection relationships of shear walls. These relationships are currently given by the empirical formulae which are defined as multi-linear relationships based on experimental results. But these formulae are applicable in limited ranges of parameters, so theoretical formulae are desirable for general purposes.

To obtain non-linear behavior, FEM is very useful and has been applied by many reseachers recently. But if obtaining only the load-deflection relationships, more simplified approaches are considered also useful. To establish these models, equilibrium equations and compatibility equations should be proposed adequately for shear walls. Then it is important to learn the distribution of stress and strain and the yielding conditions of the shear walls. To do so, the application of FEM is considered very effective.

In this paper by using FEM as a microscopic approach the resisting mechanisms are investigated and a macroscopic model is proposed for obtaining ultimate strength of shear walls by referring to the results of

FEM, where the yielding conditions are assumed.

Next a rod element method is proposed to obtain the load-deflection relationships by incremental equations while assuming the stress-strain relationships of concrete and reinforcement. This is derived from the mechanism obtained by the aforementioned macroscopic model for ultimate strength.

Finally for the design purpose a more simplified model is proposed to obtain multi-linear load-deflection relationships by calculating only the distinctive points from defined formulae.

STUDY ON MECHANISMS BY MICROSCOPIC MODEL

Microscopic Model

The analytical method using FEM is based on the average stress vs. average strain relationship proposed by Vecchio and Collins [1] with some modification for being used with FEM. This approach is very effectual for a shear wall in which many reinforcements are uniformly arranged. The details were presented in the papers [2] and [3]. The main points are described below.

−The average strain of concrete is equal to that of the reinforcement.

−The direction of principal stress for cracked concrete coincides with that of the principal strain.

−The compressive principal stress vs. strain relationship of the cracked concrete is represented by Eqs.(1) and (2) and Figure 1.

−The tensile principal stress vs. strain relationship after cracking is represented by a third-order function like Figure 1.

−A stiffness matrix is made by the principal stress vs. strain relationship with addition of an adequate shear rigidity like Eq.(3). The solution is performed by an iteration method for total stress and strain.

$$\sigma_2 = \sigma_{2max} \left\{ 2\left(\frac{\epsilon_2}{\epsilon_0}\right) - \left(\frac{\epsilon_2}{\epsilon_0}\right)^2 \right\} \tag{1}$$

$$\frac{\sigma_{2max}}{f_c'} = \frac{1}{0.8 - 0.34\frac{\epsilon_1}{\epsilon_0}} \leqq 1.0 \tag{2}$$

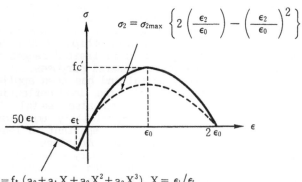

σ_1, σ_2 : Tensile and compressive principal stress

ϵ_1, ϵ_2 : Tensile and compressive principal strain

f_t, f_c': Uniaxial tensile and compressive strength

ϵ_t, ϵ_c': Strain at the tensile and compressive strength

$$\sigma_1 = f_t \left(a_0 + a_1 X + a_2 X^2 + a_3 X^3 \right) \quad X = \epsilon_1/\epsilon_t$$

Figure 1. Stress-strain relationship of concrete

$$\begin{Bmatrix} \sigma_1 \\ \sigma_2 \\ \tau \end{Bmatrix} = \begin{bmatrix} E_1 & O & O \\ & E_2 & O \\ sym & & G \end{bmatrix} \begin{Bmatrix} \epsilon_1 \\ \epsilon_2 \\ \gamma \end{Bmatrix} \tag{3}$$

E_1 : Secant modulus in tension
E_2 : Secant modulus in compression
G : Shear modulus

Separation of Mechanisms from Results of FEM

The obtained shear stresses of a shear wall are assumed in each element to be divided into those contributed by the truss mechanism of transverse reinforcement $|\tau_s$, and those contributed by the arch mechanism of concrete $|\tau_f$, from the results of FEM like Figure 2. The basic idea of these mechanisms is referred to Shohara and Kato's study [4]. This separation is useful to learn the distribution of the contributing ratio of each mechanism and to propose the macroscopic models [5].

In the truss mechanism it is defined that the equilibrium condition is obtained from shear stress $|\tau_s$, restraining force of the transverse reinforcement $|f_{SX}$, a part of concrete stress $|\kappa\sigma_1$, $|\kappa\sigma_2$, and a part of restraining force of longitudinal reinforcement $|\alpha f_{SY}$. Here the principal direction of each element is given by the analytical results of FEM. In the arch mechanism, the equilibrium condition is obtained by the shear stress $|\tau_f$, restraining forces from surrounding elements $|f_X$, $|f_Y$ and the residual restraining stresses of the concrete and the longitudinal reinforcement which are left from the truss mechanism. From these conditions the contributing ratios are calculated in each element of the FEM model.

This method was applied to many shear walls [5]. Among them the obtained results of a shear wall with a small height-to-width ratio (a/D) of 0.5 are presented in Figure 3 as an example. These studies show that in the case of a small a/D, the arch mechanism carries almost all of the shear stresses and in the case of a large a/D the truss mechanism carries some amount of the shear stresses.

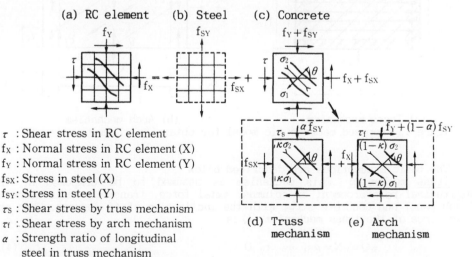

(a) RC element (b) Steel (c) Concrete

τ : Shear stress in RC element
f_X : Normal stress in RC element (X)
f_Y : Normal stress in RC element (Y)
f_{SX} : Stress in steel (X)
f_{SY} : Stress in steel (Y)
τ_S : Shear stress by truss mechanism
τ_f : Shear stress by arch mechanism
α : Strength ratio of longitudinal steel in truss mechanism
κ : Strength ratio of concrete in truss mechanism
σ_1, σ_2 : Principal stresses in concrete

(d) Truss mechanism   (e) Arch mechanism

Figure 2. Separation of stresses in a reinforced concrete element

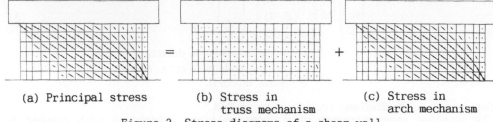

(a) Principal stress   (b) Stress in truss mechanism   (c) Stress in arch mechanism

Figure 3. Stress diagrams of a shear wall

ULTIMATE STRENGTH BY MACROSCOPIC MODEL

Analytical Method

Shohara, Minami, Shiraishi et al. proposed macroscopic models to calculate the ultimate strength of reinforced concrete shear walls by using limit analyses [4], [6] and [7]. In these models the ultimate resisiting mechanism is devided into the truss mechanism and the arch mechanism like the definition presented in the former part and the summation of the two mechanisms is considered as the ultimate strength. These models are useful for the analyses of the shear wall with typical configulations and boundary conditions. The authors also proposed a macroscopic model [8] like Figure 4 which is prominent in evaluation of the effect of the surrounding columns by considering the equilibrium conditions of a web panel and the surrounding columns in the arch mechanism, referring the basic idea proposed by Shiohara [9] and the stress flow obtained by FEM.

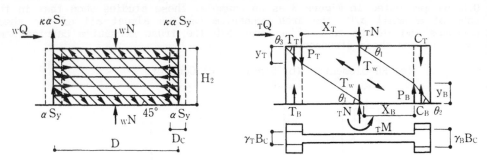

(a) Truss mechanism   (b) Arch mechanism

Figure 4. Proposed macroscopic model for obtaining ultimate strength

The basic assumptions are described below.

Truss mechanism: This mechanism is assumed to be composed of the longitudinal reinforcement of columns, axial force, transverse reinforcement of web panel and concrete strut with the inclination of 45 degrees. Then the shear force of the truss mechanism $_wQ$ is

$$_wQ = {_wN} = \rho_{WH}\ \sigma_{wy}\ t_w\ D \tag{4}$$

$_wN$: Axial force in truss mechanism
ρ_{WH} : Transverse reinforcement ratio of web panel
σ_{wy} : Yielding strength of transverse reinforcement
t_w : Thickness of web panel

Arch mechanism: This mechanism is assumed to be composed of concrete and longitudinal reinforcement of columns, longitudinal reinforcement of web panel and concrete strut which have the residual strength left from the truss mechanism. Here the basic assumptions are as follows.
  –The longitudinal reinforcement of a tensile column and the web panel is assumed as yielded.
  –The effective thickness of columns $\gamma_T B_C$, $\gamma_B B_C$ are obtained for the tensile side and compressive side respectively, with consideration for the equilibrium conditions of the compressive force of the concrete strut in the web panel, the axial forces in the column T_T, C_B and the tensile forces which are transferred by bond stress from the tensile longitudinal reinforcement in the column P_T, P_B.
  –The reduction factor of concrete strength is assumed as 1.0 in this analysis.
  From these assumptions, the shear force of arch mechanism $_T Q$ is

$$_T Q = f_C' \, t_w' \, X_B \, \sin \theta_1 \, \cos \theta_1 + f_C' \, \gamma_B \, B_C \, D_C \, \sin \theta_2 \, \cos \theta_2 \qquad (5)$$

  f_C' : Compressive strength of concrete
  t_w' : Effective thickness of web panel
  γ_B : Effective thickness ratio of compressive column

Ultimate strength of shear wall: The amount of lateral reinforcement in the web panel is varied from zero to actual value and the total shear forces summing up the truss mechanism and the arch one are calculated. Among them the maximum shear force is defined as the ultimate strength of the shear wall.

Analytical Results
The proposed macroscopic model was applied to shear walls which were studied in "the committee on RCFEM and Design Method for RC Structures" of Japan Concrete Institute by using several macroscopic models [7]. The obtained results are shown in Figure 5 as compared with the experimental ones. The proposed model gives good estimation of ultimate strength of shear walls.

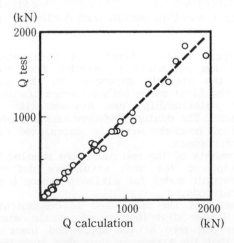

Figure 5. Comparison of ultimate strength of shear walls obtained by the proposed macroscopic model with experimental results

LOAD–DEFLECTION RELATIONSHIP BY ROD ELEMENT MODEL

Analytical Method

Several macroscopic models have been proposed to evaluate the ultimate strength of shear walls as presented in the former section but the researches concerning deflection are very few. The authors proposed a macroscopic model to calculate a load–deflection relationship by using rod elements [10]. This model would be useful to estimate the deformability which is important as the ultimate strength design becomes usable.

From the investigation of resisting mechanisms by FEM, in the case of a shear wall with a small a/D the contribution of the truss mechanism is very small and the shear force is carried mainly by the arch mechanism. Considering these characteristics a macroscopic model was proposed to represent the arch mechanism by rod elements like Figure 6. In this model under small deformations the transverse reinforcement is neglected like (a), but after the strain of the arch strut reaches the value at strength, the effect of transverse reinforcement is considered by introducing the truss strut, transverse reinforcement and hinges in the middle of columns like (b). This procedure is performed conveniently to represent the experimental trends that the strain of transverse reinforcement increases when the crush occurs in concrete and the strength reduction is observed. These phenomena mean that the contribution of the truss mechanism becomes large.

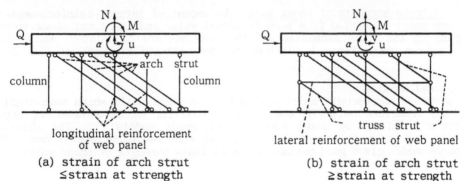

(a) strain of arch strut
≤strain at strength

(b) strain of arch strut
≥strain at strength

Figure 6. Rod element model to obtain load–deflection relationship

Here the main purpose is to estimate the relationship near ultimate strength and the resisting mechanism is modeled from the results of the macroscopic model for the ultimate strength. So the obtained deflection gives larger values at the first stage before enough cracks occur.

The stress–strain relationships are assumed like in Figure 7 for concrete and reinforcement. The dominant assumptions are presented below.
  –Columns are composed of concrete with the calculated effective thickness and longitudinal reinforcement.
  –Longitudinal reinforcements of the web panel are bundled by several rods.
  –The width and location of the arch strut are defined by the values obtained by the macroscopic model for ultimate strength and the arch strut is divided into several rods.
  –Transverse reinforcement of the web panel is concentrated at the center assuming vertically linear distribution of strain where the strain is maximum at the center and zero at the upper and lower ends. Its initial length is defined so that the stress is zero when inserted.
  –Truss struts are assumed as rigid elements. The initial length is defined by the same method as transverse reinforcement.

From these assumptions the incremental equations are given concerning the increments of forces and displacements for case (a) and case (b). The solution is controlled by the lateral displacement at the loading slab.

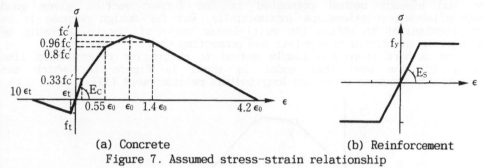

(a) Concrete   (b) Reinforcement

Figure 7. Assumed stress-strain relationship

Analytical Results

This proposed method was applied to the shear walls in which the ultimate strength was carried only by the arch mechanism in the results of the macroscopic model for the ultimate strength. Figure 9 shows the calculated result for the shear wall which was tested by Shiohara [9], comparing the envelope curve of the experimental load-deflection relationship subjected to cyclic loads. The relationship is grasped well up to the ultimate strength and furthermore the yielding load of reinforcement of tensile column is predicted by this model (Test; 1460 kN, Calculation; 1400 kN). After the peak load, its ductile behavior is obtained analytically.

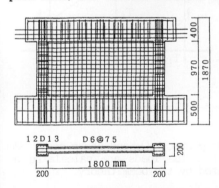

longitudinal reinforcement ratio of column	3.8	%
lateral reinforcement ratio of web panel	0.85	%
compressive strength of concrete	38.6	MPa
yielding strength of column reinforcement	364	MPa
yielding strength of web reinforcement	370	MPa

Figure 8. Analyzed shear wall (P2015-A Shiohara)

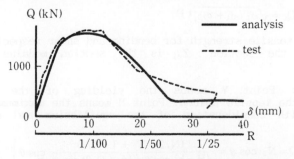

Figure 9. Load-deflection relationship

SIMPLE EVALUATION OF LOAD-DEFLECTION RELATIONSHIP

Analytical Method
The rod element method presented in the former section gives good load-deflection relationships incrementally. But for design purpose it is more convienient to obtain the multi-linear curve of the relationship by calculating the distinctive points and connecting them.

The authors propose a simple method to obtain the relationship like Figure 10. The analytical model is shown in Figure 11, which has concentrated concrete strut and longitudinal reinforcement [11].

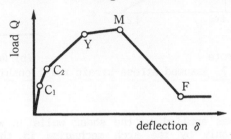

Figure 10. Proposed load-deflection relationship

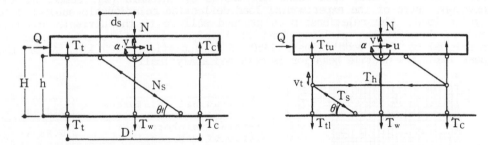

(a) Point Y, M	(b) Point F

Figure 11. Simple model to evaluate load-deflection relationship

Point $C_1|$, $C_2|$: They are defined for bending crack $_fQ_c|$ and shear crack $_sQ_c|$. Among them the smaller one is defined as $C_1|$. The loads are calculated by Eqs.(6) and (7). The deflections are calculated by the beam theory.

$$_fQ_c = (_ff_t + \sigma_0) \ Z_e/H \tag{6}$$

$$_sQ_c = \sqrt{_sf_t \ (_sf_t + \sigma_0)} | t \ D \tag{7}$$

Here $_ff_t|$, $_sf_t|$ are tensile strength for bending and shear respectively. $\sigma_0|$ is axial stress of the section. $Z_e|$ is the section modulus considering reinforcement.

Point $Y|$, $M|$: Point Y means the yielding of the longitudinal reinforcement in the tensile column. Point M means the maximum shear. They are calculated with consideration of the equilibrium and compatibility

$$Q = N_s \ \cos \theta = \frac{(N/2 + T_y + T_w/2) \ D}{(H-h) \cos \theta + (d_s + D/2) \sin \theta} \ \cos \theta \bigg| \tag{8}|$$

condition in the model of Figure 11 (a). Then the shear force Q is obtained by Eq.(8). Here N is axial load. T_y is tensile strength of column. T_w is tensile force of longitudinal reinforcement of web panel. For point Y, T_w is calculated by the compatibility condition assuming that the strain of compressive column is zero. For point M, T_w is its tensile strength. Then the horizontal displacement at the loading slab u is

$$u = \frac{l_s \ \epsilon_C + (1+2 \ d_S/D) \ v \sin \theta}{\cos \theta} \qquad (9)$$

Here ϵ_C is strain of concrete strut. For point Y it is calculated by the assumed stress strain relationship of concrete like Eq.(10). For point M it is assumed as $1.4 \epsilon_0$ in Figure 7. v is vertical displacement at the center of the loading slab and obtained by the equilibrium condition and compatibility condition in Figure 11 (a).

$$N_S = A_S \ f_C' \left\{ 2 \left(\frac{\epsilon_C}{\epsilon_0} \right) - \left(\frac{\epsilon_C}{\epsilon_0} \right)^2 \right\} \qquad (10)$$

Point F: At point F the stress of the arch strut is assumed to become zero. From this point a constant shear carried by the yielded lateral reinforcement is assumed to be maintained. These assumptions give the shear force Q by Eq.(11) where A_{WH} is area of lateral reinforcement in web and f_y is its yielding strength. The lateral displacement u is calculated by Eq.(9), where ϵ_C is substituted by the strain ϵ_u in which the stress of the concrete is zero. v is calculated by the equilibrium condition and compatibility condition in Figure 11 (b).

$$Q = A_{WH} \ f_y \qquad (11)$$

Analytical Results
The foregoing model is applied to shear walls W1 and W2 which were tested by Aoyama et al. [9][12]. W2 has the same configuration as W1 and only its material properties are different. The obtained results are shown in Figure 12. This simplified model can predict the load–deflection relationships up to the ultimate strength. As to the softening region, the general behavior is grasped for W2 but the deflection of the descending point for W1 is a little smaller. This is caused by the assumption that the arch strut is represented by one rod and therefore it gives an average deflection for a ductile shear wall like W1.

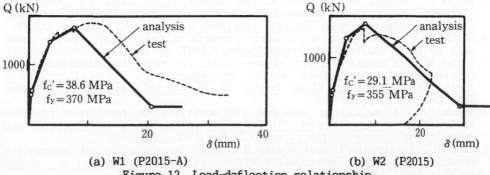

(a) W1 (P2015-A)　　　　　　　　(b) W2 (P2015)
Figure 12. Load–deflection relationship

CONCLUDING REMARKS

A microscopic model using FEM is effective to investigate the resisting mechanisms of reinforced concrete shear walls and to propose macroscopic models.

The proposed macroscopic model based on the limit analysis can estimate well the ultimate strength of shear walls. This model is prominant in evaluation of the effect of the surrounding columns.

The proposed rod element model can calculate the load-deflection relationship of shear walls incrementally by assuming the stress-strain relationship of concrete and reinforcement.

A simple analytical method is proposed which gives a multi-linear load-deflection curve of the shear wall by calculating distinctive points such as bending and shear cracking, yielding of longitudinal reinforcement in the tensile column, ultimate strength and failure strength carried by only the lateral reinforcement of the web panel. This model will be useful for design purposes.

REFERENCES

1. Vecchio F. J. and Collins M. P., The modified compression-field theory for reinforced concrete elements subjected to shear. ACI Journal, March-April 1986, pp.219-31.
2. Inoue N., Koshika N. and Suzuki N., Analysis of shear wall based on Collins panel test. In Finite Element Analysis of Reinforced Concrete Structures, ASCE, New York, 1985, pp.288-99.
3. Inoue, N., Analysis of reinforced concrete members subjected to cyclic loads. Proceedings of IABSE Colloquium Delft 1987, pp.487-502.
4. Shohara R. and Kato B., Ultimate strength of multi-storied RC shear walls. Transactions of the Architectural Institute of Japan, No.343, 1984, pp.24-31
5. Inoue, N., Shiraishi I. and Noguchi H., Verification of macroscopic models for RC walls by FEM. Structural Design, Analysis & Testing Proceedings, Structures Congress, 1989, ASCE, pp.291-300.
6. Shiraishi I., Shirai N., Murakami T. and Minami K., Macroscopic models for R/C shear walls. Structural Design, Analysis & Testing Proceedings, Structures Congress, 1989, ASCE, pp.271-80.
7. Shohara R., Shirai N. and Noguchi H., Verification of macroscopic models for R/C walls. Structural Design, Analysis & Testing Proceedings, Structures Congress, 1989, ASCE, pp.281-90.
8. Inoue N., Suzuki N., Koshika N. and Maruta M., A model for predicting ultimate strength of reinforced concrete shear walls with columns. Proceedings of the Japan Concrete Institute Vol.10 No.3, 1988, pp.373-78
9. Shiohara H., Estimation of ultimate strength of reinforced concrete shear wall by limit analysis. Concrete Journal, 1987, pp.101-14.
10. Inoue N. and Suzuki N., Rod element model for analyzing displacement of reinforced concrete shear walls. Proceedings of JCI Colloquium on Analytical Studies on Shear Design of Reinforced Concrete Structures, JCI, Dec. 1989, pp.179-86.
11. Suzuki N., A method to evaluate the ductility of reinforced concrete shear walls. Proceedings of JCI Symposium on Design of Reinforcement and Ductility of Concrete Structures, JCI, May 1990, pp.165-70.
12. Kato D., Katsumata H. and Aoyama H., Study on strength and deformation capacity of postcast shear walls without an opening. Transactions of the Architectural Institute of Japan, No.337, 1984, pp.81-88

ENERGY DISSIPATION OF REINFORCED CONCRETE SHEAR WALL PREDICTED BY NONLINEAR FEM ANALYSIS

Hajime OKAMURA
Professor
Department of Civil Engineering
University of Tokyo
Japan

Eiichi SAKATA
Research Student
Technical Research Institute
Hazama-Gumi.LTD
Japan

ABSTRACT

Based on the analysis of reversed cyclic response of various types of reinforced concrete shear walls by using the computer program called WCOMR, the effect of various parameters, such as proportion of wall, ratio of column and wall rigidity, reinforcement ratio of wall and column, steel strength, and concrete strength, on the responses, such as loading capacity, deformational capacity and energy dissipation, is discussed.

INTRODUCTION

The nonlinear FEM program WCOMR has been developed for predicting the behavior of reinforced concrete walls subjected to reversed cyclic in-plane static loads [1]. In this program reinforced concrete structures are discretized into reinforced concrete solid elements whose mechanics are denoted by the smeared crack model, and the joint elements for discrete cracks in junction planes between members with different thicknesses. Experimental verification in terms of accuracy and applicability has been systematically conducted. The ultimate capacity, deformability and hysteresis characteristics of shear walls, which are related to seismic design, have been predicted with high accuracy. Therefore, this program can be used directly for design purposes.

Based on the analytical results of WCOMR, the effects of various parameters, such as proportion of wall, ratio of column and wall rigidity, reinforcement ratio of wall and column, steel strength and concrete strength, on the responses, such as loading capacity, maximum displacement and energy dissipation, are discussed.

PARAMETRIC STUDIES

Input material properties, which are compressive strength of concrete and yield and tensile strength of reinforcements, are simpler than those required by general elastic analysis, since all the constitutive equations used are formulated by these properties [2]. Based on the dimensions of composed elements, finite element meshes are automatically generated. Twenty plate elements and twelve joint elements are used in general. The load displacement relations under monotonic loading are firstly calculated as shown in Figure 1. It takes about one hour by using 16 bit personal computer. Refering to the results, the calculations for reversed cyclic loading from the points where the tensile stress of steel at the bottom of column reaches 1000 kgf/cm² and yield strength are executed. The outputs are displacements and forces of nodal points, stresses and strains of the Gauss points in each element. The load - displacement relations, deformational shapes, stress distributions, strain distributions, crack patterns and failure modes are graphically displayed as shown in Figure 2.

Parametric calculations are conducted refering to Aoyama's test specimens P4012 (A and C series) and P2012 (B series), which have I-sectional wall surrounded by rigid beams and columns as shown in Figure 3 [3]. Cross-sectional area of columns of B series is half of that in A series. Wall height of C series is 2.5 times of that in A series. Thirty-nine cases are calculated by changing systematically the concrete strength, reinforcement ratio of wall, and applied constant compressive stress on the columns, respectively.

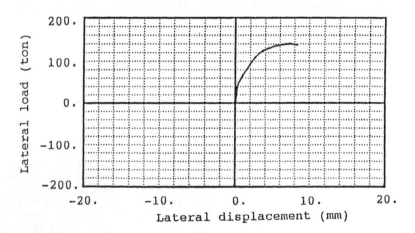

Figure 1. Graphic display of analytical result
for monotonic loading

345

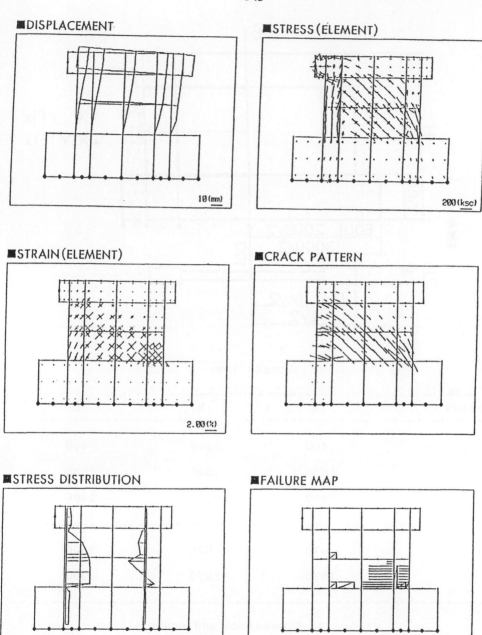

Figure 2. Graphic presentation of analytical results

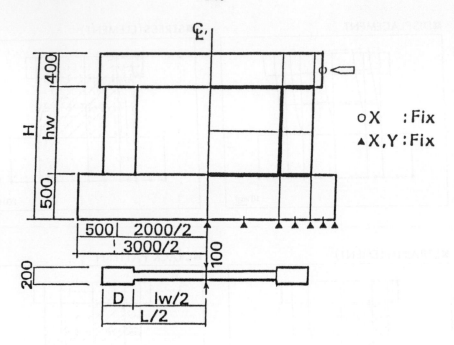

Length (mm)

Series	A	B	C
D	400	200	400
lw	1600	1800	1600
L	2400	2200	2400
hw	970	970	2300
H	1870	1870	3200

Figure 3. Dimensions and meshes

DEFORMATIONAL CAPACITY

The examles of obtained shear stress - rotational angle relations are shown in Figure 4 for 5 levels of concrete strength
in A series, where the reinforcement ratio of wall and column
are 0.0085 and 0.0190 respectively and the applied compressive
stress on each column is 15 kgf/cm². The shear stress is
defined as the lateral load divided by the product of wall
thickness and the distance between columns. The rotational
angle is defined as the lateral displacement at the top of the
wall divided by the wall height. From this kind of figures,
the deformational capacity defined in this paper as the rotational angle corresponding to the maximum loading capacity can
be obtained. Appearently the deformational capacity and the
loading capacity increase when the concrete strength increases.   This is true for all the analytical cases.
      The total displacement can be divided into three com-
ponents, bending, shear and local displacement.   The bending
displacement can be obtained in the same way as generally calculated from the experimental results, namely by doubly integrating the average curvature of the Gauss points along the
height of wall calculated on the assumption that plane sections before bending remain plane after bending.   The local
displacement can be obtained as that due to pulling out of
bars from the footing, shear displacement along the top surface of the footing, and the local sinking of the column into

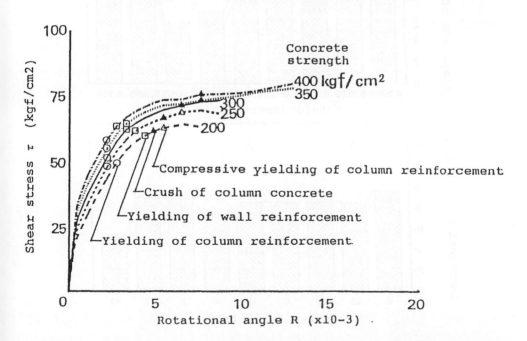

Figure 4. Effects of concrete strength on behavior under
monotonic loading ; p_w=0.0085, p_c=0.0190, σ '$_c$=15kgf/cm²

348

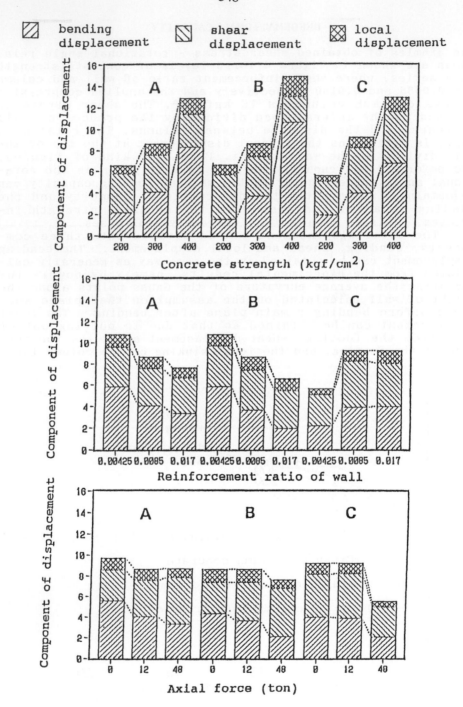

Figure 5. Effect of parameters on each component of displace
-ment

the footing. The shear displacement can be obtained as the subtraction of the bending and local displacements from the total.

The effect of various parameters on the deformational capacity can be evaluated by that on the bending displacement as shown in Figure 5, as the effect of the parameters on the shear displacement is relatively small. The bending displacement at ultimate and then the deformational capacity or the ductility of wall will increase when the concrete strength increases or the steel ratio, yield strength of reinforcement and compressive stress applied on the column decrease.

ENERGY DISSIPATION

The shapes of the nondimensional hysteresis loops for the load-displacement relations are different according to the loading level as shown in Figure 6, and to the components of displacement as shown in Figure 7. The nondimensional lateral load is defined as the ratio of the shear stress to the maximum shear stress, and the nondimensional displacement is defined as the ratio of the lateral displacement of the top of wall to the displacement corresponding to the maximum load.

The bending displacement loops have smaller residual displacement and hence have the flatter shapes not only just after cracking but also after yielding. Therefore, the energy dissipation is affected also by the amount of shear displacement even if it is not so much affected by the various parameters.

The magnitude of the bending displacement may be different in the loading direction, especially when the steel strains go deeply into the strain hardening range and the residual strains become very large. Under the displacement control test, the shear displacement should compensate for the remaining displacement. This may cause the shear failure of wall earlier than that under monotonic loading, and this may be one of the reasons why the reversed cyclic loading decreases the loading capacity or the deformational capacity of a structure compared with monotonic loading.

The nondimensional loop area is defined as the ratio of inner loop area to the triangular area made by the maximum load and the displacement, and is obtained from the nondimensional load-displacement loop as shown in Figure 8.

This nondimensional loop area or the energy dissipation near the ultimate load becomes gradually smaller when the concrete strength becomes smaller in A series although it is almost same at the lower load level as shown in Figure 9. However, in B series which has relatively heavy reinforcement (p_c = 0.038) with higher compressive stress (σ'_c = 30 kgf/cm²) acting on the smaller columns, remarkable reduction of the energy dissipation can be found when the concrete strength becomes smaller than a critical value.

Similar phenomenon can be seen in Figure 10 which shows the relation between the energy dissipation and the reinforce-

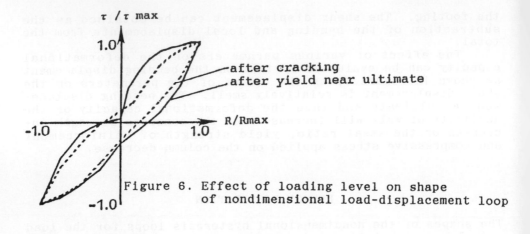

Figure 6. Effect of loading level on shape
of nondimensional load-displacement loop

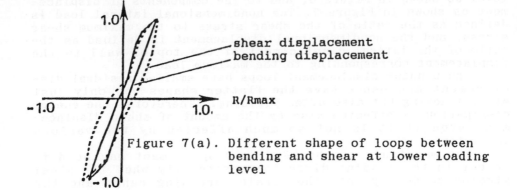

Figure 7(a). Different shape of loops between
bending and shear at lower loading
level

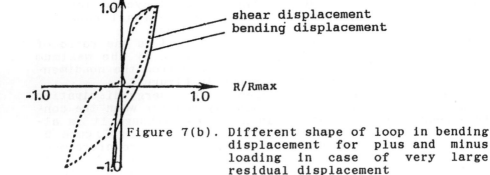

Figure 7(b). Different shape of loop in bending
displacement for plus and minus
loading in case of very large
residual displacement

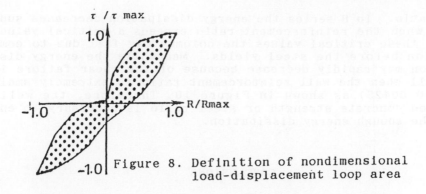

Figure 8. Definition of nondimensional
load-displacement loop area

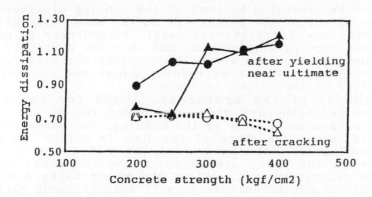

Figure 9. Critical concrete strength for remarkable reduction
of energy dissipation in weak column
$$\left(\begin{array}{l} \bigcirc, \bullet \text{ A series : } p_a = 0.019, \ \sigma'_a = 15 \text{ kgf/cm}^2 \\ \triangle, \blacktriangle \text{ B series : } p_a = 0.038, \ \sigma'_a = 30 \text{ kgf/cm}^2 \end{array} \right)$$

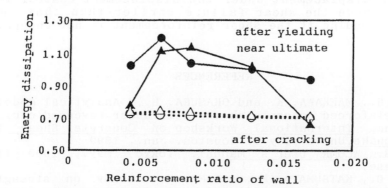

Figure 10. Upper and lower critical reinforcement ratios of
wall for remarkable reduction of energy dissipation
$$\left(\begin{array}{l} \bigcirc, \bullet \text{ A series : } p_a = 0.019, \ \sigma'_a = 15 \text{ kgf/cm}^2 \\ \triangle, \blacktriangle \text{ B series : } p_a = 0.038, \ \sigma'_a = 30 \text{ kgf/cm}^2 \end{array} \right)$$

ment ratio. In B series the energy dissipation decreases suddenly when the reinforcement ratio exceeds a critical value. Beyond these critical values the columns may fail due to compression before the steel yields. Meanwhile the energy dissipation may rapidly decrease because of the shear failure in the wall when the wall reinforcement ratio is extremely small (p_w = 0.00425) as shown in Figure 10. Therefore, the well-balanced concrete strength or steel ratio is necessary to ensure the enough energy dissipation.

CONCLUSIONS

(1) The effect of various parameters on the deformational capacity can be evaluated by that on the bending displacement, because the effect of the parameters on the shear displacement and the local one is relatively small. Accordingly it can be said that the bending displacement and then the ductility will decrease when the concrete strength of column decreases or the steel ratio, yield strength of reinforcement and compressive stress applied on the column increase.
(2) The shapes of the hysteresis loops for the load-displacement relations are different according to the components of displacement such as the bending, the shear and the local displacement. The shape of the loop is mainly dependent on the bending characteristics as the effect of various parameters on the shear displacement and the local one is found to be relatively small unless the shear failure of the wall is dominant due to extremely small reinforcement ratio in the wall.
(3) The magnitude of the bending displacement becomes different in the loading direction especially when the residual strains due to the bending displacement becomes large. The remaining displacement must be compensated by the increase of the shear displacement under the displacement control test. This may cause the shear failure earlier than that under monotonic loading if the shear reinforcement is not enough.

REFERENCES

1. SHIN, II., MAEKAWA, K. and OKAMURA, H., Analytical models for reinforced concrete shear walls under reversed cyclic loading, International Workshop on Concrete Shear in Earthquake, University of Houston, Jan., 1990
2. FORUM8, WCOMR User's Manual, Tokyo, May., 1990 (in Japanese)
3. KATO, D., KATSUMATA, H., AOYAMA, H., Study on strength and deformation capacity of postcast shear walls witout an openg, Transactions of the architectural institute of Japan, No.337, p.81, Mar., 1984

THEME III:

EFFECT OF SHEAR ON ENERGY DISSIPATION

EVALUATION OF DUCTILITY OF R/C MEMBERS AND INFLUENCE OF DUCTILITY ON INELASTIC RESPONSE BEHAVIOUR OF R/C FRAME STRUCTURES

ATSUHIKO MACHIDA and HIROSHI MUTSUYOSHI
Department of Construction Engineering, Saitama University,
Shimo-okubo, Urawa, Japan

ABSTRACT

In order to establish a reliable equation to evaluate ductility of R/C members, reversed cyclic loading tests were carried out using 33 specimens whose sectional characteristics are similar to ordinary R/C single column piers used in Japan. Based on the test results, the effects of various variables on ductility were investigated one by one. The results were summarized in a series of equations to estimate ductility quantitatively in the form of a ductility factor. It was confirmed that the ductility of R/C members derived from the proposed equation resulted in satisfactory agreement with the test results obtained by other researchers.

Moreover, in order to investigate the influence of ductility of members on inelastic response behaviour of R/C frame structures under earthquake motion, shaking table tests and pseudodynamic tests were carried out using small-scale two-story one-bay R/C bridge piers. It was observed from the tests that the inelastic behaviour of R/C frame structures depends strongly on the capacity of ductility of each member. To calculate accurately the response behaviour of R/C frame structures up to collapse, a new restoring force-displacement model which can represent ductility of each member was proposed.

INTRODUCTION

The concept of ductility has been adopted in recent seismic design codes for R/C buildings and bridge structures. It is well known that the design seismic ·forces are generally much less than the elastic response force induced by a major earthquake. However, due to lack of information on ductility evaluation, it has hardly been clarified how ductile the designed R/C structure can become during a major earthquake. This is because any reliable method to evaluate ductility of R/C members has not yet been established. There are some failure mechanisms at ultimate state in R/C columns and beams. A flexural failure mode is a typical one, but it may have no problems from the point of ductility because flexural failure generally shows ductile behaviour up to failure. On the other hand, a R/C member under reversed cyclic loading sometimes loses its load-carrying capacity, finally showing the characteristics of shear failure after the yielding of longitudinal reinforcement. Such a failure mode is very complicated and cannot be analyzed easily. The first objective in this paper is to establish a reliable equation to evaluate the ductility of such R/C members quantitatively as described above.

The influence of ductility on the overall response behaviour of a single column type structure can be well understood. However, the effect of ductility for each member on whole response behaviour of a R/C frame structure has hardly been clarified. Generally, a statically indeterminate structure such as a R/C rigid-frame structure will not collapse even if one of the members of the structure fails completely. However, the inelastic response behaviour of the structure may be influenced by the failure of such a member. The second objective is to clarify experimentally and analytically the influence of ductility in members on the inelastic response behaviour of R/C frame structures subjected to strong ground motion.

OUTLINE OF EXPERIMENT FOR DUCTILITY

The reversed cyclic loading tests were carried out using cantilever type specimens shown in Fig.1. The variables in the tests were tensile reinforcement ratio (p_t=0.59-1.66%), web reinforcement ratio(p_w=0-0.24%), compressive strength of concrete (f_c'= 128-565 kg/cm^2), shear span ratio (a/d=2.5-6), axial compressive stress(σ_o=0-30 kg/cm^2) and the number of repetitions of loading at a certain displacement amplitude(n=1-30 cycle). Table 1 and Table 2 show the experimental variables and the mechanical properties of the reinforcing bars respectively.

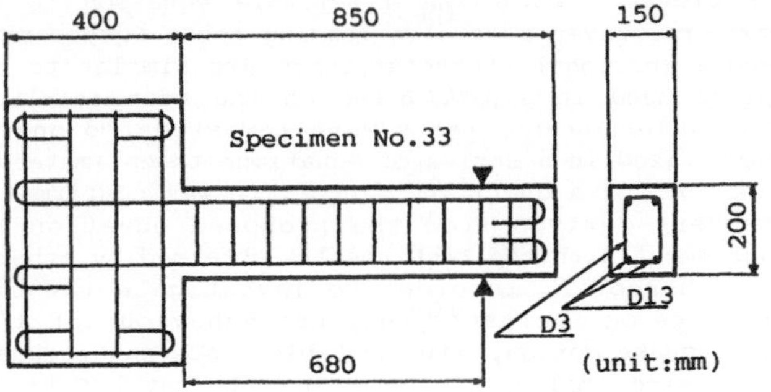

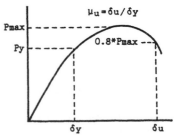

Figure 1. Dimensions of test specimens

Figure 2. Definition of ductility factor

The load was applied to the top of the specimen monotonically until the yield load, which was calculated based on the elastic theory. The measured displacement at the yield load was defined as the yield displacement (δ_y). However, when the measured strain of the main reinforcement at the bottom of the column reaches the yield strain before the yield load, the displacement at the yield strain is defined as the yield displacement. After the yield displacement, the displacement of the integral multiples of the yield displacement was applied cyclically by controlling the displacement of the specimen.

FORMULATION OF EFFECT OF VARI-
OUS FACTORS ON DUCTILITY OF R/C

Ductility factor(μ_u), that is,
the ratio of ultimate displ-
acement(δ_u) to yield displace-
ment(δ_y), was adopted as a
quantitative index of ductility
for R/C members. The yield displacement was defined as described above, and the ultimate
displacement was defined as the
limit displacement when the
load carrying-capacity decreases to 80% of the measured
maximum strength(see Fig.2).

To formulate ductility
quantitatively, one attempt was
tried at first to express the
effects of various factors on
ductility inclusively based on
the strength ratio, that is,
the ratio of shear strength to
flexural strength of R/C members. However, the results
indicated that the inclusive
expression was immoderate because the effects of various
factors on ductility were
slightly different from those
on shear strength[1]. Therefore, it was concluded that the
effect of various variables
must be investigated one by
one. In formulating, the relations between the measured ductility factor and each variable
were investigated using the
test results. In this case,
each relation was obtained by
changing only a single variable
while the other ones were kept
constant. The equation was
derived so that the best fit
for the plots could be obtained. To estimate the influence of the experimental variables on ductility factor,
standardized ductility factor

Table 1
Test variables

No.	pt	pw	a/d	σo	fc'	n	μu
1	1.06	0.12	4.00	0	203	10	5.5
2					279		4.7
3	0.59	0.12	4.00	0	406	10	≧7.1
4	0.89				357		7.1
5	1.66				338		3.9
6	1.06	0.00	4.00	0	413	10	3.6
7		0.08			400		4.9
8		0.23			318		7.5
9	1.06	0.12	3.00	0	309	10	3.8
10			5.00		389		6.5
11			6.00		363		≧4.3
12	1.06	0.12	4.00	10	294	10	4.0
13				20	301		3.4
14	0.89	0.12	3.00	0	330	10	6.5
15	1.66		5.00	0	376		4.2
16	1.06		3.00	10	307		3.5
17	0.99	0.12	4.00	10	330	10	4.5
18	0.99	0.12	4.00	0	308	10	5.6
19				5	298		5.8
20				20	321		4.3
21				30	326		4.2
22	0.99	0.12	4.00	10	335	10	——
23					254		4.0
24	0.99	0.12	4.00	10	565	10	4.3
25					140		5.1
26	0.99	0.12	4.00	10	327	1	6.0
27					323	3	4.9
28					319	30	4.3
29	0.99	0.12	2.50	10	337	10	4.3
30			5.50		348		4.6
31	0.99	0.24	4.00	10	128	10	5.4
32		0.12			128		4.4
33		0.06			128		3.5

Note:pt=tensile reinforcement ratio (%) , pw=web reinforcement ratio (%) , a/d=shear span ratio, σo=
axial compressive stress, fc'=compressive strength
of concrete (kg/cm²) , n=number of repetitions of
loads, μu=measured ductility factor

(μ_t), which is the ratio of the measured ductility factor to the ductility
factor obtained in a certain variable, was used. Thirty three test results
were used for the formulation.

Figure 3 shows the effect of the tensile reinforcement ratio on the
standardized ductility factor. The following equation which can express the
influence of only main reinforcements on ductility factor was derived from
the test results.

Table 2
Mechanical properties of reinforcing bars

Type		Yield Stress (kg/cm²)	Yield Strain (μ)	Ultimate Stress (kg/cm²)	Area of Reinforce- ment (cm²)
SD30,D10		3650	2440	5380	0.7133
SD30,D13	A	3960	2480	5470	1.267
	B	3800	2330	5500	
	C	3840	2100	5500	
SD30,D16		3580	2090	5910	1.988
SD30,D3	A	2740	2030	5690	0.06905
	B	2400	1650	4160	0.07280
	C	2540	1750	4200	

Note : Type A was used for Specimens No.1 ~16.
Type B was used for Specimens No.17~30.
Type C was used for Specimens No.31~33.

$$\beta t = \mu t - 1 = (p\,t)^{\alpha} - 1 \cdots\cdots(1)$$
$$\alpha = (-0.146 \,/\, (a/d - 2.93) - 0.978) \cdots\cdots(2) \quad (a/d \geqq 3.0)$$

where, μ_t:standardized ductility factor, p_t:tensile reinforcement ratio(%), $\beta t = 0$ when $p_t = 1\%$; α is a function depending on p_w and a/d.

The other equations were derived in almost the same manner as above for all the variables adopted in the tests. Figure 4 shows the relation between the web reinforcement ratio and μ_t. It is clear that the relation between them is almost linear. The following equation can be given.

$$\beta w = \mu t - 1 = 2.70 \,(pw - 0.1) \cdots\cdots(3)$$

where, p_w:web reinforcement ratio(%)

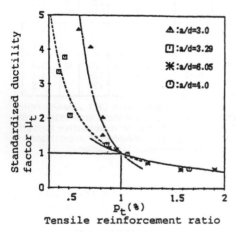

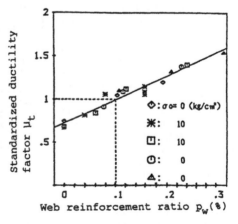

Figure 3. Standardized ductility factor and tensile rein- forcement ratio

Figure 4. Standardized ductility factor and shear rein- forcement ratio

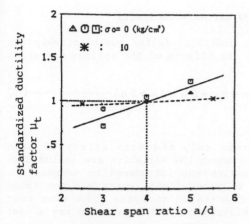

Figure 5. Standardized ductility factor and shear span ratio

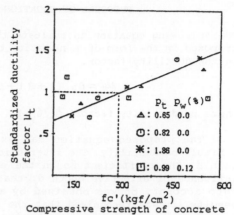

Figure 6. Standardized ductility factor and compressive strength of concrete

Figure 5 shows the relation between shear span ratio(a/d) and μ_t. The following equation was derived.

$$\beta a = \mu t - 1$$
$$= \begin{cases} (-0.0153\sigma_0 +0.175)(a/d-4.0)\cdots(4) \\ \quad (\text{where}, \sigma_0 \leqq 11. \text{ kg/cm}^2) \\ 0 \quad \cdots\cdots\cdots\cdots\cdots\cdots\cdots\cdots\cdots(5) \\ \quad (\text{where}, \sigma_0 > 11. \text{ kg/cm}^2) \end{cases}$$

Figure 6 shows the influence of compressive strength of concrete(fc') on μ_t. The result indicates that fc' has less effect on the ductility factor when the web reinforcement are arranged. Therefore, the derived equation is alternative as shown in the following equation.

$$\beta c = \mu t - 1 = \begin{cases} 0.00170(fc'-300) & (pw=0\%) \cdots(6) \\ 0 & (pw\neq0\%) \cdots(7) \end{cases}$$

fc':compressive strength of concrete (kg/cm^2)

Figure 7 shows the relation between axial compressive stress and μ_t. The equation is given as follows.

$$\beta N = \mu t - 1 = 2.18(\sigma_0 +10)^{-0.260}- 1 \cdots\cdots(8)$$

σ_0:axial compressive stress (kg/cm^2)

Figure 8 shows the influence of the number of load repetitions in a certain displacement amplitude on μ_t. The equation is shown as follows.

$$\beta n = \mu t - 1 = 1.26(n)^{-0.0990} - 1 \cdots\cdots\cdots(9)$$

n:number of load repetitions

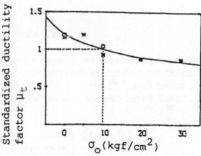

Axial compressive stress

Figure 7. Standardized ductility factor and axial compressive stress

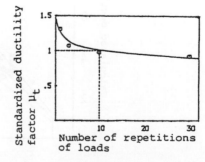

Number of repetitions of loads

Figure 8. Standardized ductility factor and number of load repetitions

PROPOSAL OF EQUATION TO ESTIMATE DUCTILITY

The following equation to evaluate the ductility factor quantitatively was proposed in the form of a summation of the effects of the various variables on the ductility factor.

$$\mu_u = \beta_0 \ (1 + \beta_t + \beta_w + \beta_c + \beta_N + \beta_a + \beta_n) \quad \cdots\cdots\cdots\cdots\cdots(10)$$

where μ_u=ductility factor(δ_u/δ_y)

The proposed equation can express only the main effect of each variable on ductility. The interactions among the variables are included in each β. The coefficient βo in the equation was introduced to express the influence of effective depth d, that is size effect, which was not taken into account. βo was obtained by a regression analysis from many test results. It was recognized that the relation between βo and 1/d was almost linear.

$$\beta_0 = 28.4 / d + 2.03 \cdots\cdots\cdots\cdots\cdots\cdots\cdots\cdots\cdots\cdots\cdots\cdots\cdots\cdots\cdots\cdots(11)$$

where d:effective depth.

EVALUATION OF PROPOSED EQUATION

The accuracy of the proposed equation was investigated using many test data including other researchers' [2], [3], [4], [5], which were not used in the formulation. Figure 9 shows the relation between the calculated ductility factors from the proposed equation and the measured ones. It is recognized

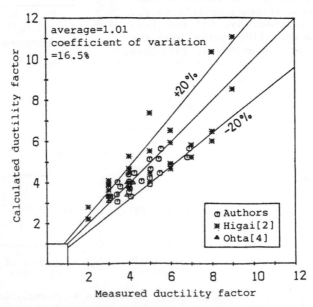

Figure 9. Comparison of calculated ductility factor
with experimental one

that the calculated values agree generally well with the experimental ones.
The average of the ratios of the experimental values to the calculated ones for
all data is 1.01 and the coefficient of
variation is 16.5%.   These values also
indicate that the proposed equation can
give satisfactory results.

OUTLINE OF EXPERIMENT FOR R/C FRAME STRUCTURES

In order to investigate the influence of
ductility of members on inelastic response behaviour in R/C frame structures
under earthquake motion, shaking table
tests and pseudodynamic tests were
carried out. The test structures are
two-story one-bay R/C frames which are
similar to typical bridge piers used for
the elevated railways of Shinkansen in
Japan. A general view of the test set-up
is shown in Fig.10. The test structures
were designed assuming that the following failure modes would occur: 1)flex-
ural failure at the bottom of the first-
level column(structures RD-1 and RP-1),
2)flexural failure in the first-level
beam(RD-3), 3)shear failure after yielding of main reinforcement in the first-
level beam(RD-4 and RP-4). To produce
the above failure modes, the tensile
reinforcement ratio and the web reinforcement ratio in the first-level beam
were changed, as shown in Table 3. In
every test, a weight of 963 kgf, which
produces an axial stress of 9.6 kgf/cm^2
in the columns, was installed at the top
of each second-column. Three structures,
RD-1, RD-3 and RD-4, were tested under
simulated earthquakes, and two structures, RP-1 and RP-4, were tested pseu-
dodynamically.
    In the simulated earthquake tests,
the first 10 seconds of EL CENTRO-NS
earthquake was repeated three times
continuously. To excite the test
structure into an inelastic range, the
original time scale was compressed by a
factor of 2 while the maximum base
acceleration was amplified to 0.8G.

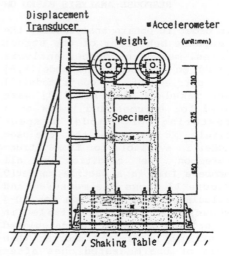

Figure 10. Test setup for simu-
           lated earthquake test

Table 3 Details of test
          structures

Common Members

Member Name	Tensile Reinforce- ment Ratio (%)	Web Reinforce- ment Ratio (%)	Relative Stiffness Ratio (*)
First -Level Column	0.75(D6X2)		1.00
Second -Level Column		0.29(D3)	1.24
Second -Level Beam	0.76(D6X2)		4.35

First-Level Beam

Speci- men Name	Tensile Reinforce- ment Ratio (%)	Web Reinforce- ment Ratio (%)	Relative Stiffness Ratio (*)
RD-1	0.85(D6X2)	0.29(D3)	1.24
RP-1			
RD-3	0.43(D3X5)	0.058(D2)	1.21
RD-4	0.73(D3X9)	0.0	1.26
RP-4			

Note (*): The stiffness of the first-level
          column is the standard value (1.0).

RESPONSE ANALYSIS BASED ON ORDINARY RESTORING FORCE MODEL

In order to obtain analytically the response behaviour of the structures, member-by-member analysis based on one component model[6,9] was carried out. Takeda's model[7] and Takeda's slip model[8] were used for columns and beams as a restoring force model respectively. It was proved that the used model is available for frame structures on the condition that all members fail in a ductile manner[9]. Figure 11 shows the measured and calculated time histories for RD-4 whose first-level beam failed in shear. The period of the calculated responses is clearly shorter than that of the measured ones after shear failure occurred(1.0 sec) in the first-level beam. This result indicates that the overall response behaviour of the structure can not be simulated accurately using the ordinary restoring force model after shear failure occurred in some members and then the load-carrying capacity of the member decreased suddenly. In a statically indeterminate structure such as a R/C rigid-frame structure, the inelastic response behaviour of the structure depends on ductility of the members even if the structure does not collapse. Therefore, the restoring force model which can effectively represent ductility for all members is required to calculate precisely the overall response behaviour.

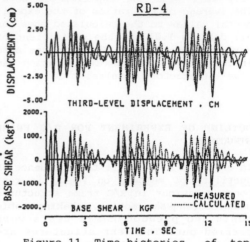

Figure 11. Time histories of top displacement and base shear obtained from test and analysis

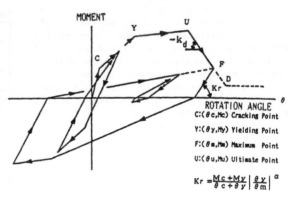

Figure 12. Proposed restoring force model

RESPONSE ANALYSIS BASED ON DUCTILITY OF MEMBERS

In order to resolve the above problem, the new restoring force model which can express the ductility for all members was proposed. Figure 12 indicates the new restoring force model, in which the decrease in the load-carrying capacity after reaching the maximum strength, was taken into consideration. The ultimate deformation (point U in Fig.12) at which the load-carrying capacity begins to decrease was determined from the proposed equation(10) for ductility previously. The slope after the point U was defined by Equation(12) derived from numerous test results. The hysteresis rule of the proposed model was the same as used in that of the ordinary model.

$$(-k_d)/k_y = 1.229(\mu_u - 1)^{-1} - 0.0539 \quad \cdots\cdots\cdots\cdots\cdots\cdots\cdots(12)$$

Using the proposed restoring force model, response analyses were carried out for all the test structures. Figure 13 shows the time histories of the top displacement obtained from the tests and analyses for structure RP-4 whose first-level beam failed in shear after the yielding of the longitudinal reinforcement. The response values and the periods of excitation obtained from the analysis agree well with those from the tests after shear failure occurred in the first-level beam(after 1.0 sec). That is, the inelastic response behaviour can be calculated accurately by using the proposed restoring force model even if the load-carrying capacity of some members decreased suddenly due to the occurrence of shear failure.

Figure 14 shows the measured and calculated base shear-displacement curves. The calculated ones were obtained from both the ordinary model and the proposed one. The inelastic behaviour cannot be represented accurately by the ordinary model after the load-carrying capacity of one member decreased

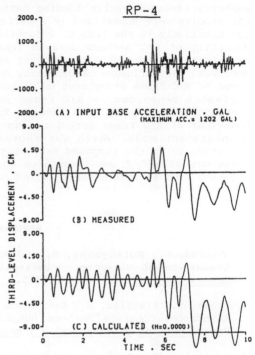

Figure 13. Time histories of top displacement obtained from test and analysis using proposed model

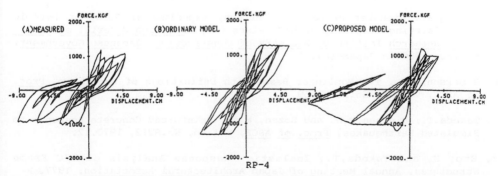

Figure 14. Base shear-displacement curves obtained from tests and analysis

due to the occurrence of shear failure. However, the overall behaviour of the structure can be obtained up to failure by the newly proposed model with satisfactory accuracy. The proposed model can be a powerful method to predict the extent of damage of each member as well as that of a structure.

CONCLUSION

In order to establish a reliable equation to evaluate the ductility of R/C members, reversed cyclic loading tests were carried out. From the tests, the results were summarized in a series of equations to estimate ductility quantitatively in the form of a ductility factor. It was confirmed that the ductility of R/C members derived from the proposed equation resulted in satisfactory agreement with the test results obtained by other researchers. Moreover, the influence of ductility of members on inelastic response behaviour of R/C frame structures under earthquake motion was investigated. The inelastic behaviour in R/C frame structures depends strongly on the ductility capacity of each member. To calculate accurately the response behaviour of R/C frame structures up to collapse, a new restoring force-displacement model which can represent ductility of each member was proposed. Using the proposed restoring force model, the inelastic response behaviour of R/C frame structures could be calculated with satisfactory accuracy even if some members failed completely.

REFERENCES

1. Toyoda, K., Mutsuyoshi, H. and Machida, A. Experimental Study on Evaluation of Ultimate Deflection of Reinforced Concrete Members. Transactions of JCI, Vol.7, 1985.

2. Higai, T., Rizkalla, S., Ben-Omran, H. and Saaday, F., Shear Failure of Reinforced Concrete Members Subjected to Large Deflection Reversals. Proceedings of Sixth Annual JCI Meeting, 1984 (in Japanese).

3. Ishibashi, T., Aoki, K. and Yoshino, S., Ductility Factor of Reinforced Concrete Members. Data of Structural Design, No.79, 1984 (in Japanese).

4. Ohta, M., Study on Seismic Resistant Design of Reinforced Concrete Piers of Single Column Type, Report of Public Works Research Institute, Ministry of Construction, Japanese Government, No.153, 1980(in Japanese).

5. Earthquake Disaster Prevention Branch, Experimental Study on Dynamic Load Carrying Capacity of R/C Piers (1), Research Material at Public Works Research Institute, Ministry of Construction, Japanese Government, No.153, 1980(in Japanese).

6. Giberson, M.F., Two Nonlinear Beams with Definitions of Ductility. Proc. of ASCE, Vol.95, No.ST2, 1969.

7. Takeda, T., Nielsen, N.N. and Sozen, M.A., Reinforced Concrete Response to Simulated Earthquakes. Proc. of ASCE, Vol.96, No.ST12, 1970.12.

8. Eto, H. and Takeda, T., Inelastic Response Analysis of R/C Frame Structures. Annual Meeting of Japan Architectural Association, 1977.

9. Machida, A., Mutsuyoshi, H. and Tsuruta, K., Inelastic Response of Reinforced Concrete Frame Structures Subjected to Earthquake Motion. Concrete Library of JSCE No.10. Dec. 1987.

A PARAMETRIC STUDY OF THE RC PANEL DUCTILITY BY USING THE ANALYTICAL MODEL

JUNICHI IZUMO
Department of Civil Engineering, Kanto Gakuin University, 4836 Mutsuura-chyo, Kanazawa-ku, Yokohama, 236, Japan

ABSTRACT

The analytical model that can accurately predict the behavior of the RC panels can also give the information on the ultimate deformations by analyzing them up to failure. In order to examine the characteristics on the ductilities of the RC panels, the existing tested specimens have been analyzed with the proposed model. Through comparison of the analytical results to the experimental ones, the ultimate deformations of the RC panels and the failure criteria have been discussed. Further, with the proposed model, the parametric study has been done to clarify the characteristics on the ductilities of the RC panels.

INTRODUCTION

In the earthquake design, it is necessary to examine the deformational ability of reinforced concrete. As the proposed model for RC panels [1] has been confirmed to describe the behavior of reinforced concrete well, it could give the information on the ultimate deformations of the RC panels by analyzing them up to failure. The aim of this study is to discuss the ultimate deformations of the RC panels with the proposed model and to clarify the characteristics on the ductilities of the RC panels through the parametric study.

FAILURE CRITERIA FOR RC PANELS

The failure of the RC panel could be categorized as: 1) Tensile failure, 2) Compressive failure. It is considered that the compressive failure will occur when the compressive stress of concrete reaches the the maximum strength of concrete. As the compressive strength of concrete will be deteriorated due to the existence of cracks, it is more appropriate to define the compressive failure by the maximum strain of concrete. Namely, the failure criterion for compression is expressed by, $\varepsilon'_{cy} \geq \varepsilon'_{cu}$, where ε'_{cy} is the compressive strain of concrete and ε'_{cu} is the compressive strain corresponding to the maximum strength. When ε'_{cu} is not given by the experiment, $\varepsilon'_{cu} = 0.00044\sqrt{f'_c}$ is used, where f'_c (MPa) is the compressive strength of concrete.

On the other hand, tensile failure will occur when the reinforcing bar reaches the limit state. What defines the limit state of the reinforcing bar in concrete? The yield point might explain the limit state of the reinforcing bar. However, examining the observed ultimate strength and the yielding load precisely, it could be judged that the RC panels have the loading capacity even after the reinforcing bars in the both directions have yielded. This means that the limit state of reinforced concrete could not be adequately explained by the yielding loads.

Analytical Model for Reinforcement

As it is considered that the analytical model for reinforcement much affects the calculated ultimate deformation of the RC panel, the model for reinforcement is discussed before examining the limit state of the RC panels. The stress of the reinforcing bar between cracks is not uniform due to the bond action.

The average stress of the reinforcing bar is obtained by integrating the stress along the reinforcing bar. To get the average stress, the simplified model has been developed by assuming that the stress distribution of the reinforcing bar could be described as a cosine function (Fig. 1). This is because the cosine curve has been judged to approximately fit the stress distribution of the reinforcing bar obtained from the uniaxial test result of reinforced concrete [2] .

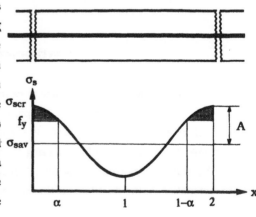

Figure 1. Stress distribution of the reinforcing bar

Therefore, the stress of the reinforcing bar is written as follows:

$$\sigma_s = A\cos\pi x + (\sigma_{scr} - A) \qquad (1)$$

where, σ_s is the stress of the reinforcing bar, A is the stress amplitude of the reinforcing bar and σ_{scr} is the stress of the reinforcing bar at the cracked plane. The average stress, σ_{sav}, is obtained by integrating the stress along the reinforcing bar between cracks.

$$\sigma_{sav} = \sigma_{scr} - A \tag{2}$$

On the other hand, the average strain of the reinforcing bar is also obtained by integrating the strain in the yielded region and the elastic region along the reinforcing bar.

$$\varepsilon_{sav} = \int_0^1 \varepsilon_s dx \tag{3}$$

where, ε_{sav} is the average strain of the reinforcing bar, ε_s is the strain given in the stress and strain relationship of the bare bar, E_s is the elastic modulus of the reinforcing bar and α indicates the location corresponding to the yield point that is expressed in Eq. (4).

$$\alpha = \frac{1}{\pi} \cos^{-1}\left(\frac{f_y - \sigma_{scr} + A}{A}\right) \tag{4}$$

The stress and strain relationship of the bare bar is given in Eq. (5) proposed by Shima [2] .

$$\sigma_s = E_s \varepsilon_s \qquad\qquad (\varepsilon_s \leq \varepsilon_y) \tag{5.1}$$
$$\sigma_s = f_y \qquad\qquad (\varepsilon_y < \varepsilon_s \leq \varepsilon_{sh}) \tag{5.2}$$
$$\sigma_s = f_y + \left\{ 1 - \exp^{(\varepsilon_{sh} - \varepsilon_s)/k} \right\} (1.01 f_u - f_y) \qquad (\varepsilon_s > \varepsilon_{sh}) \tag{5.3}$$

where, σ_s is the stress of the reinforcing bar (MPa), ε_y is the yield strain, ε_{sh} is the initial strain of strain hardening, f_y is the yield point (MPa), f_u is the tensile strength (MPa) and $K = 0.031(400/f_y)^{1/3}$.

Moreover, ε_{sh} and f_u depend on the yield point and are formulated as the following equations on the basis of the tensile test of the reinforcing bar [2], the yield points of which were at the range from 350 (MPa) to 820 (MPa).

$$\varepsilon_{sh} = 2.0 \times 10^{-3} - 1.1\varepsilon_y \qquad\qquad (\varepsilon_y \leq 2.4 \times 10^{-3}) \tag{6.1}$$
$$\varepsilon_{sh} = (14.5 - 3.1 \times 10^3 \varepsilon_y) \varepsilon_y \qquad (2.4 \times 10^{-3} < \varepsilon_y \leq 4.4 \times 10^{-3}) \tag{6.2}$$
$$\varepsilon_{sh} = \varepsilon_y \qquad\qquad (\varepsilon_y > 4.4 \times 10^{-3}) \tag{6.3}$$
$$f_u = (1.9 - 9.4 \times 10^{-4} f_y) f_y \tag{7}$$

As the average stress and average strain relationship of the reinforcing bar is expressed in the form of the implicit function, the average stress is obtained by the repeating calculation with Eq. (2) and Eq. (3), assuming the stress amplitude, A. Further, the stress of the reinforcing bar at the cracked plane is obtained from the equilibrium of the free body as shown in Fig. 2.

$$\sigma_{sxcr} = \{(\sigma_X + \sigma'_c)\cos\theta + (\tau_{XY} + \tau_{cxy})\sin\theta\}/p_x \tag{8.1}$$

$$\sigma_{sycr} = \{(\sigma_Y + \sigma'_c)\sin\theta + (\tau_{XY} - \tau_{cxy})\cos\theta\}/p_y \tag{8.2}$$

where, σ_{sxcr} is the stress of the x-reinforcement at the cracked plane, σ_{sycr} is the stress of the y-reinforcement, p_x is the x-reinforcement ratio, p_y is the y-reinforcement ratio, σ_X, σ_Y, τ_{XY} are the stresses acting on the RC panel, θ is the angle between the X-direction and the x-axis, τ_{cxy} is the shear stress along the cracked plane and σ'_c is the normal stress generated by the shear displacement those are given by the Li and Maekawa model [3]. The sign (') means that the compression is plus.

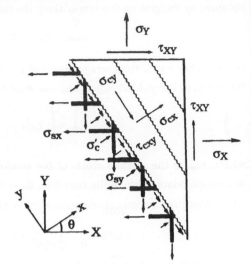

Figure 2 Stresses in RC panel.

As the yielding of the RC panels will start when the stress of the reinforcing bar at the cracked plane reaches the yield point, the criterion for yielding is given by $\sigma_{sxcr} = f_y$ or $\sigma_{sycr} = f_y$. When the cosine curve is used, the average stress always becomes the average of the maximum stress and the minimum stress. In the case that the average stress is lower than a half of the maximum stress, the minimum stress becomes minus; that is, the reinforcing bar is subjected to compression. However, such a case could not actually occur. The case that the average stress becomes lower than a half of the maximum stress will occur when the reinforcement ratio is fairy low. In such a case, the reinforcing bar is immediately yielded after cracking and the smeared cracks will not be enough formed. Consequently, the crack spacing is comparatively larger. Therefore, there seems to be a restriction in applying the cosine curve to the stress distribution of the reinforcing bar. In the case that the average stress becomes lower than a half of the maximum stress, the analysis has been done by substituting the compressive stress of the reinforcing bar for zero. However, it is necessary to develop the more accurate model for reinforcement.

Limit State of Reinforcement in Concrete

By using the proposed model [1], the limit state of the RC panel could be discussed through comparison of the analytical results with the experimental ones. The specimens that were considered to be broken due to tension have been selected from the Vecchio and Collins' experiment [4] and the Aoyagi and Yamada's [5]. The analyses have been done until the calculated stresses of the RC panel coincide with the observed maximum stresses.

The bullets as shown in Fig. 3 indicate the calculated stresses of the reinforcing bars at the cracked plane when the calculated stresses of the RC panels reach the observed maximum stresses. There is a tendency that the bullets except one are approximately studded at the range of the strain from 0.02 to 0.03. The bullet that is extremely deviated from the others has been obtained from the analysis of the Vecchio and Collins' test specimen that had the 0.45% of the y-reinforcement ratio. It is observed that this specimen is immediately yielded after cracking. As the negative stress is set at zero in this analysis, it is considered that the average stress becomes greater than that of the actual stress and consequently the stress at the cracked plane becomes the larger value. Therefore, the constitutive model for reinforcement is considered not to be applied to this case.

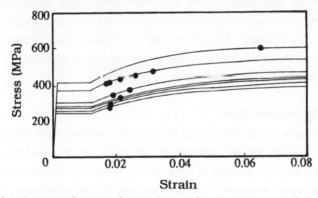

Figure 3. Calculated stress of the reinforcing bar at the ultimate strengths of the RC panels

As far as the Vecchio and Collins' experiment and the Aoyagi and Yamada's are concerned, it is considered that the limit state of the RC panel could be defined by the stress or the strain of the reinforcing bar at the cracked plane. As the strain less affects the maximum strengths of the RC panels than the stress does, the maximum strain of the reinforcing bar has been chosen to express the tensile failure for the RC panels. Considering the calculated results, the maximum strain of the reinforcing bar has been set at 0.03 for the tensile failure of the RC panels

Ultimate Strength of the RC Panels

The ultimate strengths of the RC panels have been evaluated with the 31 RC panels [4][5][6][7]. From those specimens, the one specimen which the model for reinforcement is considered not to be applied to, the two specimens that are considered to be loaded until the reinforcing bars are ruptured and the one specimen, the reinforcing bars of which are not reinforcements are ruptured and the one specimen the reinforcements of which were not yielded are excluded. The average ratios of the observed stress to the calculated ones for tensile failure and for compressive failure are 0.99 and 1.00, where the coefficients of variation are 9.7% and 10.6%, respectively. As a total, the average ratio of the observed stress to the calculated one is 0.99 and the coefficient variation is 9.8%. It is judged that the calculated ultimate strengths are good correspondence with the test results.

PARAMETRIC STUDY ON THE DUTILITIES OF RC PANELS

The parametric study has been done to examine the ductilities of the RC panels with the proposed model. The reinforcement ratio, the ratio of the y-reinforcement ratio to the x-reinforcement ratio, the angle of the x-reinforcement axis against X-axis and the combination of the stresses are selected as a parameter. For the compressive strength of concrete, the tensile strength of concrete, the yield point of the reinforcement and the elastic modulus of the reinforcement, the following values are used: f'_c=20.0 (MPa), f_t=0.27$f'_c^{2/3}$=2.0 (MPa), f_y=350 (MPa) and E_s=2.0×10^5 (MPa).

Effects of the Reinforcement Ratio

The calculated results are shown in Fig. 4 when the reinforcement ratio is varied at the range from 0.5% to 3.0%. In this analysis, the RC panel has the isotropic arrangement of the reinforcement and the loading is pure shear. When the reinforcement ratio is over 1.75%, the failure occurs due to the concrete crushing. On the other hand, the tensile failure occurs when the reinforcement ratio is less than 1.5%. The ultimate strength tends to increase, as the reinforcement ratio increases. However, the ductility is not expected even though the reinforcement ratio increases, in the case of the compressive failure.

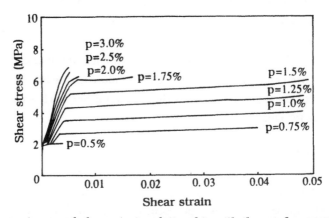

Figure 4. Shear stress and shear strain relationship with the reinforcement ratio varied

In the case of the tensile failure, the ductility will be developed as the reinforcement ratio increases. Moreover, the RC panel with the low reinforcement ratio is considered to be brittly broken as in the case of the 0.5% of the reinforcement ratio.

Effects of the Anisotropic Arrangement of Reinforcement

The calculated shear stress and shear strain relationships of the RC panels are shown in Fig. 5, when the y-reinforcement ratio is varied from 0.5% to 2.0% with the x-reinforcement ratio fixed at 2.0%. The loading is pure shear. In these cases, the shear stress along the cracked plane will be generated due to the anisotropic arrangement of the reinforcing bars.

In the case of the reinforcement ratio at the range from 0.5% to 1.0%, the reinforcing bar in the one direction is yielded and in the case of the reinforcement ratio over 1.25%, the reinforcing bars in the both directions are yielded. The ductility is considered to be developed with the increment of the y-reinforcement ratio, because the arrangement of the reinforcing bars comes closer to the isotropic arrangement and the y-reinforcement ratio also increases. However, in the case of the reinforcement ratio over 1.75%, the ductility is considered not to be expected due to the concrete crushing.

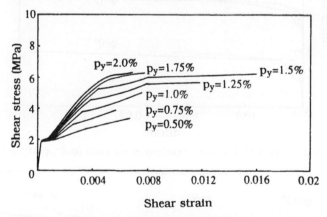

Figure 5. Shear stress and shear strain relationship with the y-reinforcement ratio varied

Effects of the Arrangement of Reinforcement

The calculated results are shown in Fig. 6 with the reinforcement ratio fixed at 1.0% while the angle of the x-reinforcement against the X-axis, φ is varied at the range from 0° to 45°. The RC panel has the isotropic arrangement of the reinforcing bars and the loading is pure shear. The differences of the ductilities could be recognized when the angle is varied. However, the clear differences of the maximum strength are not be recognized. There is a tendency that the ultimate deformation decreases, as the angle increases. This is because the stress of the x-reinforcement relatively increases compared to that of the y-reinforcement with the increment of the angle and the shear stress along the cracked plane also affects the stress of the reinforcing bar. The calculated results with the reinforcement ratio fixed at 2.0% are shown in Fig. 7. In the case of $\varphi=0°$ and $\varphi=15°$, the RC panel is broken by the concrete crushing and it results in the lower ductility differently to the case of the 1.0% of the reinforcement ratio. Further, it is observed that the ultimate deformation in the case of $\varphi=45°$ is a little more developed than that in the case of $\varphi=30°$. This is because the reinforcement ratio more affects the ultimate deformation of the RC panel than the shear stress along the cracked plane in the case of the 2.0% of the reinforcement ratio.

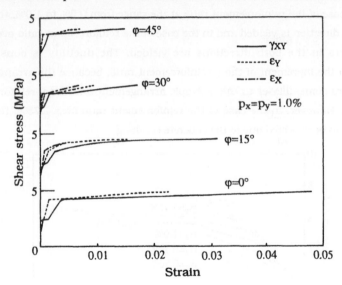

Figure 6. Calculated result with the 1.0% reinforcement ratio ($\varphi=0°$, $\varphi=15°$, $\varphi=30°$ and $\varphi=45°$)

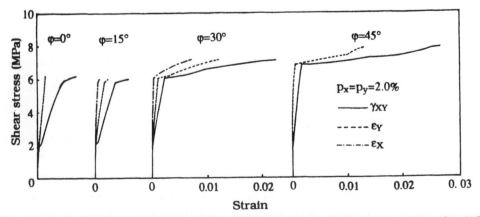

Figure 7. Calculated result with the 2.0% reinforcement ratio ($\varphi=0°$, $\varphi=15°$, $\varphi=30°$ and $\varphi=45°$)

Effects of the Combination of Stresses

The calculated results are shown in Fig. 8 when the ratio of the normal stress to the shear stress, $\eta = (\sigma_x/\tau_{xy})$ is varied from -1.0 to 1.0, while the reinforcement ratio is fixed at 1.0% and the reinforcing bars are isotropically arranged. The maximum strength of the RC panel increases, as the normal compressive stress increases. However, when the compressive normal stress is over a half of the shear stress, the ductility is judged not to be expected due to the concrete crushing. On the other hand, in the case that the RC panel is broken by the tensile force, the ductility is considered to be approximately the same regardless of the differences of the combination of the stresses. The similar tendency has been recognized in the calculated results with the reinforcement ratio fixed at 1.5%.

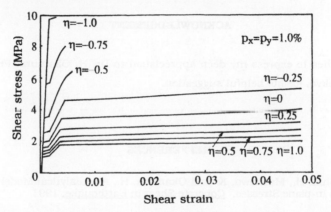

Figure 8. Calculated result when the combined stresses are applied

Through the parametric study on the ductilities, the followings are clarified.

1) The clear deferences of the deformational ability of the RC panel are recognized between the tensile failure and the compressive failure

2) The reinforcement ratio is considered to develop the ductility as far as the RC panel is broken by the tensile force.

3) The RC panel with the anisotropic arrangement of the reinforcing bars tends to have less ductility as the reinforcing bar in the one direction tends to be yielded.

4) The differences of the arrangement of the reinforcing bars affect the ductility because the reinforcing bar in the one direction tends to resist the tensile force and the stress of the reinforcing bar is affected by the shear stress along the cracked plane.

5) The effects of the combination of the stresses on the ultimate deformation are not clearly recognized in the case of the tensile failure.

CONCLUSIONS

Through the existing experiments, the limit state of the reinforcing bar has been discussed and the failure criterion for tension has been determined by the maximum strain of the reinforcing bar at the cracked plane. Further, it is confirmed that the calculated maximum strengths of the RC panels with the failure criteria are in good correspondence with the observed ones. Moreover, through the parametric study, the characteristics of the ductilities of the RC panels have been clarified.

ACKNOWLEDGEMENT

The author wishes to express my deep appreciation to Dr. H. Okamura, Professor of the University of Tokyo, for his helpful suggestion.

REFFERENCES

1. Izumo, J., Shin, H., Maekawa, K. and Okamura, H., An Analytical Model for RC panels Subjected to In-plane Stresses, Concrete Shear in Earthquake, 1991

2. Shima, H., Chou, L. and Okamura, H., Micro and Macro Models for Bond Behavior in Reinforced Concrete, Journal of the Faculty of Engineering, The University of Tokyo (B) Vol.39, No.2, 1987, pp. 133-194

3. Li, B. and Maekawa, K. and Okamura, H., Contact Density Models for Stress Transfer across Cracks in Concrete, Journal of the Faculty of Engineering, The University of Tokyo (B), Vol.40, No.1, 1989, pp. 9-52

4. Vecchio, F., J. and Collins, M., P., Response of Reinforced Concrete to In-plane Shear and Normal Stresses, Publications, No.82-03, Department of Civil Engineering, University of Toronto, Mar., 1982

5. Aoyagi, Y. and Yamada, K., Strength and Deformation Characteristics of Reinforced Concrete Shell Elements Subjected to In-plane Forces, Proc., of JSCE, No.331, Mar., 1983, pp. 167-190, Concrete Library International, JSCE, No.4, Dec., pp. 129-160

6. Sumi, K. and Kawamata, T., Mechanical Characteristics of Concrete in Reinforced Concrete Panels Subjected to In-plane Pure Shear, Concrete Journal, Vol.26, No.10, Oct, 1989, pp.97-110 (In Japanese)

7. Watanabe, F., and Muguruma, H., Ultimate Strength and Deformation of RC panels, Structural Design, Analysis and Testing, Proc., of the Sessions Related to Design, Analysis and Testing at Structure Congress '89, ASCE, 1989, pp.31-38

DUCTILE AND BRITTLE FAILURES OF
REINFORCED CONCRETE WALL ELEMENTS SUBJECT TO SHEAR

K. J. HAN, S. T. MAU, and T. T. C. HSU
Department of Civil and Environmental Engineering
University of Houston
Houston, Texas 77204-4791, USA

ABSTRACT

The deformation characteristics of reinforced concrete wall elements subjected to predominantly shear loading are investigated theoretically. The softened truss model theory forms the basis of this investigation. By tracing the load-deformation history of various wall elements under constant loading ratios, it is shown that the amount of reinforcement significantly influences the ductility of a wall element. The ductility that may be achieved also depends on whether the element is constrained or not.

GOVERNING EQUATIONS OF WALL ELEMENTS

The softened truss model theory has been well documented elsewhere [1,2]. It applies to a reinforced concrete element that has developed cracks under external loading. The concrete forms compressional struts roughly parallel to the cracks, while the reinforcing bars resist tension across the cracks, Fig. 1.

Two coordinate systems are used to characterize the element behavior: l-t and d-r systems, Fig. 2. The l-t axes are parallel to the longitudinal and the transverse steel bars, whereas the d-r axes are parallel and perpendicular to the cracks, respectively. The d-r axes are the principal axes of concrete, and considered to be the principal strain axes of the element as a whole. The orientation of the d-r system with respect to the l-t system is defined by angle α, Fig. 2.

The cracked reinforced concrete element is assumed to behave as a continuum, characterized by average strains and average stresses. The stress-strain relationships of the steel bars and the concrete are described using the two different coordinate systems. For steel bars, only axial stresses are considered, and the stress-strain ralations are expressed in the form of

$$f_l = f_l(\varepsilon_l) \tag{1}$$

and

$$f_t = f_t(\varepsilon_t) \tag{2}$$

where the ε_l and ε_t are the average strains of the element in the longitudinal and the transverse directions, respectively, and f_l and f_t are the corresponding stresses in the steel bars. The stress-strain relationships of the concrete are expressed in the principal coordinate system, namely

$$\sigma_d = \sigma_d(\varepsilon_r, \varepsilon_d) \tag{3}$$

$$\sigma_r = \sigma_r(\varepsilon_r, \varepsilon_d) \tag{4}$$

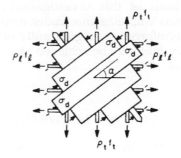

Figure 1: Truss Model

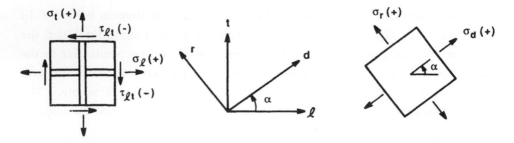

Figure 2: Definitions of Stresses and Coordinate Systems

where σ_d and σ_r are the principal stresses in the concrete, and ε_r and ε_d are the principal strains of the element.

Since the material laws of the steel and the concrete are defined in two different coordinate systems, it is necessary to relate the strains of the element in the two systems. This is achieved by a coordinate transformation of strains, i.e.,

$$\varepsilon_l = \varepsilon_d \cos^2\alpha + \varepsilon_r \sin^2\alpha \tag{5}$$

$$\varepsilon_t = \varepsilon_d \sin^2\alpha + \varepsilon_r \cos^2\alpha \tag{6}$$

$$\gamma = 2(\varepsilon_d - \varepsilon_r) \sin\alpha \cos\alpha \tag{7}$$

where γ is the shearing strain of the element in the l-t system.

After a similar transformation of the concrete stresses, the equations of equilibrium of the element in the l-t system are obtained as

$$\sigma_l = \sigma_d \cos^2\alpha + \sigma_r \sin^2\alpha + \rho_l f_l \tag{8}$$

$$\sigma_t = \sigma_d \sin^2\alpha + \sigma_r \cos^2\alpha + \rho_t f_t \tag{9}$$

$$\tau = (\sigma_d - \sigma_r) \sin\alpha \cos\alpha \tag{10}$$

where ρ_l and ρ_t are the reinforcement ratios in the longitudinal and transverse directions, respectively, and σ_l, σ_t, and τ are the normal and the shearing stresses of the element in the l-t system. These three stresses may be considered as the externally applied loads on the element, and are termed element stresses.

The above ten equations contain fifteen variables: five strains, ε_l, ε_t, ε_d, ε_r, and γ, seven stresses, f_l, f_t, σ_d, σ_r, σ_l, σ_t, and τ, two reinforcement ratios, ρ_l and ρ_t, and angle α. If the two reinforcement ratios and the two constants that define the proportionality of the three element stresses are given, the load-deformation history of the element may be obtained by solving the ten equations for incrementally varying magnitudes of element stresses.

In the following sections, the steel bars are assumed to behave as a linear elastic-perfectly plastic material, characterized by the Young's modulus of steel, E_s, and the yielding stress f_y. The compressive stress-strain relation of the concrete struts is assumed to obey the softened concrete model suggested by Vecchio and Collins [3],

$$\sigma_d = \zeta f_c' \left[2\left(\frac{\varepsilon_d}{\zeta \varepsilon_o}\right) - \left(\frac{\varepsilon_d}{\zeta \varepsilon_o}\right)^2 \right] \qquad \text{if } \varepsilon_d / \zeta \varepsilon_o \leq 1 \qquad (11a)$$

$$\sigma_d = \zeta f_c' \left[1 - \left(\frac{\varepsilon_d / \zeta \varepsilon_o - 1}{2/\zeta - 1}\right)^2 \right] \qquad \text{if } \varepsilon_d / \zeta \varepsilon_o > 1 \qquad (11b)$$

where ε_o is the compressive strain at the maximum stress of concrete under uni-axial compression, and ζ is a softening coefficient that is related to the principal strains in the concrete by

$$\zeta = \frac{1}{\sqrt{0.7 - (\varepsilon_r / \varepsilon_d)}} \qquad (12)$$

In this study, ε_o is assumed to be -0.002. The tensile stress in concrete is ignored, i.e.

$$\sigma_r = 0 \qquad (13)$$

ELEMENT BEHAVIOR - GENERAL

The behavior of reinforced concrete wall elements under shear may be best illustrated using the load-deformation curves. Fig. 3 shows three such curves that are plotted in terms of the applied shearing stress and the corresponding shearing strain. These curves are obtained by analyzing three elements with different amounts of steel under an identical loading condition. The applied shear is acting in an inclined angle with respect to the l-t axes, Fig. 4. The element stresses in the l-t coordinate system, Fig. 2, are obtained through a coordinate transformation. The reinforcement ratios used in the three elements are indicated in Fig. 5. The curves in Fig. 3 reveal three different types of failure modes, namely under-reinforced plastic, under-reinforced, and over-reinforced modes. Element 1 has relatively small amounts of steel in both directions. In this case, both reinforcements yield before the concrete struts reach their maximum capacity. Thereafter, the load-carrying capacity of the element remain constant until, after a large deformation, the concrete strength starts to decrease. The failure mode is very ductile, similar to the plastic failure of steel frames.

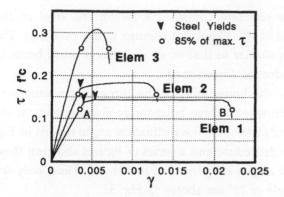

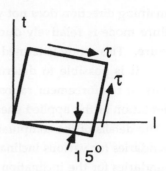

Figure 3: Load-Deformation Curves of Three Element Types - General

Figure 4: Orientation of the Applied Shear

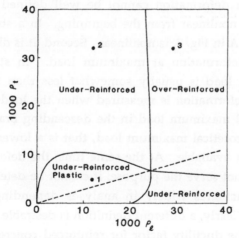

Figure 5: Boundaries of the Three Types of Failure Modes - General

This mode is termed under-reinforced plastic mode. Element 3 has a relatively large amount of steel. In this case, the compressive stress in the concrete struts reaches the maximum value before any of the steel yields; afterwards, the load-carrying capacity drops sharply. The failure mode is brittle, and the pattern of the element behavier is similar to that of an over-reinforced beam in flexure. This mode is termed over-reinforced mode. Inbetween the two modes, there is a failure mode that is characterized by the yielding of steel in only one direction, followed by the peaking of the concrete compresive stress; the steel in the remaining direction does not yield at all. Element 2 belongs to this category. The failure mode is relatively ductile, similar to that of an under-reinforced beam in flexure. This mode is termed under-reinforced mode.

It is possible to determine the boundaries of the three failure modes in terms of reinforcement ratios. However, these boundaries also depend on the orientation of the applied shearing stress, i.e., the inclination angle shown in Fig. 4. The details of the computation procedure and a series of figures showing these boundaries for various inclination angles are presented in Ref. [4]. Here, only the boundaries for the inclination angle of 15° are shown in Fig. 5.

For steel structures, the ductility factor is defined as the ratio of the total deformation at maximum load to the elastic-limit deformation [5]. If the same definition is applied to reinforced concrete wall elements, two difficulties arise. First, the elastic-limit deformation cannot be well defined because the load-deformation curve is nonlinear from the beginning. In a strict sense, even the initial path, such as OA in Fig. 3, is nonlinear. Second, it is difficult to determine precisely the total deformation at maximum load. In steel structures, the theoretical maximum load is usually somewhat less than the true maximum load, and the total deformation is measured when the load-deformation curve crosses the theoretical maximum load in the descending branch. In reinforced concrete walls the theoretical maximum load, that is a lower bound of the true maximum load, is not available. At the same time, the deformation at the true maximum load does not serve the purpose; it cannot be determined precisely in experiments, and fluctuates widely in analyses depending on the assumed material properties. Clearly, a different definition is desirable.

In this study the ductility factor for reinforced concrete wall elements is defined as the ratio of the total deformation at 85% of the maximum load in the descending branch to the deformation at the same load in the ascending branch. This definition may be written as

$$D.F. = \gamma_B / \gamma_A \qquad (14)$$

where A and B are as indicated in Fig. 3. From the numerical analysis, the shearing strains at the 85% of the maximum load in the ascending and the descending branches of Element 1 are obtained as 0.00354 and 0.0221, respectively. The corresponding values for Element 2 are 0.00328 and 0.0129; and for Element

3, 0.00354 and 0.00699. Thus, the ductility factors are computed as 6.2 for Element 1, 3.9 for Element 2, and 2.0 for Element 3.

The theoretical load-deformation curves cannot be expected to continue indefinitely. At a certain point on the curve, the compressive strain in concrete will reach a value at which the concrete will disintegrate. If the magnitude of the load at this limiting strain is greater than 85% of the maximum load, the shearing strain at this point should be substituted for γ_B in Eq. 14 to compute the ductility factor. For the lack of enough experimental data, this limiting strain is assumed to be $\varepsilon_d = -0.003$ in this study. It turns out that this limitation does not apply to any of the three elements considered here. For each element, the curve in Fig. 3 is terminated when the compressive strain in the concrete struts reaches -0.003, and the load magnitude at this point is less than 85% of the maximum.

ELEMENT BEHAVIOR - CONSTRAINED

For low-rise shearwalls, the wall elements may be assumed to be constrained in the horizontal direction by a rigid foundation, i.e.,

$$\varepsilon_l = 0 \tag{15}$$

Once this assumption is made, Eq. 1 and the amount of the longitudinal steel become irrelevant. Also, the expression for the softening coefficient, Eq. 12, is greatly simplified by the application of Eq. 5. For a typical shear loading, σ_t may be assumed to be 0, and the load-deformation history may be traced by solving the remaining equations. The details of the solutions are presented in Ref. [6] and will not be repeated here. It suffices to state that this theoretical analysis compares well with the experimental results on low-rise shearwalls in terms of the overall load-deformation history.

In a subsequent study [7], it is further delineated that there are three different failure modes. When a relatively small amount of transverse steel is used, the steel bars yield first, but the overall shear-carrying capacity continues to rise until the concrete capacity peaks. Beyond this point the shear carrying capacity decreases gradually. This phenomenon is similar to that of a beam in flexure with a ductile failure mode. Thus, this mode of failure is termed under-reinforced mode. When an excessive amount of steel is used, on the other hand, the shear-carrying capacity reaches the maximum before the steel yields and decreases sharply afterwards. This type of failure is brittle and may be classified as over-reinforced mode. Inbetween the two modes, a balanced mode, for the lack of a better term, exists that is characterized by the yielding of steel preceded by the peaking of the concrete stress. The differences in the overall load-deformation behaviors of these three failure modes are illustrated in Fig. 6.

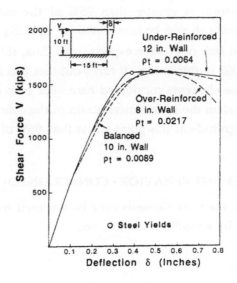

Figure 6: Load-Deflection Curves of Shearwalls - Constrained

Since the failure mode depends on the amount of reinforcement, an effort has been made to find the boundaries of the failure modes in terms of some measure of reinforcement. It turns out that the boundaries depend on two factors. They are the reinforcement index, ω, defined in Eq. 16, and the relative stiffness between the steel and concrete, η, defined in Eq. 17.

$$\omega = \frac{p_t f_y}{f'_c} \tag{16}$$

$$\eta = \frac{p_t E_s \varepsilon_0}{f'_c} \tag{17}$$

The boundaries that separate the three modes of failure are shown in Fig. 7. Also shown in the figure are two straight lines marked by $\omega/\eta=0.6$ and $\omega/\eta=1.4$. These two lines demarcate the region that covers the range of yielding stresses of steel from 40 ksi to 80 ksi. Thus, the shaded area in the figure represents the area of practical interest.

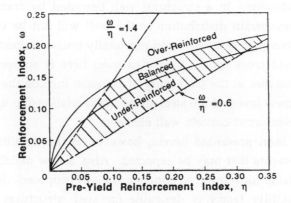

Figure 7: Boundaries of the Three Types of Failure Modes - Constrained

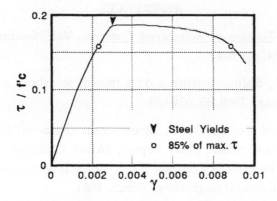

Figure 8: Load-Deformation Curve of the Under-Reinforced Element

The load-deformation curve of the under-reinforced element in Fig. 6 is replotted in Fig. 8 in terms of shearing stress and shearing strain. The terminal point of the plot corresponds to ε_d=-0.003. The shearing strains at the two points where the shearing stress is 85% of the maximum are 0.00232 and 0.00889. Thus, using Eq. 14, a ductility factor of 3.8 may be expected in this case.

DISCUSSION AND CONCLUSIONS

The above theoretical results are obtained under various simplifying assumptions. It should be emphasized that the results are for element behavior, not structural behavior. In a structural wall bounded by frame or boundary elements, the stress-strain distribution in the wall will not be uniform and the strain is neither completely constrained nor totally unconstrained.

The softened truss model theory as used here is somewhat simplified. The ongoing researches at the University of Houston include the modification of the softened concrete law and the steel stress-strain relationship to better describe the behavior of reinforced concete wall elements.

What has been presented herein, however, is the ductility of reinforced concrete wall elements that may be expected. Also, a new definition of ductility factor that is suitable for reinforced concrete walls is proposed. Specifically what value of the ductility factor is desirable for such structures remains to be determined through experiences and researches.

REFERENCES

1. Hsu, T. T. C., Torsion of Reinforced Concrete, Van Nostrand Reinhold Co., New York, 1984, pp. 544.

2. Hsu, T. T. C., Softened truss model theory for shear and torsion, ACI Structural Journal, 1988, 85, 624-635.

3. Vecchio, F. and Collins, M. P., Stress-strain characterisitics of reinforced concrete in pure shear. Final Report, IABSE Colloquium on Advanced Mechanics of Reinforced Concrete (Delft, 1981). Internatioanl Association for Bridges and Structural Enginering, Zurich, 1981.

4. Han, K. J. and Mau, S. T., Membrane behavior of R/C shell element and limit of reinforcement, J. Structural Engineering, ASCE, 1988, 114, 425-444.

5. Manual of Steel Construction, Load and Resistance Factor Design, 1st edition, American Institution of Steel Construction, Inc., Chicago, 1986, p. 6-212.

6. Hsu, T. T. C. and Mo, Y. L., Softening of concrete in low-rise shearwalls, ACI Journal, 1985, 82, 883-889.

7. Mau, S. T. and Hsu, T. T. C., Shear design and analysis of low-rise structural walls, ACI Journal, 1986, 83, 306-315.

ANALYTICAL APPROACH TO DYNAMIC HYSTERESIS DAMPING OF R/C IN-PLANE STRUCTURES SUBJECTED TO BASE ACCELERATION

KOICHI MAEKAWA
Division of Structural Eng. & Construction
Asian Institute of Technology
GPO Box 2754, Bangkok, Thailand

CHONGMIN SONG
Institute of Hydraulics and Energy
Swiss Federal Institute of Technology
CH-1015 Lausanne, Switzerland

ABSTRACT

The relation of the dynamic restoring force and displacement of R/C shear walls is a target to be analytically approached. The dynamic finite element analysis program [WCOMD] was developed to get an answer on whether "the characteristics regarding the dynamic restoring force and the hysteresis damping of R/C in-plane structures be the same as the static one or not". Both smeared and discrete crack R/C elements were implemented in [WCOMD] for analysis of multi-story framed shear walls. The static and dynamic analysis results manifested that the horizontal restoring force versus displacement relation in seismic base acceleration differs from the static one with the result of the higher energy absorption and hysteresis damping, but reduced ductility. It is analytically clarified that the higher dynamic hysteresis damping arises from the secondly induced inertia forces due to the two-dimensional extent of the structure and the path-dependent material nonlinearity of R/C constituent elements.

INTRODUCTION

Most of studies on dynamic analysis of R/C structures have been concerned with linear shaped members and their composites with lumped masses as a whole. The nonlinear restoring force-displacement relation of ingredient members has been directly or indirectly implemented in the dynamic equations of motion in terms of equivalent damping [C] and stiffness [K] as,

$$[M]\{\ddot{X}\} + [C]\{\dot{X}\} + [K]\{X\} = - [M][I]\{\ddot{X}_g\} \tag{1}$$

$$[I] = \begin{bmatrix} 1 & 0 & \cdots & 1 & 0 \\ 0 & 1 & \cdots & 0 & 1 \end{bmatrix}^T \quad \{\ddot{X}_g\} = \{\ddot{X}_{gx}, \ddot{X}_{gy}\}^T$$

where, [M], $\{X\}$ and $\ddot{X}_{gi}$ are global mass matrix, structural response nodal vector and ground base acceleration in i-th direction. It has been usual to install the characteristics of static stiffness and hysteresis damping obtained by static loading tests in Eq. (1). The whole system is modeled with reduced degree of freedom with several numbers of lumped mass. As a matter of fact, the energy absorption and restoring forces of beams and columns were experimentally clarified to be very similar to those under static conditions [1]. The point of discussion of this paper is laid on whether this static and dynamic consistency would be applicable to the framed shear walls or not, and if not, what sort of mechanisms cause discrepancy between dynamic and static hysteresis damping and stiffness in R/C in-plane structures.

Analytical Approach and Background

In-plane shear walls in a frame system with beams and columns (See Figure 1) have been simplified as resisting units against horizontal shear forces. For equivalent viscous damping and stiffness as mechanical parameters of a member, single story framed shear walls have been tested under horizontal static forces. Here, two dimensional walls are supposed in Eq. (1) to be one-dimensional element having the same horizontal shear characteristics as the static tests.

However, the shear walls are capable of carrying not only horizontal but vertical and flexural actions with two dimensional extent of stress field. These secondary forces affect the horizontal shear resisting mechanism in turn because of its material nonlinear property. For mechanics of structural walls in static and dynamic shear, the authors deem that the fully two-dimensional nonlinear analysis is indispensable without any simplification in the member levels, that is, initiation rooted in the multi-axial constitutive model of constituent materials as follows.

$$\{R\} = -[M][I]\{\ddot{X}_g\} - [M]\{\ddot{X}\} - \iiint_V [B]^T \{\sigma(\varepsilon,t)\} dV = 0 \tag{2}$$

where, the third term in Eq. (2) represents the internal force vector, and $\{R\}$ is the residual force vector to be zero at convergence. Up to now, few studies have been made on the in-plane nonlinear dynamic process of whole structures due to the difficulty of accurately predicting spatially averaged stresses $S(e,t)$ in R/C domains subjected to reversed cyclic loads. The constitutive model for two-way cracked R/C plates is to be modeled for solving Eq. (2) under reversed cyclic shear in dynamics. The reinforced concrete constitutive model, which was developed in the University of Tokyo, was adopted in the dynamic equation (2) of motion.

If the constitutive model used would be applicable to reversed cyclic stress path and arbitrary specified rate of stress, the hysteresis, viscous damping effects and associated energy absorption on the seismic loads are automatically taken into account in the internal force term of Eq. (2). This paper is to investigate the dynamic shear resisting mechanism of wall elements in the framed shear wall structures. Here, the general nonlinear dynamic program [WCOMD] played a major role on clarifying the difference of dynamic shear from the static one.

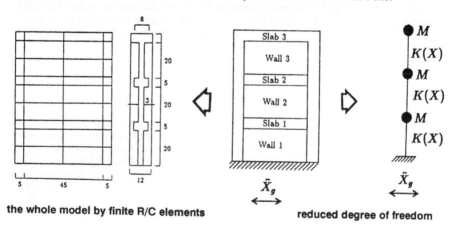

the whole model by finite R/C elements   reduced degree of freedom

Figure 1.   Multi-story in-plane shear wall and structural modeling

GENERAL FORMULATIONS

To solve Eq. (2) with respect to the nodal displacement $\{X\}$, the Newmark scheme based on the direct integration is employed. At time $t + dt$, the governing equation (2) takes the form of

$$\{R\}_{n+1} = \{F\}_{n+1} - [M]\{\ddot{X}\}_{n+1} - \{P\}_{n+1} = 0 \tag{3}$$

$$\{F\}_{n+1} = -[M][\,I\,]\{Xg\}_{n+1}, \quad \{P\}_{n+1} = \iiint_V [B]^T \{\sigma(\varepsilon, t)\} dV$$

where, $\{R\}_{n+1}$, $\{F\}_{n+1}$, $\{\ddot{X}\}_{n+1}$ and $\{P\}_{n+1}$ are error vector, applied force, acceleration and internal force vectors at step $n+1$ when time is $t + dt$. In the Newmark scheme, the acceleration vector is transformed to be function of previous nodal vector and up-dated one. We look for solution $\{X\}_{n+1}$ to satisfy the following equations as,

$$\{X\}_{n+1} = \{X_p\}_{n+1} + \Delta t^2 \beta \{\ddot{X}\}_{n+1}$$

$$\{\dot{X}\}_{n+1} = \{\dot{X}_p\}_{n+1} + \Delta t \gamma \{\ddot{X}\}_{n+1} \tag{4}$$

$$\{X_p\}_{n+1} = \{X\}_n + \Delta t \{\dot{X}\}_n + \Delta t^2 (0.5 - \beta)\{\ddot{X}\}_n$$

$$\{\dot{X}_p\}_{n+1} = \{\dot{X}\}_n + \Delta t (1 - \gamma)\{\ddot{X}\}_n$$

Note that $\{X_p\}_{n+1}$ and $\{\dot{X}_p\}_{n+1}$ are predictor values of displacement and velocity. Parameter B and r are the free ones which control the accuracy and stability of solution. If $\gamma \geq 0.5$ and $\beta = 0.25(\gamma + 0.5)^2$, the Newmark scheme is unconditionally stable for consistent finite elements. The time step can be chosen according to requirement of precision. The simultaneous equations from (2) to (4) are transformed to an effective static problem which can be solved by using Newton iterative scheme. The algorithm of this method is summarized as below.

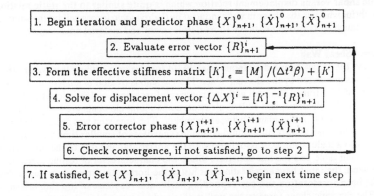

1. Begin iteration and predictor phase $\{X\}_{n+1}^0$, $\{\dot{X}\}_{n+1}^0$, $\{\ddot{X}\}_{n+1}^0$

2. Evaluate error vector $\{R\}_{n+1}^i$

3. Form the effective stiffness matrix $[K]_e = [M]/(\Delta t^2 \beta) + [K]$

4. Solve for displacement vector $\{\Delta X\}^i = [K]_e^{-1}\{R\}_{n+1}^i$

5. Error corrector phase $\{X\}_{n+1}^{i+1}$, $\{\dot{X}\}_{n+1}^{i+1}$, $\{\ddot{X}\}_{n+1}^{i+1}$

6. Check convergence, if not satisfied, go to step 2

7. If satisfied, Set $\{X\}_{n+1}$, $\{\dot{X}\}_{n+1}$, $\{\ddot{X}\}_{n+1}$, begin next time step

The stiffness matrix [K] of Step 3 above is used only to adjust displacement increments when the norm of the error vector denoted by Eq. (3) would not be satisfactorily small. Its suitability affects the speed of convergence but has no influence on the accuracy of solution, because the internal stresses are directly computed based upon the assumed total strain in each iterative loop. According to this procedure, which is not the linear integration but the step-by-step nonlinear one, time step partition is related only to the accuracy of acceleration and path-dependency, and free from the accumulation of error on equilibrium requirement.

Constitutive Model of R/C Element

If we would access to dynamics of R/C based on Eq. (2), the constitutive law for R/C must possess the ability to predict spatially averaged stresses of nonlinear domain including cracks for any given strain-time history. The authors adopted the smeared crack model developed in Concrete Laboratory of University of Tokyo, which consists of the following cyclic material models, (1) Compression model for cracked concrete [Maekawa, Okamura], (2) Tension stiffness model [Shima, Chou, Okamura], (3) Crack shear transfer model [Li, Maekawa] and (4) Reinforcing bar model [Shin, Kato]. The detail is discussed in the accompanying paper by Izumo et al.

Shin et al. [11] installed this model to the finite element program [WCOMR] for analysis of shear walls subjected to static reversed cyclic shear. In order to represent the localized displacement between members, such as junction plane between footing and R/C walls, the joint element was formed by combining the model of (1), (2) and (3). This is the same strategy as the above static analysis program [12].

The time-dependent model for stresses in Eq. (2) is essential to take into account the viscous damping, but at present, under development. Tentatively, we used the path-dependent R/C elements mentioned above. This means that the viscous damping effect corresponding to the loading speed on which the path-dependent constitutive laws are based are taken into account, and that the higher loading rate and corresponding variety of viscous energy absorption be neglected. This assumption seems reasonable so that the viscous damping of materials has a minor role compared with the hysteresis one. Accordingly, the authors' interest is to investigate the dynamic structural energy dissipation for the framed shear wall.

DYNAMIC SHEAR RESISTANCE IN FRAMED SHEAR WALLS

In the static analysis and experiments, the horizontal shear force applied to the walls through the rigid beam-slabs is identified according to the static equilibrium. Similarly, the dynamic shear force applied to the walls is easily computed by transforming the acceleration response of mass into the inertia forces. It was reported that R/C slender columns with top heavy weight exhibit the dynamic shear versus displacement relation which is quite similar to the static relation [1]. But it is doubtful whether the shear in walls induced by the dynamic forces will be the same as the static one, because of its two dimensional stress development and material nonlinearity.

Single Framed Shear Wall

Figure 2 shows the target adopted by Inoue et al. [9,10] for dynamic loading tests where the base acceleration was generated by the shaking table. The top slab is treated in analysis as a rigid body with the mass of 17.5 tons. The wall with isotropic reinforcement of 1% is idealized by 9 R/C smeared crack finite elements, and 13 R/C elements and 16 joint elements are assigned to frames surrounding walls and junction planes between members. The time step was specified to be 0.005 second so that the analytical results would coincide regardless of smaller time step. In the test, the vertical base acceleration was associated with the horizontal one due to the interaction between the shaking table and the test structure. Since the characteristics of testing machine could not be simulated, the input base acceleration was just in the horizontal direction in this analysis.

The base acceleration and corresponding horizontal response of top slab are shown in Figure 3. The dynamic shear force to the wall is obtained by multiplying the response acceleration by the mass of rigid slab. The dynamic shear versus displacement relation is shown in Figure 4 accompanied by the static analysis result. Figure 5 also shows the experimental static and dynamic shear of the wall subjected to the base acceleration [9]. It can be concluded that the dynamic restoring shear force and energy absorption capacity are indeed different from the static ones even though the same shear displacement is attained. The dynamic energy absorption and corresponding hysteresis damping are observed to be greater than the static one.

Let us discuss the mechanism which causes different dynamic and static restoring shear forces. It should be noted that the secondary inertia forces take place, which are excited by the primary horizontal shear and stiffness change due to higher nonlinearity of elements. The vertical and rotational inertia forces are obtained by timing the mass or the internal moment of the top rigid slab with corresponding accelerations. The larger shear displacement gives rise to the greater moment and vertical forces with higher frequency as shown in Figure 6. It can be said that the secondary inertia forces, which are completely zero in the static load, affects the horizontal shear in earthquake, because the behaviors of cracked R/C walls are highly nonlinear.

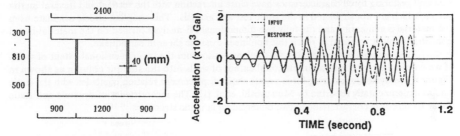

Figure 2. Single framed shear wall.   Figure 3. Response and input acceleration.

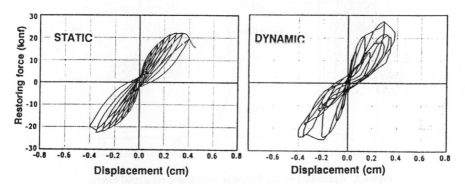

Figure 4. Dynamic and static shear restoring force versus displacement relations (analysis).

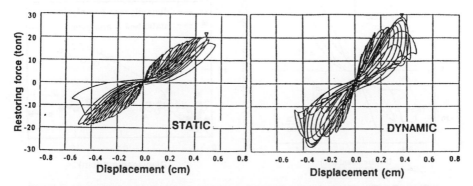

Figure 5. Dynamic and static shear restoring force displacement relations (Experiment [9]).

The static and dynamic internal loops are shown in Figure 7, where approximately double the equivalent viscous damping coefficient is achieved in dynamics. The increased damping factors are also experimentally reported by Inoue, et al. [10] as shown in Figure 7 including analytical results. The difference of the static from dynamic damping coefficients is also fairly simulated. If the material would be perfectly linear, the dynamic restoring shear is independent on the secondary inertia components because of its perfect superimposition of stresses and displacement as checked by Song [6]. Then, it is concluded that the dynamic shear cannot be fully investigated just with the static cyclic loading tests, because the horizontal shear energy absorption and restoring force characteristics have close interaction with the vertical and flexural inertia forces, which are not simulated in the static test and analysis. This coupling effects originate from the material nonlinearity, and the fully two dimensional FE analysis formed on the material-based constitutive law is essential to reach this nonlinear dynamic shear in earthquake.

The associated inertia moment originates also from the two dimensional extent of walls. If we would narrow the two dimensional extent of walls into the slender column as shown in Figure 8, the dynamic restoring force versus shear displacement relation coincides with the static one as experimentally reported by Mutsuyoshi, et al. [1]. The inconsistency of dynamic and static shear hysteresis is characteristic to two dimensional in-plane structures.

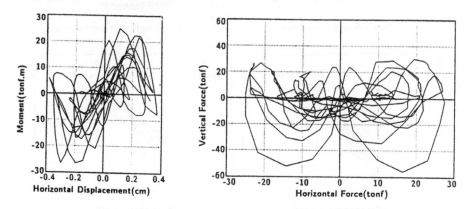

Figure 6. Secondary inertia forces in vertical and rotational forces.

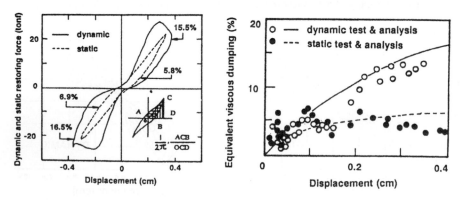

Figure 7. Dynamic and static energy absorption and equivalent viscous damping.

The dynamic restoring force and hysteresis damping is affected also by the dynamic base acceleration. Compared with the dynamic response shown in Figure 4 where the frequency of input base acceleration is kept constant but the amplitude is getting larger, the constant amplitude but varying frequency of base shear gives rise to a little bit distorted internal loop as shown in Figure 9. The induced secondary inertia forces exhibit different response to the horizontal shear. The dynamic hysteresis damping is not unique to members nor structures.

Multi-Story Framed Shear Walls

The three-story framed shear wall is targeted as shown in Figure 1. It was used in studying the ductility by static tests [5]. The dynamic shear resistance in each layer is of great interest from seismic design view points. The static and dynamic response of discretized structure shown in Figure 1 was computed. The analysis of the whole structure concerned is to examine the dynamic shear of each R/C wall in base acceleration by comparing those dynamics with the static shear restoring force-displacement relation as an isolated structural element. Similarly, the horizontal ground acceleration was input and the mass was concentrated to the three slabs with 13.2 tons. The static horizontal force was applied to the lowest story slab. The input and response acceleration in each story appear to be complicated (See Figure 10), but by multiplying the mass with the acceleration, we get the dynamic shear applied to each shear wall as shown in Figures 11 and 12.

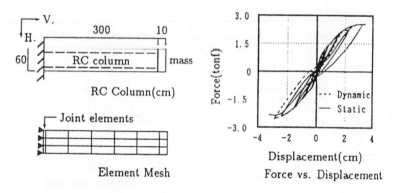

Figure 8. Effect of the two dimensional extent of the objective (beam dimension).

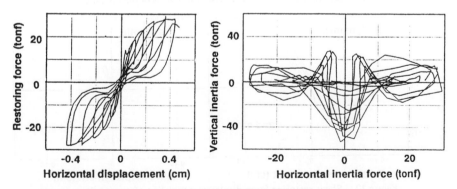

Figure 9. Dynamic hysteresis damping and stiffness in different base acceleration.

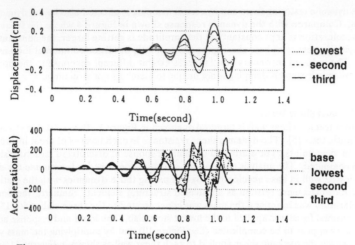

Figure 10. Input and response acceleration and displacement in each story.

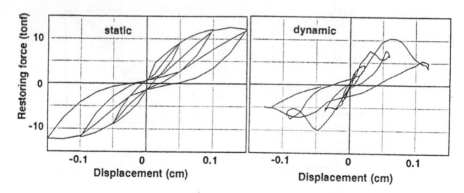

Figure 11. Dynamic and static shear hysteresis curves of the lowest story.

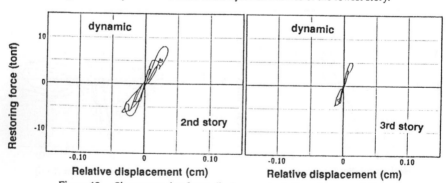

Figure 12. Shear restoring force-displacement relation of upper stories.

The dynamic and static restoring force characteristics of the first story wall are shown in Figure 11. In this case, the dynamic shear force to the first story is the sum of the inertia forces of three slabs. The shear force provoked by the base acceleration is also far from the static one. The reduced capacity and ductility of shear in earthquake are found. The dynamic shear force-displacement relations of the second and third stories are shown in Figure 12. The shear displacement is the relative horizontal one between slabs. The effect of the rigid body rotation of each slab is offset. Due to the lower dynamic shear force transferred, the nonlinearity is not remarkable. So far as the shear stiffness and hysteresis damping under the lower shear forces, there is very little difference among three walls both in static and dynamic loads.

The different shear resisting mechanism is caused by the secondary inertia forces which result in the different stress distribution and deformation as shown in Figure 13. In case of the higher static load, the more uniform stress field in the first story wall is produced rather than the dynamic case. This is related to the different ductility and capacity of shear between static and dynamic loads. For dynamic shear of nonlinear R/C in-plane structures, the whole domain of two dimensional extent is to be fully discretized by use of the R/C finite elements where the path-dependent constitutive laws of R/C are installed.

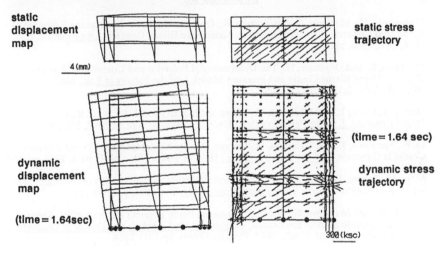

Figure 13.   Dynamic and static deformational mode and stress distribution.

CONCLUSION

Two dimensional nonlinear dynamic analysis program was developed based on the smeared crack R/C model and the discrete one. The whole domain of in-plane R/C shear walls were directly idealized with R/C smeared and discrete crack models. The shear resisting mechanism of R/C walls within the whole structural system was investigated under dynamic and static loads. Based on the fully two dimensional analyses of framed shear walls, the followings were clarified.

The dynamic hysteresis damping and restoring shear stiffness of R/C walls is generally different from the static one, and affected by structural geometry, material nonlinearity and the sort of base acceleration. This means that tested static shear characteristics of walls, such as the equivalent viscous damping, cannot be directly used but are to be factored for spatially simplified system with lumped mass subjected to dynamic base acceleration.

This sort of dynamic and static inconsistency is caused by the secondly excited inertia forces and the material nonlinearity of R/C members. If the in-plane walls would be perfectly linear, the horizontal shear behavior is both linear and independent on other inertia forces due to the linear superimposition. The accompanied rotational inertia arises from the two dimensional extent of the whole structures and mass distribution. The two dimensional whole model of the objective is indispensable to fully take into account the nonlinear coupling between R/C dynamic shear nonlinearity and excited inertia forces applied to the wall itself in turn.

ACKNOWLEDGMENT

The authors express their gratitude to the Ministry of Education of JAPAN for Grant-in-Aid for Scientific Research No.01420034. The second author is grateful to the Japanese Government for financing his scholarship in the University of Tokyo, and the first author to JICA for financing his research in AIT as a secondment from the Government of Japan.

REFERENCES

1. Mutsuyoshi, H., Machida, A. and Tsuruta, K., Dynamic Nonlinear Earthquake Response of Reinforced Concrete Structures Based on Strain Rate Effect, Concrete Library of JSCE, No.8, Dec., 1986, pp.101-115.

2. Maekawa, K. and Okamura, H., The Deformational Behavior and Constitutive Equation of Concrete Using Elasto-Plastic and Fracture Model, J. of the Faculty of Eng., Univ. of Tokyo (B), Vol.37, No.2, 1983, pp.253-328.

3. Shima, H., Chou, L. and Okamura, H., Micro and Macro Models for Bond Behavior in Reinforced Concrete, J. of the Faculty of Eng., Univ. of Tokyo(B), Vol.39, No.2, 1987, pp.133-194.

4. Li, B., Maekawa, K. and Okamura, H., Contact Density Model for Stress Transfer across Cracks in Concrete, J. of the Faculty of Eng., Univ. of Tokyo (B), Vol.40, No.1, 1989, pp.9-52.

5. Maki, M., Takagi, H. and Kanoh, Y., Experimental Study on the Ductility of the Multi-storied Reinforced Concrete Shear Wall, Transactions of AIJ, Oct, 1984, pp.1829-1830.

6. Song, C. M. and Maekawa, K., Nonlinear Dynamic Finite Element Analysis of RC Shear Walls Subjected to Seismic Excitations, Computer Aided Analysis and Design of Concrete Structures, Proc. 2nd Int. Conf., Vol.2, Pineridge Press, pp.1213-1224.

7. Izumo, J., Shin, H., Maekawa, K. and Okamura, H., An Analytical Model for RC Panels subjected to In-Plane Stresses, Int. Workshop on Concrete Shear in Earthquake, Elsevier Ltd., 1991.

8. Okamura, H. and Maekawa, K., Non-linear Analysis and Constitutive Models of Reinforced Concrete, Computer Aided Analysis and Design of Concrete Structures, Proc. 2nd Int. Conf., Vol.2, Pineridge Press, pp.831-850.

9. Maruta, M., Shibata, A., Kubo, T., Inoue, N. and Nakamura,M., Model Test for Evaluation of Seismic Behavior of Reactor Buildings, PART-9 Damping Characteristics Test (Comparison Between Dynamic Tests and Static tests), Transaction of AIJ, Oct., 1990 (in Japanese).

10. Inoue, N., Shibata, A., Kubo, T., Nagashima, T. and Akino,K., Model Test for Evaluation of Seismic Behavior of Reactor Buildings, PART-10 Discussion on Damping Performance, Transaction of AIJ, Oct., 1990 (in Japanese).

11. Shin, H., Maekawa, K. and Okamura, H., Analytical Models for Reinforced Concrete Shear Walls Under Reversed Cyclic Loading, Int. Workshop on Concrete Shear in Earthquake, Elsevier, 1991.

THEME IV:

DESIGN FOR EARTHQUAKE RESISTANCE

SEISMIC DESIGN STRATEGIES FOR SHEAR
RESISTANCE IN REINFORCED CONCRETE STRUCTURES

THOMAS PAULAY
Department of Civil Engineering
University of Canterbury
Christchurch, New Zealand

ABSTRACT

A brief review of various features of shear mechanisms in reinforced concrete building structures, to be utilized when ductile response resulting from earthquake actions is to be expected, is presented. With few exceptions, shear mechanisms are not considered to be suitable for energy dissipation in a seismic environment. For this reason, elastic response in shear is desirable. An outline of a deterministic design philosophy and some examples of its application to potentially plastic regions in beams, walls and joints, to preserve adequate shear resistance under extremely adverse conditions, is also given.

INTRODUCTION

It is now well established that potential plastic regions in reinforced concrete ductile structures that may be subjected to severe earthquakes, must be clearly identified. There is a certain amount of freedom involved in the selection by the designer of such plastic regions, commonly referred to as plastic hinges. Hence, as part of a deterministic seismic design strategy, the designer can usually make a clear decision as to the position of plastic regions within the structural system which will then result in a well-behaved structure.

Desirable Hysteretic Response
Ductility, that is the ability for inelastic deformations without significant loss of strength, enables structures designed to resist moderate intensity of lateral forces to survive without collapse the effects of large earthquakes. To this end, the designer's aim will be to ensure optimum hysteretic response associated with good energy-dissipating capacity and acceptable rates of stiffness and strength degradation. A desirable response, similar to that achievable in steel structures, is shown by the full line curves in Fig. 1. With well

detailed plastic hinges in beams, columns and structural walls this is achievable.

However, reliance must be placed on flexural response quantified by the attainable maximum curvature ductility. As a general rule, limitations on maximum curvature are set by the strain capacity of the concrete in compression. When necessary, as in the case of columns and occasionally in walls, this strain capacity can be adequately increased by suitable confinement of the compressed concrete. Design procedures leading to such desirable hysteretic response are now well established.

The Role of Shear in Inelastic Response
The role of shear mechanisms in the ductile response of structures, designed for earthquake resistance, is as yet not clearly defined. The amount of research directed to the reliable prediction of shear resistance based on shear failure mechanisms, rather than shear strength necessary to maintain flexural mechanisms in a seismic environment, underscores this observation. Some structural designers hold the view that the seismic design strategy adopted should aim to prevent a shear failure under any circumstances. Considerations supporting this approach include:

(1)   While a diagonal tension failure in a web reinforced member may be ductile, the mechanism cannot be used for efficient energy dissipation under reversed cyclic shear. Transverse shear reinforcement (stirrups) can contribute to shear strength only in tension.

(2)   Upon shear reversal, a new compression field will need to develop. This is only possible if diagonal cracks which, as a result of extensive previous yielding of the shear reinforcement, may have become wide, close. In the process of crack closure, greatly reduced shear resistance is offered by the structure, and this results in a *"pinched"* hysteretic response, such as that shown by the dashed curves in Fig. 1.

(3)   Cumulative tensile strains in stirrups will seriously reduce the compression strength of the concrete, diagonally cracked at least in two directions, necessary to sustain the diagonal compression field.

THE PHILOSOPHY OF CAPACITY DESIGN

Conclusions derived from a comparison of the hysteretic responses of flexural and shear mechanisms of potential plastic regions, lead to the need to establish a desirable and quantifiable hierarchy in failure modes. Strong column-weak beam relations in ductile frames are well established examples. Similarly it is postulated that if the shear strength of a plastic region is in excess of the shear demand arising from the development of plastic hinges, the undesirable effect of shear (Fig. 1) can be eliminated or at least minimised. The prediction of the shear demand depends on the ability of the designer to predict the maximum flexural strength of plastic hinges when subjected to the expected curvature ductility demands. The prediction of earthquake induced shear demand, as part of an established capacity design procedure [1] is obtained from:

$$V_i - \omega_v \phi_o V_E \qquad (1)$$

where V_E = the shear derived by analysis for the specified intensity of lateral forces on the structure.

ϕ_o = flexural overstrength factor which quantifies the largest expected flexural strength in plastic hinges in relation to the moments M_E derived from the actions of specified lateral forces.

ω_v = a dynamic shear magnification factor, applicable to situations in which there is no unique static relationship between the flexural strength and shear, as in the case of cantilever walls. For beams $\omega = 1.0$.

Plastic regions of elements designed to resist this shear can be expected to maintain flexural strength under reversed cyclic seismic displacements without detrimental interference by inelastic shear mechanisms.

LIMITATIONS ON MAXIMUM SHEAR STRENGTH

The maximum attainable shear strength is controlled, as is well known, by the compression strength of the concrete when it is also subjected to transverse tensile strains. To this end codes usually impose an upper limit on dependable shear strength, generally in terms of shear stresses v, for example $v \leq 10\sqrt{f_c'}$ (psi) [1] or $v < 0.2f_c' \leq 6$ MPa [2]. However, in plastic hinge regions in beams and walls further conservatism to prevent shear compression failures is warranted because of the concentration of diagonal and flexural compression forces and the effects of reversed cyclic inelastic displacements. The issue is more critical in structural walls, which in this context are large beams with thin webs.

The effect of imposed inelastic strains, corresponding with the expected curvature ductility capacity in a plastic hinge, should be taken into account when attempting to set an upper limit on dependable shear strength. This may be achieved by limiting the expected ductility demand on the structure. This, however, involves the use of larger lateral design forces, which in turn may affect strength requirements for many, if not all, other components of a structure. Alternatively the detrimental effect of large and repeated curvature ductility demands on the (diagonal) compression strength of the concrete may be recognized by relying on shear strength which is less than that used in situations controlled by gravity loads and possibly by wind forces. The advantage of this approach is that the largest value of the appropriate overall ductility capacity for a structural system may still be chosen to determine the required intensity of lateral design forces. Subsequently only those areas of plastic regions would need to be strengthened where diagonal compression, as a result of high shear, is considered to be critical. A corresponding approach, similar to that used for the last 10 years in New Zealand, takes the form

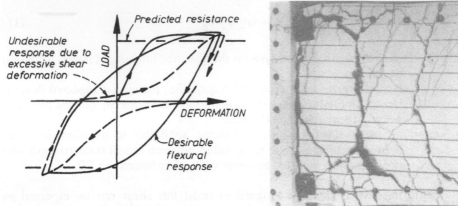

Figure 1   A comparison of hysteretic responses

Figure 2   Shear displacements along interconnecting flexural cracks across a plastic hinge of a beam.

$$v_{max} = \left(\frac{0.32\ \phi_o}{\mu_\Delta} + 0.03\right)f_c' < 0.16f_c' < 6\ MPa \tag{2}$$

where μ_Δ estimates the displacement ductility capacity of the structure which in turn determines the reduction of lateral design forces in comparison with those corresponding with elastic seismic response, and ϕ_o is the flexural overstrength factor, defined with (Eq. (1)). Its significance in Eq. (2) is that, in addition to the maximum probable strength of constituent materials, it also allows for the development of strengths possibly in excess of that which was required. Excess strength is often inevitable because of rounding up of dimensions and quantities of flexural tension reinforcement, or because load combinations, other than those involving earthquake forces, may have governed critical strength requirements. If excess strength is available during inelastic seismic response, Eq. (2) assumes that a proportional reduction in ductility demand will ensue.

For example when a very ductile coupled structural wall system is used with $\mu_\Delta = 6$, and a near optimum design for strength resulted in $\phi_o = 1.5$, Eq. (2) gives $v_{max} = 0.11\ f_c'$, a value considerably less than that envisaged by most codes. On the other hand the maximum shear in a structure with some reserve strength, so that $\phi_o = 1.9$, designed for restricted ductility capacity $\mu_\Delta = 3.5$, will be limited to $v_{max} = 0.15\ f_c'$.

The restriction of this procedure manifests itself for example in increased web thickness, but only in the plastic hinge region of cantilever walls in multi-storey buildings.

SLIDING SHEAR IN PLASTIC HINGES IN BEAMS

As a consequence of curvature ductility demands in plastic hinges of beams, interconnected wide flexural cracks may lead to a potential sliding shear plane. When the flexural compression reinforcement in the section is relatively large, or perhaps equal to that of the tension reinforcement, the role of the concrete in transmitting compression diminishes and a major fraction of a moment is transmitted by steel forces. At this stage shear transfer across a potential sliding plane, as seen in Fig. 2, must rely primarily on dowel resistance of flexural bars and on rather limited contribution of shear friction by interlocking aggregate particles. Sliding shear deformations can be large enough to reduce significantly energy-dissipation capacity in plastic hinges of beams. A control of sliding shear may be achieved with the use of diagonal bars placed across the plastic hinge region in one or both directions as shown in Fig. 3.

In frames designed for large system ductility capacity ($\mu_\Delta \approx 6$), reduction of energy dissipation will increase with the increased magnitude of the shear force, quantified by the shear stress v_i, and the ratio r of the negative to the positive shear at the plastic hinge. The ratio r quantifies the amplitudes of shear reversals.

It will be appreciated that diagonal steel will contribute to sliding shear resistance in both tension and compression, as indicated in Fig. 3, and that it can respond in a very ductile manner. This concept has been incorporated into seismic design practice in New Zealand [1] where the use of diagonal reinforcement of the type shown in Fig. 3 is recommended whenever the total shear stress, derived from considerations of the overstrengths of both plastic hinges in a member, in accordance with Eq. (1), exceeds

$$v_i - 0.3 \ (2+r) \ \sqrt{f_c'} \ (MPa) \qquad or \qquad v_i - 3.6(2+r) \ \sqrt{f_c'} \ (psi) \qquad (3)$$

where r is the ratio of the smaller shear force to the larger one encountered during reversals, and is thus always negative. When r > -0.2 the use of diagonal reinforcement is not considered to be necessary. When the maximum shear stress exceeds the value given by Eq. (3), diagonal reinforcement is necessary because the loss of energy dissipation resulting from uncontrolled sliding displacements is considered excessive. The threshold set by Eq. (3) is based on experimental observations. The diagonal steel so provided (Fig. 3), taking into account yield strength in tension and/or compression, as appropriate, should resist a shear force

$$V_{di} - 0.7\left(\frac{v_i}{\sqrt{f_c'}} + 0.4\right) (-r) \ V_i \qquad (4)$$

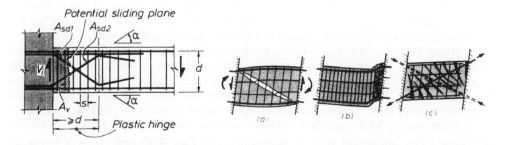

Figure 3   Control of slicing shear in the plastic hinge by diagonal reinforcement

Figure 4   Mechanisms of shear resistance in coupling beams

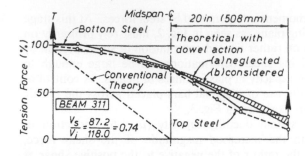

For example in a spandrel beam with r = 0.9 and a maximum shear stress of $v_i = 0.6\sqrt{f_c}$ (MPa), diagonal reinforcement should be provided to resist a shear force of $V_{di} = 0.7(0.6+0.4)(-0.9) V_i = 0.63 V_i$, i.e. 63% of the maximum shear force derived from Eq. (1). In most situations much smaller amounts of diagonal shear reinforcement will be required. Using the example of Fig. 3 the area of diagonal steel provided must be such that

$$A_{sd1} + A_{sd2} \geq \frac{V_{di}}{f_y \sin \alpha} \tag{5}$$

COUPLING BEAMS OF STRUCTURAL WALLS

As in all reinforced concrete squat elements, shear in relatively deep coupling beams, connecting structural walls in multi-storey buildings, leads to special features. Such beams may be subjected to exceptionally large ductility demands. When designed in accordance with some existing code requirements [2] they will fail in diagonal tension, as shown in Fig. 4(a), as often observed in earthquakes. When adequate stirrup reinforcement is provided to ensure that it can not yield when the flexural overstrength of the coupling beam is developed at both ends (Eq. (1)), improved ductility will result, but eventually a sliding shear failure, as shown in Fig. 4(b), may occur.

Elementary analysis [4] as well as experimental evidence (Fig. 5) shows that the conventional flexural reinforcement in such deep beams may be in tension over the full span and that shear is primarily transmitted by one single diagonal concrete compression strut. Beam 311, quoted in Fig. 5 with a span to depth ratio of 1.29, was reinforced with parallel top and bottom bars in equal quantities (1.58%). Stirrup reinforcement, satisfying code [2] requirements, and based on the traditional 45° model, was provided to resist 74% of the maximum shear force when the beam was subjected to equal moments at each end. While this mechanism is quite satisfactory when wind induced shear forces are to be transmitted between coupled walls, it is less suited for seismic situations. The horizontal bars in beams seen in Figs. 4(a) and (b) will never be subjected to compression but rather to incrementally increasing inelastic tensile strains. This leads to progressive elongation of the beam, to eventual deterioration of the diagonal compression strut and hence to inferior energy dissipation manifested by a hysteretic response, such as shown by the dashed curves in Fig. 1.

In such situations it is again advantageous to follow the trend in structural behaviour with suitable detailing, for example the use of diagonal bars, as in Fig. 4(c). It should be appreciated that once a group of diagonal bars has yielded these bars will be able to carry close to the full diagonal compression force upon load reversal. Thus, with cyclic reversed loading the contribution of the concrete to diagonal compression will eventually diminish. Because both moments and shear forces are then transmitted by diagonally placed steel, optimum hysteretic response will ensue.

Squat Walls

Over the last two decades squat walls have received considerable attention in many countries. The dominance of shear in influencing the overall response of such walls is well established. However, in the context of their role in the seismic resistance of low-rise buildings, the *"shear problem"* has often been overrated. Squat walls have significant flexural capacity, even when close to minimum vertical reinforcement is used, while gravity loads assigned to such walls in real buildings are generally small. As a consequence it is often difficult to provide adequate foundations to transmit overturning moments which are associated with even moderate shear demands in squat walls. In such cases seismic response is more likely to be governed by rocking of the wall on its foundation than by flexural or shear strength. Critical situations for squat cantilever walls arise therefore only when massive foundations or tension piles are present or where significant gravity loads are also to be carried. These conditions may exist when squat walls are used in the lower storeys of multi-storey buildings where they are required to resist the major part of the seismic shear, normally assigned by ductile frames.

When squat walls are expected to exhibit dependable ductility, the designer's attention will need to be focused on the mode of energy dissipation. As stated earlier, it is preferable to avoid shear mechanisms involving potential diagonal tension or compression failures or the yielding of horizontal (shear) reinforcement. This may be readily achieved by the application of a capacity design philosophy, whereby the dependable shear strength of a wall is made at least equal to the shear force associated with the flexural overstrength developed at the base of the wall (Fig. 6(a)) thereby significant ductility can be developed as a result of yielding of the vertical reinforcement without interference by a diagonal tension failure. It is, however, also well established that in such situations a sliding shear failure along the base, or in its close vicinity, may be induced after few displacement reversals with only moderate displacement ductility.

Mechanisms of shear friction, including dowel resistance of the vertical bars, will be mobilized along the base section, which normally coincides with a construction joint (Fig 6(a)). Because of small gravity loads or squat walls, the flexural compression zone, which is to transfer by shear friction the bulk of the shear force to the foundations, will be rather small. As the major portion of the (vertical) flexural reinforcement will have yielded, wide cracks will be present and shear transfer by aggregate interlock over the cracked region will diminish. The erosion of crack faces after a few inelastic load reversals will reduce the ability of the flexural compression zone to transfer the necessary shear, and sliding may occur. This results in progressive loss of stiffness, strength and hence energy dissipation of significant magnitudes.

The provision of additional dowels at the base and other innovative techniques have recently been developed to improve sliding shear resistance in squat walls.

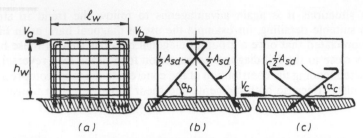

Figure 6   Modes of sliding shear resistance in squat walls

Effective and relatively simple solutions are suggested in Figs. 6(b) and (c), which show how diagonal reinforcement may be used. These bars, generally of larger size than those used in the conventional mesh, but few in numbers, are placed approximately in the middle of the wall thickness. As in the case shown in Fig. 3, these bars can contribute to shear resistance in both tension and compression. The single most important feature of this detailing is that shear along the potential sliding plane is transferred in a ductile manner, with little deterioration under progressive seismic displacement cycles [5].

The arrangement shown in Fig. 6(b) has the disadvantage that it will increase the flexural resistance of the wall thereby an additional shear force, shown as V_b, will be generated. However, the aim in design should be to allow the vertical wall reinforcement to determine the flexural strength, and hence the intensity of lateral forces that could be applied, and to use diagonal bars only to resist an appropriate fraction of the shear force, shown V_a in Fig. 6(a).

Figure 6(c) shows the optimum solution. Although the diagonal bars may increase the flexural resistance of the wall, because the vertical tension components can operate on a reduced internal moment arm, this contribution will immediately diminish upon sliding displacement. At this stage one set of bars will be subjected to compression so that the vertical components of the diagonal forces will cancel each other.

A design procedure based on this concept must rely to some extent on engineering judgement when aiming to improve energy dissipation and to enable acceptable ductile response to occur. Neither of these properties can be uniquely quantified. The major parameters which need to be considered in the design strategy are, however:

(i)   The desirable level of displacement ductility capacity of the squat wall to be used. Fully ductile response will seldom be necessary.

(ii)   The fundamental period of the structure, which is indicative of the number of displacement reversals a squat wall may encounter over the duration of an earthquake.

(iii)   The flexural overstrength of the wall, as designed. Excess strength, which may be available and indeed often unavoidable, will reduce ductility demand.

(iv)   The contribution of dowel mechanisms [4] of the vertical wall reinforcement along the potential sliding plane.

(v)   The frictional shear capacity of the precracked compressed concrete in the flexural compression region of the wall section with due allowance for likely strength degradation due to reversed cyclic and multidirectional straining of the concrete.

(vi)   The configuration of the flexural compression zone of the wall section. Flanged wall sections possess inherently large flexural strength while developing a very small neutral axis depth. Consequently they represent greater hazard in terms of sliding shear failure than comparable squat walls with rectangular cross section.

JOINTS BETWEEN REINFORCED CONCRETE MEMBERS

Significant research efforts over the last two decades addressed issues relevant to joints, particularly between beams and columns of reinforced concrete ductile frames. Design recommendations have been incorporated into various codes [1,2], although international consensus on several important aspects of joint behaviour is still lacking.

Joint shear forces are readily derived from first principles. However, as yet there is no general agreement with regard to mechanisms that are involved in shear transfer within a joint core. At one hand mechanisms have been postulated [4] as models on which joint design may be based, while another approach [2] prefers to rely on empirical findings supporting the role of joint hoop reinforcement in providing confinement to and preserving the integrity of the concrete in a joint core.

The assumption of infinitely rigid joints, often assumed in the analysis of frames subjected to lateral forces, is untenable. Joints relying primarily on the mechanism which enable efficient transfer of forces from reinforcement to the concrete in the joint core, necessitating considerable amounts of joint shear reinforcement, account typically for 20% of the interstorey deflection in frames. Mechanisms relying for shear transfer primarily on a single diagonal strut across the joint encourage large tensile strains to develop in beam reinforcement within the joint. Such joints may account for 50% of the interstorey deflection in frames subjected to lateral forces.

A detailed review of the state of the art of joint behaviour, recently studied in a United States, New Zealand, Japan and China cooperative reseach project, is beyond the scope of this summary.

CONCLUSIONS

Issues relevant to the shear resistance in reinforced concrete structures, expected to respond in a ductile manner to severe earthquake induced motions, were briefly reviewed.

It was postulated that shear mechanisms in which concrete needs to contribute to diagonal compression, are not suitable for energy dissipation. Therefore the hierarchy of mechanisms in the structural system should be such that shear mechanisms can operate essentially within the elastic domain.

When establishing a suitable shear resisting mechanism, the skill of the designer to predict the ability of such mechanisms to sustain the maximum strength of plastic hinges in which energy dissipation occurs by means of curvature ductility, is of considerable importance.

A conservative approach to the compression strength of concrete within alternating diagonal compression fields should be adopted because of large multidirectional tensile strains to which the compressed concrete may be repeatedly subjected.

When large nominal shear stresses need be sustained in plastic hinges of beams and particularly in squat walls, mechanisms of sliding shear, unique to inelastic seismic response, should be addressed. Diagonal reinforcement, to sustain a significant fraction of the shear in these regions will contribute to greatly increased energy dissipation.

In deep and relatively short spandrel and coupling beams, shear resistance by means of diagonal reinforcement alone will ensure optimal seismic response.

When capacity design procedures ensure that plastic hinges in frame members or walls are developed only in predetermined locations, the design strategies outlined above are applicable to these regions. Shear resistance in all other locations may be based on mechanisms relevant to structures subjected to gravity loads and wind forces.

REFERENCES

1. NZS 3101:1982, Code of Practice for the Design of Concrete Structures, Standards Association of New Zealand, 127 p.

2. ACI 318-89, Building Code Requirements for Reinforced Concrete, American Concrete Institute, Detroit, 1989, 353 p.

3. Paulay, T. and Bull, I.N., Shear Effects on Plastic Hinges of Earthquake Resisting Reinforced Concrete Frames, Comité Euro-International du Béton, Bulletin d'Information No. 132, April 1979, pp.165-172.

4. Park, R. and Paulay, T., Reinforced Concrete Structures, John Wiley & Sons, New York, 1975, 769 p.

5. Paulay, T., Priestley, M.J.N. and Synge, A.T., Ductility in Earthquake Resisting Shear Walls, ACI Journal, V.79, No. 4, July-Aug. 1982, pp.257-269.

DESIGN PHILOSOPHY FOR SHEAR IN EARTHQUAKE RESISTANCE IN JAPAN

HIROYUKI AOYAMA
Professor of Structural Engineering
Department of Architecture, University of Tokyo
Bunkyo-ku, Tokyo 113, Japan

ABSTRACT

In accordance with the basic philosophy of the new AIJ Design Guideline to clearly recognize the collapse mechanism of the structure to be designed, a new shear design method was developed in which deformation capacity of members was explicitly considered. The paper introduces the basic philosophy of the Guideline, as well as the basic concept of its shear provisions, derivation of design equations for non-ductile and ductile members, and confirmation by experiments. One of the most remarkable features of shear provisions is the design equation for ductile members, i.e., shear design to ensure required inelastic deformation capacity, which was achieved by relating concrete strength and inclination angle of diagonal strut in the truss mechanism with the required deformation capacity.

INTRODUCTION

In 1988, "Design Guideline for Earthquake Resistant Reinforced Concrete Buildings Based on Ultimate Strength Concept" [1] was published from the Architectural Institute of Japan (A.I.J.). It was not intended to replace the current Reinforced Concrete Design Standard, as the new Design Guideline has a very limited scope of application, and furthermore it is as yet in the proposal stage and cannot be actually used in practice. Nevertheless, the release of the Design Guideline, after a long study by various committees involved, was enthusiastically greeted by researchers and practicing engineers, as it marked an onset of a new and rational method of earthquake-resistant design of reinforced concrete buildings.

The basic philosophy of the Design Guideline is the clear recognition by structural engineer of the collapse mechanism of the building structure to be designed. The engineer first determines a possible and desirable collapse mechanism of the structure and location of plastic hinges. Secondly reinforcement at these plastic hinges is determined so that the collapse load would exceed the postulated design loading. This stage is called hinge mechanism design. Finally structural elements other than plastic hinges are designed in such a way that no yielding nor premature failure would occur in these elements before the hinge mechanism of the structure is attained. This stage is called hinge mechanism guarantee design.

The above-mentioned basic philosophy had been fostered in the Reinforced Concrete Committee of A. I. J. under the direction of the writer. In 1983-86, a joint research project on "the Development of Seismic Design of Reinforced Concrete Buildings Depending on the Ductility of Structures" was organized by the Grant in Aid from the Ministry of Education with the writer as the representative researcher. The report of this joint research project [2], issued in 1986, formed a basic draft of the work by the Seismic Design Subcommittee (chairman: Prof. T. Okada, Univ. of Tokyo) of the Reinforced Concrete Committee (chairman: Prof. Y. Kanoh, Meiji Univ.).

Among various items that had been developed in the report [2], the shear provisions were a completely new proposal along the recent trend to observe macroscopic shear resistance mechanism in the member. It was the work by Dr. T. Ichinose of Nagoya Institute of Technology [3]. After it was presented to the committees of A.I.J., it was subjected to intense discussions, and was refined and improved into its final form in the Chapter 6 of the Guideline, through the effort of a working group (chairman: Prof. F. Watanabe, Kyoto Univ.).

BASIC CONCEPT

Shear design of reinforced concrete members have been carried out on the basis of empirical equations, such as ACI 326 or Arakawa's equations. However works by Nielsen [4], Thürlimann [5], Collins [6], Shobara [7], and Minami [8] opened the way for shear design based on the equilibrium of two-dimensional stress distribution. Shear design approach of the Guideline can be positioned on the extension of these previous works.

The Guideline accepted the concept of truss mechanism and arch mechanism from the previous works. In case of a simple beam subjected to a uniform load, two load-carrying mechanisms are illustrated in Fig. 1. The truss mechanism depends on the amount of web reinforcement; if no web reinforcement is provided, this mechanism is non-existent. Diagonal compression members of the truss mechanism and compression arch in the arch mechanism both consist of concrete; hence the compressive strength of concrete is shared by two mechanisms. As the amount of web reinforcement increases, the share for arch mechanism has to be reduced.

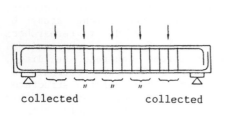

collected   collected

(a) Resisting Reinforcement

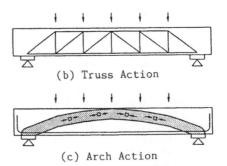

(b) Truss Action

(c) Arch Action

Figure 1. Truss Action and Arch Action in a Simple Beam

One of the most important difference of the Guideline approach and above-mentioned previous works is that the Guideline considers inelastic deformation capacity of member to be designed. It was an original idea of Ichinose [2, 3], and became one of the most remarkable feature of the Guideline. Shear design to ensure required inelastic deformation capacity was achieved by relating concrete strength and inclination angle of diagonal strut in the truss mechanism with the required deformation capacity.

TRUSS MECHANISM AND ARCH MECHANISM

The concept of truss mechanism and arch mechanism is explained for a short member with a rectangular section subjected to end moments and shear with or without axial load as shown in Fig. 2 (a). Here the member length L corresponds to the clear span length, and both member ends are assumed to be connected to member-to-member joints which are sufficiently strong and rigid. Fig. 2 (b) and (c) illustrate associated truss and arch mechanisms, respectively. For the time being, it is assumed that the member is not expected to have any inelastic deformation capacity.

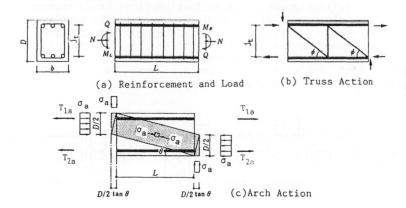

(a) Reinforcement and Load   (b) Truss Action

(c) Arch Action

Figure 2. Truss Action and Arch Action in a Member Subjected to
Bending Moment and Shear Force

The simplified member configuration, shown in Fig. 2, corresponds more to columns than beams. In case of beams having vertical loads in the span, a part of shear produced by vertical load may be carried by a curved arch mechanism as shown in Fig. 1 (c). Considering a curved arch mechanism, however, leads to a cumbersome equation derivation. Replacing a curved arch by a linear arch is a safe-side approximation. Moreover, the effect of vertical load is relatively small in Japanese buildings where in general the effect of earthquake load is predominant. Hence members are to be designed for maximum shear in the clear span.

First, let us consider the arch mechanism shown in Fig. 2 (c). As shown, the depth of compression strut is taken to be D/2. The reason for this assumption is (1) we assume yield strength of axial reinforcement to be infinitely large as we are designing for shear after the completion of design for flexure, and (2) we derive design equations following the lower bound theorem of the theory of plasticity. Hence, we can maximize the shear carried by the arch action corresponding to the same stress σ_a in the compression strut, which can be achieved by putting the depth equal to $D/2$. By geometrical relation in Fig. 2 (c), we obtain for shear carried by the arch action,

$$V_a = b \frac{D}{2} \sigma_a \tan\theta \tag{1}$$

where

$$\tan\theta = \frac{\sqrt{L^2 + D^2} - L}{D} \fallingdotseq \frac{D}{2L} \tag{2}$$

V_a : shear carried by arch mechanism (N)
b : width of the section (m)
D : total depth of the section (mm)
σ_a : average stress in compression strut (MPa)
θ : angle of arch action
L : member length (mm)

Secondly, we consider the truss mechanism shown in Fig. 2 (b). The height of truss j_t is taken to be the distance between centroids of upper and lower axial reinforcement. Noting that vertical truss members actually represent web reinforcement of the member, we may better idealize the truss mechanism as shown in Fig. 3 (b). This may be regarded as the superposition of many trusses of Fig. 2 (b) by shifting one a little after another. Then, diagonal truss force, carried by concrete, is uniformly distributed with angle ϕ. Web reinforcement is to be uniformly distributed, and is carrying its yield stress of σ_{wy}.

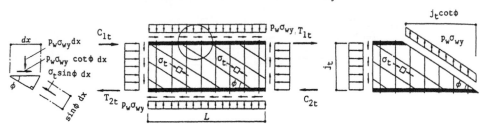

(a) Equilibrium near   (b) Truss Action   (c) Vertical Equilibrium
    the Top Bars

Figure 3. Equilibrium in the Truss Action

Considering the equilibrium of a small portion along the upper chord shown in Fig. 3 (a), we obtain

$$\sigma_t = \frac{p_w \sigma_{wy}}{\sin^2\phi} = p_w \sigma_{wy} (1 + \cot^2\phi) \tag{3}$$

where σ_t : average diagonal stress in compression (MPa)
p_w : web reinforcement ratio
σ_{wy} : yield strength of web reinforcement (MPa) to be taken not more than 25 times concrete strength
θ : angle of truss action.

As to the shear carried by the truss action, equilibrium of a free body shown in Fig. 3 (c) leads to

$$V_t = b\, j_t\, p_w\, \sigma_{wy} \cot\phi \tag{4}$$

where V_t : shear carried by the truss mechanism (N)
  j_t : distance between centroids of axial reinforcement (mm).

Total shear carried by the member is expressed by

$$V_u = V_t + V_a = b\, j_t\, p_w\, \sigma_{wy} \cot\phi + b\, \frac{D}{2}\, \sigma_a \tan\theta \tag{5}$$

where V_u : total shear at ultimate (N)

in which the truss angle ϕ and arch stress σ_a are yet to be determined.

PROVISIONS FOR NON-DUCTILE MEMBERS

The ultimate shear V_u is attained when the web reinforcement stress reaches the yield point σ_{wy} and concrete stress reaches its maximum capacity. For simplicity we ignore the difference of ϕ and θ, and simply add σ_t in eq. (3) and σ_a in eq. (1) to obtain resultant concrete stress. Furthermore, we assume that the compressive strength in web concrete is reduced from cylinder strength σ_B by a factor ν_0 because of diagonal cracking.

$$\sigma_t + \sigma_a = \nu_0 \sigma_B \tag{6}$$

Nielsen's proposal [4] from shear tests of T-beams is adopted here for the reduction factor ν_0.

$$\nu_0 = 0.7 - \frac{\sigma_B}{200} \tag{7}$$

where ν_0 : strength reduction factor for web concrete of non-ductile member
  σ_B : cylinder strength of concrete (MPa).

Equation (7) gives increasing effective strength $\nu_0\sigma_B$ up to σ_B equal to 70 MPa at which point $\nu_0\sigma_B$ reaches the maximum of 24.5 MPa.
  Introducing a factor β as the ratio of truss stress σ_t of eq. (3) to the effective strength $\nu_0\sigma_B$,

$$\beta = \frac{\sigma_t}{\nu_0\sigma_B} = \frac{p_w\sigma_{wy}(1+\cot^2\phi)}{\nu_0\sigma_B} \tag{8}$$

we can rewrite eq. (5) as follows.

$$V_u = b\, j_t\, p_w\, \sigma_{wy} \cot\phi + b\frac{D}{2}(1-\beta)\nu_0\sigma_B \tan\theta \tag{9}$$

where the truss angle ϕ, or $\cot\phi$, is the only one unknown quantity.
  As to the quantity $\cot\phi$, some of the previous works [7, 8] fixed the value of $\cot\phi$ to be 1 (or $\phi = 45°$). However the Guideline allows the value

of cotϕ to be greater than 1, considering aggregate interlock through inclined cracks. Its upperbound is taken to be 2 (or ϕ = 26.5°) according to the proposal by Thurlimann [5], considering the possible limitation of aggregate interlock with increasing crack width. Hence

$$\cot\phi \ \leqq \ 2 \tag{10}$$

From eq. (6), σ_t must not be greater than $\nu_o\sigma_B$. Thus from eq. (3) we obtain

$$\cot\phi \ \leqq \ \sqrt{\frac{\nu_o\sigma_B}{p_w\sigma_{wy}} - 1} \tag{11}$$

Equation (9) is a quadratic equation of cotϕ which gives increasing ultimate shear V_u within the range of cotϕ shown below.

$$\cot\phi \ \leqq \ \frac{j_t}{D\tan\theta} \tag{12}$$

As stipulated by the lower bound theorem of the theory of plasticity, cotϕ may be taken to be the largest value within the range of eqs. (10), (11) and (12), in order to maximize the ultimate shear V_u of eq. (9).

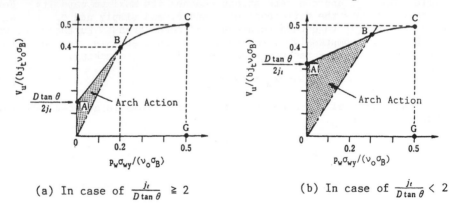

(a) In case of $\dfrac{j_t}{D\tan\theta} \geqq 2$ (b) In case of $\dfrac{j_t}{D\tan\theta} < 2$

Figure 4. Relationship between Shear Reinforcement and Shear Strength in a Non ductile Member

Figure 4 (a) illustrates the relationship between normalized ultimate shear vs. normalized web reinforcement, for members in which eq. (10) controls over eq. (12), or $j_t/(D\tan\theta) \geqq 2$ (this means that the member has an ordinary length). Eq. (9) can be rewritten as follows

$$\frac{V_u}{b\,j_t\,\nu_o\,\sigma_B} = \frac{p_w\sigma_{wy}}{\nu_o\sigma_B}\cot\phi + (1-\beta)\frac{D\tan\theta}{2\,j_t} \tag{13}$$

When $p_w\sigma_{wy}$ = 0, β = 0, hence point A in Fig. 4 (a) is obtained (which cannot be greater than 0.25 from the above-mentioned member length limitation). Between points A and B, eq. (10) governs. Hence cotϕ = 2.0 and β < 1, and ultimate shear consists of truss mechanism of line OB and arch mechanism of shaded zone OAB. Beyond the point B eq. (11) governs and β = 0. The second term in the right hand side of eq. (13) disappears and the first term gives a circle with center at point G. In this range shear is carried by truss mechanism only. The upper bound for shear reinforcement is given by the point C, or $p_w \sigma_{wy} = \nu_o \sigma_B/2$.

Fig. 4 (b) illustrates a similar relationship for member in which eq. (12) controls over eq. (10), or $j_t/(D\tan\theta) < 2$ (in other words, a very short member).

PROVISIONS FOR DUCTILE MEMBERS

For members where yield hinges are expected to form at one or both ends, the range within a distance 1.5 times the total depth from the critical section is called hinge zone. A criss-cross set of inclined cracks is expected to form in hinge zones. In order to take the possible effect of such cracks into account, two modifications are made to the previous case of non-ductile members.

First, effective strength of web concrete $\nu\sigma_B$ is lowered by following equations.

$$\nu = (1 - 15R_p)\,\nu_o \qquad \text{for} \qquad R_p \leqq 0.05 \qquad (14\text{-}1)$$

$$\nu = \nu_o/4 \qquad \text{for} \qquad R_p > 0.05 \qquad (14\text{-}2)$$

where ν : strength reduction factor for web concrete of ductile member
R_p : expected maximum hinge rotation angle.

Secondly, upper limit of truss angle $\cot\phi$ in eq. (10) is reduced as follows.

$$\cot\phi = \lambda \qquad (15\text{-}1)$$

where $\qquad \lambda = 2 - 50R_p \qquad \text{for} \qquad R_p \leqq 0.02 \qquad (15\text{-}2)$

$$\lambda = 1 \qquad \text{for} \qquad R_p > 0.02 \qquad (15\text{-}3)$$

: upper limit value of $\cot\phi$.

Assuming that the expected maximum hinge rotation is equal to those recommended in the commentary to Chapter 4 of the Guideline, we obtain following values for ν and $\cot\phi$.

Columns : $Rp = 1/67$

$= 0.775\nu_o$, $\cot\phi = 1.25$

Beams : $Rp = 1/50$

$= 0.7\nu_o$, $\cot\phi = 1$

Beams connected to walls : $Rp = 1/40$

$= 0.625\nu_o$, $\cot\phi = 1$

Reduction of $\cot\phi$ value corresponds to the increase of truss angle ϕ within hinge zones. For a member having hinge zones at both ends, we are in fact assuming a truss mechanism as shown in Fig. 5 (a). Reduced cot in the hinge zones correspond to the reduction of aggregate interlock due to large rotation of hinge zones. Beyond hinge zones, cotϕ value gradually increases to a constant value as determined by eqs. (10), (11) and (12).

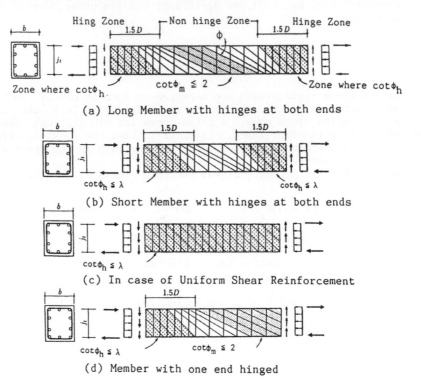

(a) Long Member with hinges at both ends

(b) Short Member with hinges at both ends

(c) In case of Uniform Shear Reinforcement

(d) Member with one end hinged

Figure 5. Truss Action in Members with Hinges at Both Ends

When the member length is shorter, we might have a truss mechanism shown in Fig. 5 (b), where the central zone with a constant cotϕ value disappears. In such a case the member should better be designed assuming a truss mechanism as shown in Fig. 5 (c). Fig. 5 (d) illustrates a member with one end hinged, which can be treated as in case of Fig. 5 (a).

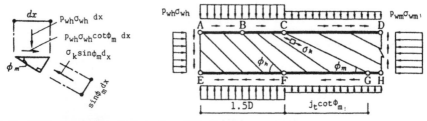

(a) Equilibrium near Point C   (b) Equilibrium in Plane Stress

Figure 6. Truss Action Near the Hinge Zone

The ultimate shear in case of uniform web reinforcement as in Fig. 5 (c) can be obtained by replacing ν_o in eqs. (8), (9), (11) by ν in eq. (14), and replacing eq. (10) by eq. (15).

If a model as in Fig. 5 (a) or (d) is assumed, that is, truss angle is to be varied within hinge zones and outside, we assume a truss model in Fig. 6 in and near the hinge zone. Suffixes h and m denote hinge zone and member zone, respectively. Considering equilibrium at point C, we obtain

$$\sigma_t = p_{wh}\sigma_{wh}(1+\cot^2\phi_m) \tag{16}$$

This stress is the largest both in hinge zone and outside. From the condition that $\sigma_t \leq \nu\sigma_B$, we obtain the following equation,

$$\cot\phi_m \leq \sqrt{\frac{\nu\sigma_B}{p_{wh}\sigma_{wh}} - 1} \tag{17}$$

which is to replace eq. (11).

The equilibrium of vertical forces between sections BF and CG leads to

$$p_{wh}\sigma_{wh}\cot\phi_h = p_{wm}\sigma_{wm}\cot\phi_m \tag{18}$$

A similar discussion as in case of eqs. (8) and (9) gives the following equations,

$$\beta = \frac{p_{wh}\sigma_{wh}(1+\cot^2\phi_m)}{\nu\sigma_B} \tag{19}$$

$$V_u = b\, j_t\, p_{wh}\sigma_{wh}\cot\phi_h + b\,\frac{D}{2}(1-\beta)\,\nu\sigma_B\tan\theta \tag{20-1}$$

$$V_u = b\, j_t\, p_{wm}\sigma_{wm}\cot\phi_m + b\,\frac{D}{2}(1-\beta)\,\nu\sigma_B\tan\theta \tag{20-2}$$

In the above equations, stress σ_{wh} and σ_{wm} are undetermined. Because the entire derivation is based on the lower bound theorem of the theory of plasticity, ignoring or liberating some of constraints always leads to a solution on the safe side. Hence it is permissible to assume that

$$\sigma_{wh} = \sigma_{wm} = \sigma_{wy} \tag{21}$$

and assume $\cot\phi_m$ and $\cot\phi_h$ in the following way.

First, $\cot\phi_m$ is chosen as the maximum value within constraints of eqs. (10), (12) and (17). Then $\cot\phi_h$ is chosen as the maximum value within constraints of eqs. (12), (15) and (22).

$$\cot\phi_h \leq \cot\phi_m \tag{22}$$

Ultimate shears in the hinge zone and outside are calculated from eqs. (20-1) and (20-2), respectively, where β is evaluated by eq. (19), and the assumption of eq. (21) is used. As the equilibrium in eq. (17) is ignored in the above procedure, smaller of the two ultimate shear values gives a safe side evaluation of the ultimate shear of the member.

CONFIRMATION BY EXPERIMENT

The commentary to the Guideline gives a comparison of measured vs. calculated shear strength of numerous test data of reinforced concrete members. The data bank included specimens with the ranges of concrete strength 16.5–62.9 MPa, tensile reinforcement ratio 0.39–3.21%, web reinforcement ratio 0–2.44%, web reinforcement yield strength 253–1470 MPa, product of the above two 0–19.1 MPa, and axial stress ratio to concrete strength 0–0.732.

Fig. 7 shows relationship between V_{max}/V_f and V_u/V_f where V_{max} is measured maximum shear force, V_f is shear force associated with calculated maximum flexural strength, and V_n is calculated shear strength by eq. (9) for non-ductile member. When V_u/V_f is greater than 1.0, the member is supposed to fail in flexure. The bilinear straight line in Fig. 7 thus represents theoretical prediction. The comparison is generally satisfactory from practical point of view, regardless of the fact that eq. (8) does not consider effect of axial forces.

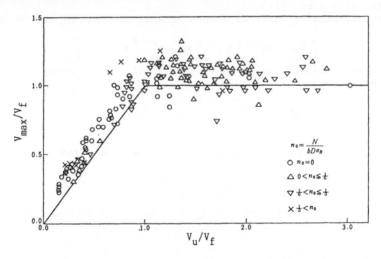

Figure 7. Comparison between Theoretical Shear Strength
and Experimental Data

The Guideline also gives a comparison for deformation capacity. Since the provisions for ductile members give reduced shear strength with the increase of hinge rotation angle, it is possible to evaluate theoretical hinge rotation angle at which shear strength equals the flexural strength. Adding the deflection angle at yielding to the above, we obtain calculated deformation capacity of test specimens. Fig. 8 shows the comparison of measured vs. calculated values of deformation capacity where measured value was determined at the point where the strength decreased to 80% of the measured maximum. Several points of underestimation correspond to specimens under high compressive stress, which mostly failed in flexural compression. Except for those, it can be concluded that the design method of the Guideline generally provide sufficient deformation capacity of yield hinges.

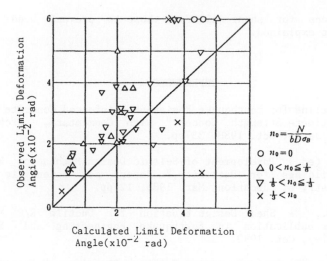

Figure 8. Comparison of Measured vs Calculated Values
of Deformation Capacity

ALTERNATE METHOD

The commentary of the Guideline includes an alternate method of shear design, called method B, as contrasted to the method A which is the method in the Guideline text as explained above. This method B was developed by Dr. K. Minami [9]. It is very similar to the method A except for the following.

(1) The truss angle ϕ is taken to be 45° at all times.

(2) The strength reduction factor for web concrete is a function of shear span ratio.

(3) The design for shear in hinge zone is carried out by expressing shear vs. flexural strength ratio as a function of expected maximum hinge rotation angle.

CONCLUSION

The shear provisions in the Design Guideline are developed to cover the following three phases of hinge mechanism guarantee design: first, design to ensure shear strength of columns, beams, and shear walls; secondly, design to ensure plastic deformation capacity of yield hinges subjected to shear in addition to flexure; thirdly, design to prevent bond splitting failure of columns and beams. In this paper, detailed explanation was given to (1) design to ensure shear strength of non-ductile columns and beams, and (2) design to ensure plastic deformation capacity of ductile members with yield hinges. By comparing with existing experimental data, the adequacy of these provisions was established. Due to limitation of space, provisions for shear walls and those to prevent bond splitting failure were not explained.

REFERENCES

1. "Design Guideline for Earthquake Resistant Reinforced Concrete Buildings Based on Ultimate Strength Concept" with Commentary, Architectural Institute of Japan, Oct. 1988, 337pp.

2. Aoyama, H. (ed.) "Development of Seismic Design Method of Reinforced Concrete Buildings for Ductility," Report for Scientific Research Grants, Ministry of Education, Mar. 1986, 121pp.

3. Ichinose, T., "A Shear Design Equation for Ductile R/C Members," presented for publication in Earthquake Engineering and Structural Dynamics, Wiley, Oct. 1990.

4. Nielsen, M.P., "Limit Analysis and Concrete Plasticity," Prentice Hall, 1984, 420pp.

5. Thürlimann, B., "Plastic Analysis of Reinforced Concrete Beams," Introductory Report of IABSE Colloquium, Copenhagen, 1979, pp. 71-90.

6. Collins, M.P. and Mitchell, D., "Shear Design - the 1984 Canadian Code," Journal of ACI, Vol.83, No.6, 1986, pp.925-933.

7. Shobara, R. and Kato, B., "Ultimate Strength of Reinforced Concrete Members by Compression Field Theory," Architectural Inst. Japan Annual Convention, Oct. 1978, pp.1731-1732.

8. Wakabayashi, M. and Minami, K., "Shear Strength of Concrete Structural Members," Annual Report, Kyoto Univ. Disaster Prevention Res. Inst. No.24B-1, Apr. 1981, pp.245-277.

9. Minami, K. and Kuramoto, H., "A Shear Design Equation for Reinforced Concrete Members," Proc. Japan Concrete Inst. Vol.9, No.2, June 1987, pp.347-352.

EVALUATION FOR EARTHQUAKE-RESISTANT DESIGN
OF REINFORCED CONCRETE STRUCTURES IN SHEAR

SHOJI IKEDA
Department of Civil Engineering
Yokohama National University, Yokohama City, Japan

ABSTRACT

Earthquake-resistant design of structures should be performed to fulfill required serviceability of the structure after the earthquake, as well as required safety during the earthquake. Since required serviceability after the earthquake can be varied according to the type and importance of the structure, a rational seismic design should be required to consider an acceptable level of damage due to the earthquake. This paper describes the following matters such as the concept of the rule of earthquake-resistant design, verification of the design rule by using pseudo-dynamic tests, execution of shear failure of reinforced concrete column specimens under actual earthquake acceleration records by the pseudo-dynamic test, and simulation of visual seismic failure by using a video-computer recording system.

INTRODUCTION

The level of structural damage can be quantified relating to the plastic deformation or residual deformation of the structure. Therefore, it is essential to obtain seismic behavior of reinforced concrete structures during earthquakes in order to predict for the design.

Construction cost can be reduced significantly if some degree of damage by a severe earthquake is accepted on the basis of avoiding loss of lives. Therefore, the seismic design considering ultimate limit state is supposed to be preferable[1,2]. However, in seismic zones, severe earthquakes occur definitely within a certain period and the important functions of a concrete structure must be maintained during and after the earthquake. Hence, structures must be designed to fulfill the required safety during the earthquake and serviceability and performance after the earthquake. This concept of the seismic design of concrete structures was proposed by the author.

The Committee on Concrete in the Japan Society of Civil Engineers (JSCE) stipulated a method of seismic design of concrete structures in the Specification of Reinforced Concrete Design in 1986 in which the above-mentioned design concept was adopted[3,4]. Here, the basic concept and background of the new design method relating to the Specification of JSCE are described. The results of pseudo-dynamic test are shown to clarify the appropriateness of the design method. Furthermore, a visual simulation system for real time earthquake failure is stated using pseudo-dynamic test.

CONCEPT OF SEISMIC DESIGN

Concrete structures should be designed corresponding both with safety during earthquakes and with serviceability and performance of the structures after earthquakes. Acceptable degrees of damage of stuctures due to earthquakes differ from the type and importance of structures. Although required integrity of the structure during and after the earthquake depends on the type of the structures, most of them are expected to be useful after earthquakes. Therefore, it is of primary importance to determine the required performance of the particular structure after the earthquake in the design. In this context, safe and economical structures can be designed by determining the degree of damage due to the earthquake in the level of a certain limit state. In general, the structures which allow a small damage are designed against high response loading, while the structures which allow damage without loss of safety are designed against rather low response loading in the same earthquake due to energy absorption in the plastic deformation.

Design Earthquake
The magnitude and intensity of a design earthquake should, in principle, be determined according to the type of the structure and characteristics of the construction site. In general, a design earthquake can be determined as the one the magnitude of which might be the largest in the life of the structure.

The useful life of concrete structures may be about fifty years. Then, anticipated horizontal seismic coefficient at the base of the structures during the life of structures might be 0.2 through previous data and experiences. In addition, regional effect of the seismic activity, the effect of ground condition, the magnification due to dynamic response, the importance of the structures, etc. must be included in a design earthquake loading.

In the case of a highly severe earthquake which is much stronger than the design earthquake, the structure can hold sufficient reserve strength if adequate ductility in the plastic stages would be maintained due to suitable design details and suitable structural planning. Therefore, it is possible to avoid collapse of the structures even under stronger earthquakes than anticipated in the design.

Limit States
The limit state under a design earthquake should be determined in relation to necessary strength and deformability which are based on safety and serviceability during and after the earthquake, respectively.

The acceptable level of damage can be varied in accordance with the type and importance of the structure. When a reinforced concrete structure suffers significant plastic deformation, residual deformation and large crack opening are remained in the structure after the earthquake, and eventually the functions of the structure are impaired.

Though it is not easy to relate the level of damage to the level of serviceability after the earthquake, this is an essential point of seismic design. Referring to previous examples of damage to concrete structures caused by earthquakes and to experimental results, the acceptable levels of damage may be classified as sound condition, light damage, medium damage and significant damage.

From the results of reversed cyclic tests on reinforced concrete specimens, $1\delta_y$ (here, δ_y is yield displacement) displacement shows quite sound condition because of elastic range. Under $2\delta_y$ displacement, the width of crack is not large and structural integrity is kept in general. Therefore, $2\delta_y$ displacement may refer to light damage. Under $3\delta_y$ displacement, diagonal crack can be seen at the base of columns and deterioration in concrete can be seen during the repetition of reversed loading. Hence, this is classified as medium damage. In this case, the structure can be used by repairing the portion of damage at appropriate time after the earthquake and can be served with adequate inspection. In the case of $4\delta_y$ displacement, significant residual deformation and large crack opening may be seen after the earthquake and it may be necessary to repair or strengthen in the earliest possible time after the earthquake. From the above-mentioned relationship between the maximum elasto-plastic displacement and the level of the damage, the classification as shown in Table 1 may be acceptable to quantify the level of damage.

By determining the maximum acceptable plastic deformation in the design stage, necessary strength of the elements of the structure can be determined quantitatively for the design earthquake.

The acceptable level of damage may be "light damage" for the usual civil engineering structures considering the public safety, economy, period of design life, and long term serviceability of the structure.

INFLUENCES OF EARTHQUAKE

Seismic Coefficient

A seismic inertial force is obtained by multiplying seismic coefficient to the weight of the structure and its surcharges. Seismic coefficients should be determined in relation to the characteristics of construction site, the response characteristics of the structures, etc.

In general, seismic coefficients are to be obtained by equation (1).

$$k_h = \nu_1 \nu_2 \nu_3 \nu_4 \nu_5 k_0 \qquad (1)$$

where k_h: horizontal seismic coefficient
k_0: basic horizontal seismic coefficient, 0.2 in general
ν_1: zone modification factor, 0.7 to 1.0 in general
ν_2: ground modification factor, 0.9 to 1.2 in general
ν_3: response modification factor according to the natural period of the structure, 0.5 to 2.0 in general
ν_4: limit state modification factor, 0.4 to 1.0 in general, where $\nu_3 \cdot \nu_4 \geqq 0.5$
ν_5: modification factor due to the contribution of non-calculated elements, 0.7 to 1.0 in general.

In the case of seismic earth pressure, ν_3, ν_4, ν_5 are 1.0. The

TABLE 1
Classification of level of damage

Maximum response displacement	Level of damage
1 δ_y	Sound condition
2 δ_y	Light damage
3 δ_y	Medium damage
4 δ_y	Significant damage
δ_y : yield displacement	

TABLE 2
Modification factor of seismic coefficient

modification factor	classification	value
regional modification factor (ν_1)	high seismic zone	1.0
	medium seismic zone	0.85
	low seismic zone	0.7
ground modification factor (ν_2)	class 1: rock	0.9
	class 2: diluvium	1.0
	class 3: alluvium without soft ground	1.1
	class 4: soft ground	1.2
response modification factor (ν_3)	Fig. 1	0.5～2.0
limit state modification factor (ν_4)	sound condition	1.0
	light damage	0.7
	medium damage	0.55
	significant damage	0.4
modification factor due to effect of non-calculated members (ν_5)	no effect of non-calculated members	1.0
	medium effect of non-calculated members	0.85
	high effect of non-calculated members	0.7

coefficient for seismic liquid pressure is the same value as the one for the vessel.

The modification factors ν_1 to ν_5 can be determined referring to Table 2. The modification factors ν_1 and ν_2 are referred to, for example, the Seismic Design Standard of Highway Bridge of Japan Road Association.

The modification factor ν_3 is the factor which expresses the effect of dynamic response of the structure according to the natural period of the structure and the ground condition. The value of ν_3 can be obtained by Fig. 1 which was determined by judging the response spectra of previous severe earthquakes.

The modification factor ν_4 is the factor which can reduce the earthquake force according to the serviceability of the structure after the earthquake. The values given in Table 2 was obtained from the elasto-plastic response analysis using various severe earthquakes such as El Centro, Taft and so on. If the energy conservation rule is applied to obtain the values of ν_4, the values of 1.0, 0.58, 0.45, 0.38 are obtained for $1\delta_y$, $2\delta_y$, $3\delta_y$, $4\delta_y$ displacements respectively using the equation of $1/\sqrt{2\mu-1}$ where μ is a ductility factor. These values are lower than the values in Table 2.

Complex and large-scale structures should be verified by dynamic analysis in which design earthquake input motions should be determined according to the characteristics of the ground and the structure.

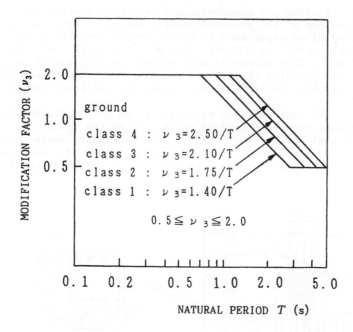

Figure 1. Response modification factor (ν_3) vs. natural period of structure.

Analysis of Loading Effect

To evaluate appropriate loading effects, adequate structural models should be assumed according to the structural system. Linear static analysis can be used to calculate the loading effects using static seismic coefficient modified by various factors. The natural period of the structure can be obtained by using total cross sectional inertia, i.e., neglecting the influence of cracking of concrete.

When dynamic analysis is applied, it is important to use an adequate damping factor. Since approximate damping factor of reinforced concrete structures is 0.05 to 0.10, it is recommended to use a damping factor of 0.07 considering slight inelastic behavior of reinforced concrete structures.

Ductility

Adequate design details should be provided to avoid collapse of the structure under earthquakes stronger than anticipated. Longitudinal reinforcement ratio, web reinforcement ratio, shear arm ratio, and axial compressive stress ratio are major factors influencing the ductility of reinforced concrete members during earthquakes.

EVALUATION BY EXPERIMENT

Reversed cyclic loading and pseudo-dynamic test were carried out to clarify the appropriateness of the proposed design method. The pseudo-dynamic test is a method of an experiment in which dynamic response of the specimen under a seismic wave is obtained in combination with the analysis of computer on line. This method is also called "On line hybrid experiment". This method can predict the dynamic response of structures under any digitalized earthquake motion without knowing the nonlinear behavior of the structures[5].

Fig.2 is the test specimen which was fixed to the test floor. A constant vertical force was applied at the top of the column to simulate dead load in actual structures. Fig.3 shows the basic characteristics of the specimens under reversed cyclic loading and observed cracks are shown in fig.4. From these figures, the classification of damage level was

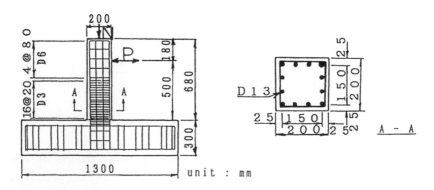

Figure 2. Test specimen.

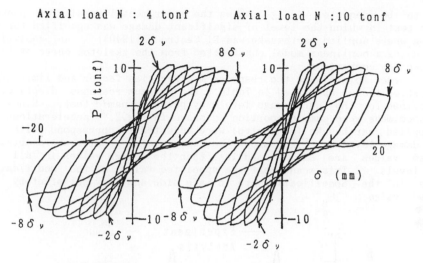

Axial load N : 4 tonf   Axial load N :10 tonf

Figure 3. Lateral load P vs. displacement δ under constant axial loads.

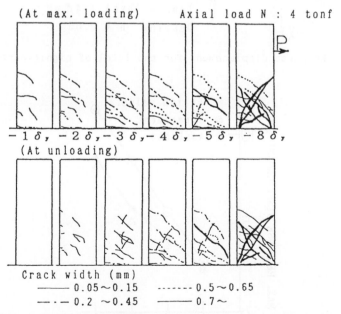

(At max. loading)   Axial load N : 4 tonf

$-1\delta, -2\delta, -3\delta, -4\delta, -5\delta, -8\delta,$

(At unloading)

Crack width (mm)
——— 0.05～0.15   ------- 0.5～0.65
—·— 0.2 ～0.45   ——— 0.7～

Figure 4. Cracks in the columns under reversed cyclic loading at various displacements.

found to be adequate.   Fig.5 shows the experimental results of pseudo-dynamic test in which the level of significant damage was specified for the specimen where applied earthquake was El Centro NS(1940).   An analytical result using a nonlinear model constructed from the skeleton curve of Fig.3 was shown for comparison.

In order to evaluate the response modification factor and limit state modification factor specified in Table 2 the maximum response displacement was obtained for the four damage levels by pseudo-dynamic test.   Since the test specimens were same proportion as shown in Fig.2,   accelerations of the applied earthquake (El Centro) were adjusted to correspond with the design damage levels.   Fig.6 shows the experimental results where obtained response values are similar and less than the design values in all the damage levels.   This means that the specified values of the modification factors in the Specification are in safe side and are satisfactory for intended design.

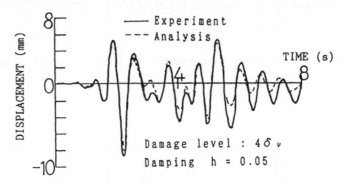

Figure 5. Response displacement for the level of significant damage.

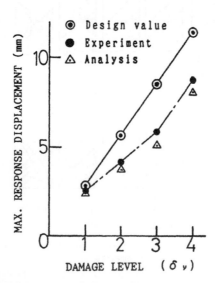

Figure 6.   Max. response displacement vs. adopted damage level.

Simulated Dynamic Response Visualization

It will serve greatly for seismic design and research to observe ongoing failing phenomena of concrete structures under a severe earthquake. Though the shake table test can be applied for this purpose, it is not only costly but also has difficulties such as adjusting appropriate time scale for the specimen size according to the change of mechanical properties of materials due to the influence of strain velocity.

The pseudo-dynamic test is less costly and has little difficulty for any size of the specimen to apply seismic response. However, it takes much more time to apply seismic displacement than the actual seismic movement. Therefore the dynamic response behavior cannot be observed in the pseudo-dynamic test in the real time basis. A simulated dynamic response visualization system was successfully developed by utilizing computer controlled video tape recorder in the pseudo-dynamic test[6].

This system operates such that a video tape recorder records for 0.2 second at the every 0.1 second of the seismic acceleration record which is applied for the pseudo-dynamic test. Then, simulated dynamic response of the specimen can be visualized by playing back the tape.

This system is quite useful to realize the ongoing seismic failure of the structure under actually recorded seismic acceleration in virtually real time behavior. Evaluation of shear failure of reinforced concrete columns under an earthquake can be done by using this system in the visual manner.

Fig.7 shows an example of the response behavior of a specimen under the El Centro NS(1940) earthquake motion reproduced by a videoprinter, where the amount of shear reinforcement in the specimen is one half of the required value.

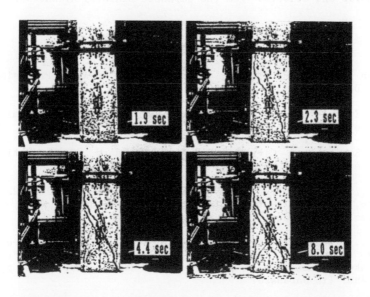

Figure 7. Visualization of response of a reinforced concrete column specimen under El Centro (NS) earthquake at the time(second) shown in each photograph

CONCLUDING REMARKS

A new and rational seismic design method was proposed and was adopted in the Specification of Reinforced Concrete Design of the Japan Society of Civil Engineers. The design method was ascertained by experimental and analytical studies including pseudo-dynamic test. The special features of the proposed design method and the experimental results are as follows:

(1) Seismic design should be performed to fulfill required serviceability after the design earthquake as well as required safety during the earthquake.

(2) High magnification factor due to response was introduced according to actual observation in the earthquakes.

(3) A reduction factor that depends on the acceptable level of damages remaining in the structure after the earthquake was introduced. The name of this factor is limit state modification factor.

(4) Pseudo-dynamic test clarified the appropriateness of the values of design magnification factors and design limit state modification factors specified in the Specification.

(5) A simulated dynamic response visualization system was successfully developed by utilizing computer controlled video tape recorder in the pseudo-dynamic test.

ACKNOWLEDGEMENT

Part of this study was supported by Grant-in Aid for Scientific Research (B) No.61460159 and 01460174 from the Japanese Ministry of Education.

REFERENCES

1. American Concrete Institute, Appendix A - Special Provisions for Seismic Design, ACI 318-83, 1983.

2. Comite Euro-International du Beton, CEB Model Code for Seismic Design of Concrete Structures, 1985.

3. Japan Society of Civil Engineers, Standard Specification of Reinforced Concrete Design, Oct. 1986. (in Japanese)

4. Ikeda,S., Machida,A., Commentary on Seismic Design of Reinforced Concrete, Concrete Library Vol.61, Oct. 1986, Japan Society of Civil Engineers. (in Japanese)

5. Ikeda,S., Tsubaki,T., Yamaguchi,T., Seismic limit state of reinforced concrete structures and evaluation of serviceability after the earthquake, Proceedings of the 43rd Annual Conference, Vol.5, pp.44-45, JSCE, 1988. (in Japanese)

6. Ikeda, S., Yamaguchi, T., Visual simulation of earthquake response from pseudo-dynamic test, Proceedings of the 45th Annual Conference, Vol.5, pp.648-649, JSCE, 1990. (in Japanese)

STRENGTH AND DUCTILITY DESIGN OF RC MEMBERS SUBJECTED TO COMBINED BENDING AND SHEAR

FUMIO WATANABE [1] and TOSHIKATSU ICHINOSE [2]

1) Associate Professor, Department of Architecture, Kyoto University, Kyoto, Japan
2) Associate Professor, Department of Architecture, Nagoya Institute of Technology, Nagoya, Japan

ABSTRACT

A consistent shear design method for reinforced concrete members subjected to combined bending and shear is proposed. In modeling of shear-resisting mechanism, superimposition of truss and strut action is introduced based on the lower bound theory. For ductile members, maximum allowable value of total diagonal stress of concrete due to truss and strut action is suppressed to avoid the degradation of diagonally compressed concrete under cyclic loading in post yield range. Checking procedure for splitting bond failure along longitudinal reinforcement is also proposed.

INTRODUCTION

In recent years, the capacity design method for reinforced concrete ductile frames was proposed by Thomas Paulay [1] and its philosophy was adopted by the New Zealand Design Code [2]. The draft of design guidelines based on similar method was also prepared by AIJ (Architectural Institute of Japan) Task-committee for Seismic Design of Reinforced Concrete [3] in 1988.

In the capacity design method, it is of primary importance to provide the required strength or strength and ductility to constituent members based on the intended collapse mechanism. In this paper, the strength and ductility design method of reinforced concrete members subjected to combined bending and shear is proposed. The proposed design method was adopted by the AIJ Guidelines mentioned above.

The shear design method described in this study is fundamentally based on the lower bound theory. Superimposition of truss and strut action is introduced in the modeling of shear resisting mechanism. In determining the stress at yield point for concrete, the effectiveness factor proposed by M. P. Nielsen [4] is used. The stress at yield point for shear reinforcement is limited less than 25 times of compressive strength of concrete. This restriction is to allow the high-strength shear reinforcement to be used in harmony with the corresponding high-strength concrete, in order to derive full efficiency from it. To satisfy the ductility demand for the plastic hinging regions, inclined compressive stress in concrete generated by truss and strut actions is suppressed less than a particular value to avoid the degradation of diagonally compressed concrete due to load cycling in the post-yield range. For hinge regions, the permissible minimum angle of concrete compressive stress in truss action is also restricted. A checking procedure to avoid the bond-splitting failure along longitudinal reinforcements is described in the last chapter.

In this study, the shear design method for members without any intermediate loading, such as dead load and live load, is discussed so that for a member subjected to intermediate loading, design shear force is to be maximum shear force in a member.

SHEAR DESIGN METHOD FOR NON-DUCTILE MEMBERS

Basic concept in modeling

The composed materials, i.e., concrete and reinforcing bar are assumed to be rigid-plastic materials. In determining the stress at yield point for concrete, σ_{cy}, the concept of effective compressive strength proposed by M. P. Nielsen [4] is introduced as given by Eqs.1.

$$\sigma_{cy} = v_o f'_c \quad , \quad v_o = 0.7 - \frac{f'_c}{200} \tag{1}$$

where, σ_{cy}=stress at yield point for concrete, v_o=effectiveness factor, f'_c=compressive strength of concrete in Mpa. In this study, Eqs. 1 is considered to be applicable to the concretes from 18 to 60 Mpa in compressive strength. For higher strength concrete, this equation seems to be unsuitable because σ_{cy} shows maximum value when f'_c=70Mpa.

The stress at yield point for shear reinforcement is given by the stress at actual yielding of it. However, it is restricted less than 25 times of the compressive strength of concrete. When actual yielding stress of shear reinforcement exceeds 25 times of the compressive strength of concrete, the yield strength used in the design equations is to be 25 times of compressive strength of concrete as

$$\sigma_{wy} = \sigma_y \quad \sigma_y \leq 25f'_c \tag{2}$$

$$\sigma_{wy} = 25f'_c \quad \sigma_y > 25f'_c \tag{3}$$

where, σ_y=actual stress at yielding of shear reinforcement, σ_{wy}=stress at yield point to be used in the design equations for shear.

As the equilibrium condition, only the balance to the transverse external shear is considered. It means that the longitudinal reinforcement should have enough strength to balance the bending moment and axial force at any section. Dowel action of longitudinal reinforcements is ignored. Conditions at ultimate limit state are: (1) total diagonal compressive stress in concrete generated by combined truss and strut actions reaches the effective compressive strength of concrete, $v_o f'_c$, defined by Eq. 1, and (2) stress in shear reinforcement reaches the stress at yield point, σ_{wy}, defined by Eq. 2 or Eq. 3.

Truss action

Assumed analogous truss model is shown in Fig.1.a. Assumed stress conditions in the truss model are: (1)homogeneous, uniaxial compressive stress of concrete, σ_t, with angle of ϕ to the member axis, (2) stress at yield point of shear reinforcement, σ_{wy}, which is uniformly and densely distributed along member axis, and (3) linearly varying stresses of longitudinal reinforcement, which is represented by the upper and lower stringers. Bond strength of longitudinal reinforcement is assumed to be large enough to satisfy the equilibrium of an infinitesimal stringer element (see Fig.1.b).

From the equilibrium condition of a free body in Fig.1.a., the shear force carried by truss action is given as

$$V_t = b \, j_t \, p_w \, \sigma_{wy} \cot\phi \tag{4}$$

where, b=width of beam or column section, j_t=distance between upper and lower longitudinal reinforcements, p_w= shear reinforcement ratio (=$\Sigma A_w/bs$), A_w=total sectional area of a set of shear reinforcement, s= spacing of a set of shear reinforcement, σ_{wy}=stress at yield point of shear reinforcement, ϕ= inclination of compressive stress of concrete to member axis.

From the equilibrium of an infinitesimal stringer element in Fig.1.b., the compressive stress in concrete, σ_t, is given as

$$\sigma_t = p_w \sigma_{wy} (1+\cot^2\phi) \quad (5)$$

In Eq. 5, $\cot\phi$ must satisfy the following inequality because σ_t cannot be larger than the effective compressive strength of concrete ($\sigma_t \leq \sigma_{cy} = v_o f'_c$).

$$\cot\phi \leq \sqrt{\frac{v_o f'_c}{p_w \sigma_{wy}} - 1} \quad (6)$$

Thurlimann [5] proposed that $\cot\phi$ should not exceed 2 to prevent the excessive transverse strain due to the loss of aggregate interlocking. This restriction for the value of $\cot\phi$ is also introduced to the design equations in this study.

$$\cot\phi \leq 2 \quad (7)$$

Strut action
The stress conditions and geometry of the strut action are shown in Fig.2. Assumed stress conditions are : (1) uniaxial stress ,σ_s, in the concrete strut BCEF with an angle of θ to the member axis , and (2) uniform stresses of the longitudinal reinforcements.

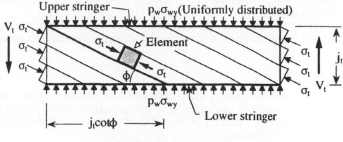

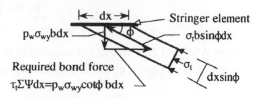

a) Analogous truss model

b) Equilibrium of an infinitesimal stringer element

Figure 1 Assumed analogous truss model

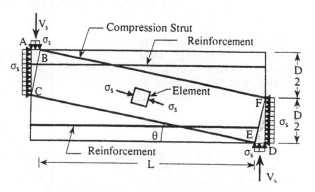

Figure 2 Stress conditions and geometry of assumed strut model

Lower bound theory permits us to assume the geometrical condition providing maximum load carrying capacity so that the depth of the compressive strut , AC and DF in Fig.2 becomes a half of the total section depth, D/2, (Ref.4). Geometry of Fig.2 gives the following equation,

$$\tan\theta = \frac{\sqrt{L^2 + D^2} - L}{D} = \sqrt{\left(\frac{L}{D}\right)^2 + 1} - \frac{L}{D} \quad (8)$$

where, θ=angle of compression strut to member axis, L:clear span length of member, D:total section depth.

At the ultimate state, the total compressive stress of concrete generated by the truss and strut actions should be equal to the effective compressive strength of concrete, $v_o f'_c$, as

$$v_o f'_c = \sigma_t + \sigma_s \tag{9}$$

In the above equation, the difference of the direction of the compressive stresses in the strut and truss action is neglected as a conservative simplification for the design purpose. By substituting Eq. 5 into Eq. 9, the following equation for the compressive stress allotted to the strut action is obtained.

$$\sigma_s = (1 - \beta)v_o f'_c \tag{10}$$

where,

$$\beta = \frac{\sigma_t}{v_o f'_c} = \frac{p_w \sigma_{wy}(1 + \cot^2\phi)}{v_o f'_c} \tag{11}$$

The triangles ABC and DEF in Fig.2 is assumed to be under bi-axial hydrostatic pressure, σ_s, so that the shear force carried by the strut action is given by the following equation.

$$V_s = 0.5bD(1 - \beta)v_o f'_c \tan\theta \tag{12}$$

Shear strength equation
The preceding discussion leads to a following equation for shear strength of a member based on superimposition of truss and strut action.

$$V_u = V_t + V_s = b\,j_t\,p_w\,\sigma_{wy}\cot\phi + 0.5bD(1 - \beta)v_o f'_c \tan\theta \tag{13}$$

Equation 13 is a quadratic equation concerning $\cot\phi$ because β includes $\cot^2\phi$. V_u increases with the increase of $\cot\phi$ in the following range.

$$\cot\phi \le j_t/(D\tan\theta) \tag{14}$$

The lower bound theory permits us to select the maximum V_u as the shear strength so that the largest $\cot\phi$ within Ineqs. 6, 7 and 14 can be selected. In this study, Ineq.14 is conservatively ignored for the simplicity. The relationship between the amount of shear reinforcement, $p_w\sigma_{wy}$, and the shear strength of a member ,V_u/bj_t, given by Eq.13, is indicated by the solid line ABC in Fig.3. The point A has the coordinates $(0, v_o f'_c D\tan\theta/2j_t)$

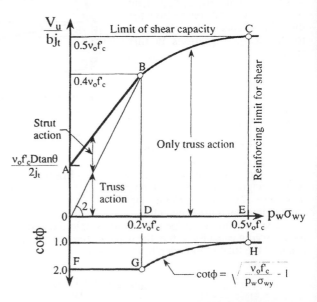

Figure 3 Relationship between V_u and $p_w\sigma_{wy}$

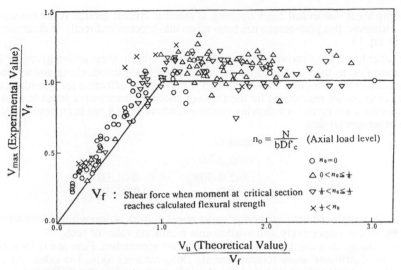

Figure 4   Relationship between theoretical V_u and experimentally obtained shear strength V_{max}

and it depends on the value of L/D because tanθ includes L/D as given by Eq.8. Up to the point B with the coordinates $(0.2v_o f_c, 0.4v_o f_c)$, the contribution of strut action exists as shown by triangle OAB, where cotφ is 2. Beyond the point B, shear force is resisted only by the truss action and cotφ is decreasing with the increase of $p_w \sigma_{wy}$ as given by Ineq.6. The line BC is a circular arc around the point 0, because cotφ is determined by Ineq.6. For $p_w \sigma_{wy} > 0.5 v_o f_c$, Eq. 13 gives the shear strength less than that at the point C, because in Eq.13 the yielding of shear reinforcement is assumed. However, the lower bound theory permits us to select the same shear strength as the point C, allowing that the stress of shear reinforcement is less than the yield strength and the compressive stress of concrete, σ_t, is $v_o f_c$ with angle of 45 degree.

In Fig.4, comparison between the shear strength of non-ductile members calculated by Eq. 13 and that obtained from shear tests on beams and columns in the past. Y axis indicates the experimentally obtained shear strength and X axis indicates the calculated shear strength where both axis are normalized by the shear force at which the bending moment at critical section could reach the theoretical flexural capacity. From this figure, it is seen that the shear strength equation proposed in this study can be applied for the design purpose.

SHEAR DESIGN METHOD FOR DUCTILE MEMBERS

Basic Concepts in Modeling
In this study, the effective compressive strength of concrete in hinge region is reduced in proportion to the required inelastic hinge rotation angle as

$$v f'_c = (1.0 - 15R_p) v_o f_c \qquad 0 < R_p \leq 0.05$$
$$= 0.25 v_o f'_c \qquad 0.05 < R_p \qquad (15)$$

where, vf'_c = effective compressive strength of concrete for hinge region, R_p = required inelastic hinge rotation angle in radian. This treatment is allow for the degradation of diagonally compressed concrete due to diagonal cracking in two directions under reversed loading and crack widening due to the progress of plastic elongation of flexural reinforcements in post yield range. Sliding shear due to full crack opening at member critical section is not considered in this study. However, this phenomena has been taken into account indirectly in determining the value of v in Eq. 15.

On the other hand, the maximum allowable value of $\cot\phi$ in truss action given by Ineq. 7 should be modified to take into account the loss of aggregate interlocking action along the cracked surfaces in hinge region. However, Ineq. 7 may be still effective for non-hinge region. From above discussions, restrictions for the angle of concrete compressive stress in truss action of ductile members are given by following equations, where Eq. 18 has been determined based on the past experimental data.

$$\cot\phi_h \leq \lambda \tag{16}$$

$$\cot\phi_m \leq 2.0 \tag{17}$$

$$\begin{aligned}\lambda &= 2.0 - 50R_p \qquad & 0 < R_p \leq 0.02 \\ &= 1.0 & 0.02 < R_p \end{aligned} \tag{18}$$

where, ϕ_h and ϕ_m =angles of concrete compressive stress in truss action within hinge region and outside hinge region, respectively, and λ =allowable maximum value of $\cot\phi_h$.

In the shear design of a ductile member, there are two approaches. First one is for a member with uniformly distributed shear reinforcements along member axis. The other one is for a member with non-uniformly distributed shear reinforcements, i.e., the different amounts of shear reinforcements are provided for the hinge region and for the non-hinge region in a member, respectively. Shear strength of a ductile member with uniformly distributed shear reinforcement can be simply obtained by replacing the effective strength $v_o f'_c$ with $v f'_c$ and Ineq.7 ($\cot\phi \leq 2$) with Ineq.16($\cot\phi_h \leq \lambda$). In the following, we discuss the shear design method for ductile members with non-uniform shear reinforcement.

Truss action

Assumed analogous truss and its stress conditions near the hinge region in a ductile member with non-uniform shear reinforcement is shown in Fig.5, where the length of design hinge region is assumed to be 1.5 times of the total section depth. The shear reinforcement ratio within hinge region is denoted by p_{wh} and that for outside of hinge region by p_{wm}, respectively. It is assumed that the stresses of the shear reinforcement gives the vertical binding stresses, $p_{wh}\sigma_{wy}$ and $p_{wm}\sigma_{wy}$ to upper and lower stringers within hinge and non-hinge regions, respectively. The inclinations of the compressive stresses in truss action are designated by ϕ_h and ϕ_m for regions ABFE and CDHG, respectively as shown in Fig.5.

The principal compressive stress of concrete with angle ϕ_m just inside the point C is denoted by σ_k as shown in Fig.5. From the equilibrium condition in the vertical direction of a infinitesimal element of upper stringer with a length of dx just inside the point C, following equation is obtained.

$$p_{wh}\,\sigma_{wy}\,dx = \sigma_k\,dx\,\sin^2\phi_m \tag{19}$$

From Eq.19, we obtain

$$\sigma_k = p_{wh}\,\sigma_{wy}\,(1 + \cot^2\phi_m) \tag{20}$$

The principal compressive stress of concrete varies in the region BCGF due to the change of the angle of concrete compressive stress in truss action and the above stress , σ_k , is the maximum compressive stress of concrete in this region. The concrete stresses in the region ABFE and CDHG are smaller than σ_k. The requirement $\sigma_k \leq v f'_c$ gives the following equation, where the effective compressive strength of concrete $v f'_c$ is given by Eq. 15 because the point C is considered to be inside the hinge region.

$$\cot\phi_m \leq \sqrt{\frac{v f'_c}{p_{wh}\sigma_{wy}} - 1} \tag{21}$$

From the equilibrium of a free body obtained by cutting the analogous truss along the line BF in Fig.5, the shear force carried by the truss action is given as

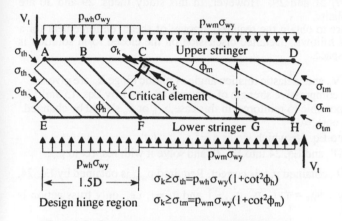

$$V_t = b \, j_t p_{wh} \sigma_{wy} \cot\phi_h \quad (22)$$

Shear force transferred along the line CG is also obtained by similar way as

$$V_t = b j_t p_{wm} \sigma_{wy} \cot\phi_m \quad (23)$$

Equations 22 and 23 require the following condition.

$$p_{wm} = p_{wh} \frac{\cot\phi_h}{\cot\phi_m} \quad (24)$$

$$\sigma_k \geq \sigma_{th} = p_{wh}\sigma_{wy}(1+\cot^2\phi_h)$$

$$\sigma_k \geq \sigma_{tm} = p_{wm}\sigma_{wy}(1+\cot^2\phi_m)$$

Figure 5 Geometry and stress conditions of truss action near hinge region

Strut Action

Compressive stress of strut action may pass over the point C in Fig.5. The concrete stress allotted to strut action in a ductile member, σ_s, is therefore the remainder of the effective strength of concrete in hinge region minus the compressive stress at the point C, σ_k, as follows.

$$\sigma_s = (1 - \beta_1) \, v \, f'_c \quad (25)$$

where,

$$\beta_1 = \frac{p_{wh}\,\sigma_{wy}(1 + \cot^2\phi_m)}{v\,f'_c} \quad (26)$$

The shear force carried by the strut action, V_s, is obtained by replacing $v_o f'_c$ by $v f'_c$ and β by β_1 in Eq.12.

$$V_s = 0.5bD(1 - \beta_1)vf'_c\tan\theta \quad (27)$$

Shear strength equation

The preceding discussion leads to the following equation for shear strength of a ductile member as the sum of the contribution of the truss and strut actions.

$$V_u = V_t + V_s = b \, j_t \, p_{wm} \, \sigma_{wy} \cot\phi_m + 0.5bD(1 - \beta_1)vf'_c\tan\theta \quad (28)$$

Equation 28 is a quadratic equation concerning $\cot\phi_m$. The value of V_u increases with the increase of $\cot\phi_m$ in the following range.

$$\cot\phi_m \leq \left[\frac{p_{wm}}{p_{wh}}\right]\frac{j_t}{D\tan\theta} \quad (29)$$

Substituting of equation 24 into inequality 29 gives the following.

$$\cot\phi_h \leq \left[\frac{p_{wm}}{p_{wh}}\right]^2 \frac{j_t}{D\tan\theta} \quad (30)$$

The lower bound theory permits us to select the maximum V_u as the shear strength. Therefore, we can select the largest $\cot\phi_h$ within Ineqs. 16 and 30. We can also select the largest $\cot\phi_m$ within Ineqs. 17, 21 and 29. However, in this study Ineqs. 29 and 30 are conservatively ignored for simplicity.

A practical design procedure to obtain the required amount of shear reinforcement for a ductile member is indicated as follows. Detailed discussion on the closed form solution is omitted due to the limitation of space.

Step 1: Required shear strength V_u and inelastic rotation angle of plastic hinge R_p are given.
Step 2: Calculation of effective compressive strength of concrete vf'_c and λ by Eqs. 15 and 18.
Step 3: If $V_u > bj_t vf'_c/2$, it is impossible to reinforce. In this case, b, j_t or f'_c should be changed.
Step 4: Assume $\cot\phi_m = 2.0$ and $\cot\phi_h = \lambda$.
Step 5: Substitute $\cot\phi_m = 2.0$ into Eq. 26, which yields $\beta_1 = 5p_{wh}\sigma_{wy}/vf'_c$.
Step 6: Substitute $\beta_1 = 5p_{wh}\sigma_{wy}/vf'_c$ and Eq. 24 into Eq. 28 and solve it with respect to p_{wh}.
Step 7: If $\sqrt{vf'_c/p_{wh}\sigma_{wy} - 1} \geq 2.0$, obtained p_{wh} is correct. Required p_{wm} is obtained by Eq. 24.

If not, $\cot\phi_m$ is given by $\cot\phi_m = \sqrt{vf'_c/p_{wh}\sigma_{wy} - 1}$ and $\beta_1 = 1$, i.e., only truss action is effective. Follow step 8.

Step 8: If $\cot\phi_m = \sqrt{vf'_c/p_{wh}\sigma_{wy} - 1} < \lambda$, follow step 10.

Step 9: Substitute $\cot\phi_m = \sqrt{vf'_c/p_{wh}\sigma_{wy} - 1}$, $\beta_1 = 1$ and Eq. 24 into Eq. 28 and solve it with respect to p_{wh}. Required p_{wm} is obtained by Eq. 24.

Step 10: For this case, $\cot\phi_h = \cot\phi_m = \sqrt{vf'_c/p_{wh}\sigma_{wy} - 1}$ and $p_{wh} = p_{wm}$. Required p_{wh} and p_{wm} are obtained by Eq. 28, where $\beta_1 = 0$.

PREVENTION OF SPLITTING BOND FAILURE ALONG LONGITUDINAL REINFORCEMENT

In beams or columns with large number of longitudinal deformed bars in the extreme layer of a section, particularly when closely spaced, splitting bond failure along longitudinal reinforcements likely to occur due to the circumferential or transverse tensile stresses induced by wedge action of the protrusions of deformed bars. Such failure is not favorable because it might lead to abrupt reduction in load carrying capacity due to the loss of truss action in a member. To avoid the splitting bond failure, the estimated maximum bond stress of longitudinal reinforcements, i.e., design bond stress, τ_d, should not be greater than the splitting bond strength of them, τ_{bu}, as

$$\tau_d \leq \tau_{bu} \tag{31}$$

Design bond stress
In the analogous truss shown in Fig. 1, bond action of longitudinal reinforcement is required to balance against the diagonal compression force of concrete strut and the tensile force of shear reinforcement. The design bond stress to assure the truss action is given by Eq. 32 from the equilibrium condition of an infinitesimal element of upper and lower stringers as shown in Fig.1.b.

$$\tau_t = bp_{wt}\sigma_{wy}\cot\phi/\Sigma\psi \tag{32}$$

where, τ_t=design bond stress due to truss action, p_{wt}=required shear reinforcement ratio in the design for shear, σ_{wy}=yield strength of shear reinforcement, ϕ=angle of compressive stress of concrete in truss action to member axis, which is used in shear strength design of a member, $\Sigma\psi$=total perimeter of longitudinal reinforcements arranged in upper or lower layer of a section.

As an alternative method, design bond stress is also obtained by considering the force difference of a longitudinal reinforcement between both member end sections computed by the section analysis for bending moment and axial force as

$$\tau_f = \frac{d_b \Delta\sigma}{4(L-d)} \tag{33}$$

where, τ_f=design bond stress based on section analysis, d_b=bar diameter, $\Delta\sigma$=the stress difference of a longitudinal reinforcement between both member end sections, d=effective depth of longitudinal reinforcement.

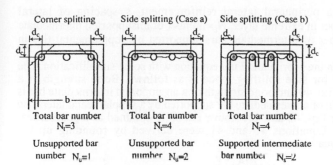

Corner splitting   Side splitting (Case a)   Side splitting (Case b)

Total bar number N_t=3   Total bar number N_t=4   Total bar number N_t=4

Unsupported bar number N_u=1   Unsupported bar number N_u=2   Supported intermediate bar number N_s=2

Figure 6 Typical splitting mode

The stresses of longitudinal reinforcements at the member end section is calculated by section analysis based on plain remain plain hypothesis, however for plastic hinge regions the stresses should be over strength of the longitudinal reinforcement. A denominator of (L-d) in equation 33 is allow for that the stress of longitudinal reinforcement does not change so much at member end region in a length d.

For non-ductile members or ductile members with a plastic hinge at either member end, design bond stress might be whichever smaller of equation 32 and 33.

For ductile members with plastic hinges at both member ends, design bond stress is to be given by equation 33.

Splitting bond Strength
Several equations to estimate the splitting bond strength of deformed bars was proposed in the past. In this study, the original equations proposed by Morita and Fujii [6] are fundamentally used, however some simplifications and modifications were added for the simplicity.

Based on the past studies, bond strength of a deformed bar , τ_{bu}, is given by Eq. 34 as the total of the contribution of concrete, τ_{co}, and the contribution of lateral reinforcement, τ_{st}. However, for the upper horizontal longitudinal reinforcements in a beam section the value of τ_{bu} computed by Eq. 34 have to be reduced to 80 %.

$$\tau_{bu} = \tau_{co} + \tau_{st} \tag{34}$$

Based on the Morita and Fujii equation, the contribution of concrete, τ_{co}, is given as

$$\tau_{co} = 0.313(0.4b_i + 0.5)\sqrt{f_c} \tag{35}$$

where, the value of b_i depends on the splitting mode of cover concrete (see Fig.6) and is given as whichever smaller value of equation 36 or 37,

For corner splitting mode $b_{ci} = (2\sqrt{2}d_c - d_b)/d_b$ (36)

For side splitting mode $b_{si} = (b - \Sigma d_b)/\Sigma d_b$ (37)

where, d_c=cover thickness of a corner bar measured from center of it, f'_c=compressive strength of concrete in Mpa, d_b=bar diameter and b=section width.

Contribution of lateral reinforcement , τ_{st} is given by Eq. 38 or 39 for each splitting mode where the bar size of supplemental tie is assumed to be equal to the peripheral lateral reinforcement in this study,

For corner splitting mode ($b_i=b_{ci}<b_{si}$) $\qquad \tau_{st}=0.313\dfrac{50A'_w\sqrt{f_c}}{sd_b}$ (38)

For side splitting mode ($b_i=b_{si}\leq b_{ci}$) $\qquad \tau_{st}=0.313\dfrac{\left(\dfrac{40}{N_t}+\dfrac{10N_u}{N_t}+\dfrac{30N_s}{N_t}\right)A'_w\sqrt{f_c}}{d_b s}$ (39)

where, A'_w=sectional area of a peripheral lateral reinforcement, s=spacing of lateral reinforcement, N_t=total bar number in extreme layer, N_u=number of unsupported intermediate bars in extreme layer, N_s=number of intermediate bars supported by supplemental ties in extreme layer.

Equation 38 was obtained from the Morita and Fujii equation with some simplification. The derivation procedure of Eq. 39 for side splitting mode is as follows. Bond strength of a supported corner bar is given by Eq. 40 and bond strength of a unsupported intermediate bar is given by Eq. 41, and bond strength of a intermediate bar supported by a supplemental lateral tie is given by Eq. 42 as the total of Eqs. 40 and 41. Above Eq. 39 is obtained as a geometrical mean of Eqs. 40, 41 and 42. Equations 40 and 41 were derived by rounding up the coefficients in Morita and Fujii equations.

$$\tau_{st,1}=0.313\dfrac{20A'_w\sqrt{f_c}}{d_b s}$$ (40)

$$\tau_{st,2}=0.313\dfrac{10A'_w\sqrt{f_c}}{d_b s}$$ (41)

$$\tau_{st,3}=\tau_{st,1}+\tau_{st,2}=0.313\dfrac{30A'_w\sqrt{f_c}}{d_b s}$$ (42)

ACKNOWLEDGEMENTS

Design equations proposed in this study were formed by the AIJ Working Group on Shear chaired by F. Watanabe. Authors deeply acknowledge to the members of the group.

REFERENCES

1)Paulay T. ; Capacity Design of Reinforced Concrete Ductile Frames, Workshop on Earthquake-resistant Reinforced Concrete Building Constructions, University of California, Berkeley, July, 1977

2)Code of Practice for THE DESIGN OF CONCRETE STRUCTURES, NZS3101, 1982

3)Design Guidelines for Earthquake Resistant Reinforced Concrete Buildings Based on Ultimate Strength Concept, Architectural Institute of Japan, 1988

4)Nielsen M. P. ; LIMIT ANALYSIS AND CONCRETE PLASTICITY, Prentice-Hall, 1984

5)Thurlimann B. ; Plastic Analysis of Reinforced Concrete Beams, Introductory Report of IABSE Colloquium, Copenhagen, 1979, pp.71-90

6)Fujii S. and Morita S. ; Splitting Bond Strength of Deformed Bars, Transactions of the Architectural Institute of Japan, No.319, Sept, 1982, pp.47-55

DESIGN OF EARTHQUAKE RESISTANT SHEARWALLS TO PREVENT SHEAR FAILURE

S. K. GHOSH and VYTENIS P. MARKEVICIUS
Portland Cement Association
5420 Old Orchard Road
Skokie, Illinois 60077, USA

ABSTRACT

In earthquake-resistant design of structures, to ensure inelastic deformability, prevention of premature shear failure of members becomes very important. The current code-prescribed design procedures for earthquake-resistant shearwalls against shear needs improvement with respect to both required strength and available strength. This paper is concerned with required strength, and aims at establishing the design values of dynamic shears at the bases of concrete shearwalls in buildings subject to seismic ground motion.

INTRODUCTION

Shearwalls in multistory buildings are slender, and behave essentially as vertical cantilever beams. Isolated shearwalls or individual walls connected to frames will tend to yield first at the base where the moment is the greatest. In earthquake-resistant design, to assure inelastic deformability, it becomes necessary to ensure that no brittle shear (diagonal tension) failure would develop prior to or simultaneously with the development of flexural hinging at the base of a wall. Prevention of premature shear failure through proper design against shear thus assumes primary importance. Proper shear design for a structural element involves satisfaction of the inequality:

$$V_u \leq \phi V_n$$

where V_u is the required shear strength at a critical section, V_n is the nominal shear strength of that section, and ϕ is a strength reduction factor.

In the case of beams, the required shear strength is determined by assuming that moments of opposite sign corresponding to probable strength act at the joint faces and that the member is loaded with the tributary gravity load along the span. Similarly, for a column, the required shear strength is determined from the consideration of maximum developable moments, consistent with the axial force on the column, occurring at the column ends. In the case of shearwalls, a similar design condition is not readily established. This is because the magnitude of the shear at the base of a wall (or at any level above) is dependant not only on the maximum developable flexural strength at the section, but on the location of the resultant of the earthquake-induced horizontal forces that are distributed along the height of the wall. The current seismic codes thus prescribe that the required shear strength, V_u, for a shearwall be obtained from an elastic analysis of the structure containing the wall under code-prescribed factored static lateral seismic forces. To compensate for the inadequacy of the approach, a reduced strength reduction factor is prescribed for the design of shearwalls against shear [1].

The aim of this research is to establish design values of dynamic shears (required shear strengths) at the bases of concrete shearwalls in buildings subject to seismic ground motion. The objectives are accomplished primarily through an extensive parametric study in the course of which a large number of dynamic inelastic response history analyses were carried out on isolated shearwalls of varying heights, initial fundamental periods, and flexural strengths, under earthquake ground motions of varying intensities and frequency contents. The large volume of numerical results was systematically processed and synthesized, leading to conclusions with potentially valuable design implications.

ANALYTICAL INVESTIGATION

In multi-degree-of-freedom systems subject to variable reversing loads (e.g., seismic excitation), the relationship between moment and shear at a given section varies due to the effect of the different modes of vibration. Figure 1a, which is a composite representation of the variation with time of the moment and shear at the base of an isolated shearwall subject to earthquake input motion, demonstrates clearly that the moment and shear are not in phase. The M/V ratio varies continuously during the motion, as indicated in Fig. 1b. The various moment-shear combinations to which the critical section at the base is subjected during the motion may be plotted as in Fig. 1c. The envelope of these plotted points contains the critical moment-shear combinations that must be considered in design. It can be reasoned that for design purposes the critical portion of the envelope is taken into account if the following moment-shear combinations are considered:

a. The maximum shear V_{max}, and the corresponding moment M_V (point A, Fig. 1c).

b. The maximum moment M_{max}, and the corresponding shear V_M (point B, Fig. 1c).

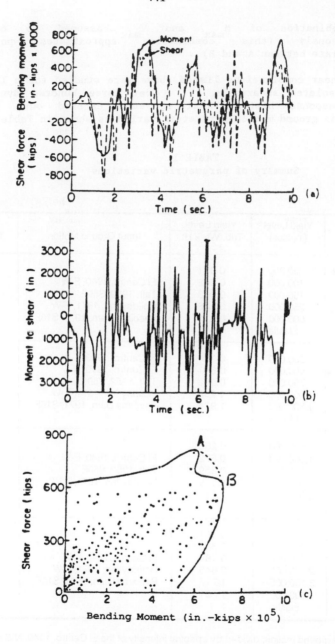

Figure 1.   Results of inelastic dynamic analysis of an isolated
shearwall: (a) variations of bending moment and shear
force with time, (b) variations of moment-to-shear ratio
with time, (c) bending moment vs. shear force plot.

c. The combination of M_{max} and V_{max} assumed to occur simultaneously (this combination approximates points intermediate between A and B).

The moment-shear combinations listed above were studied for a large number of isolated shearwalls [2]. Over three hundred dynamic inelastic response history analyses were carried out, with structural and ground motion parameters varied as shown in Table 1.

TABLE 1
Summary of parametric variations

No. of Stories	Period (sec.)	Yield Level (in.-kips)	Yield Level Trib. Wt. × Ht.	Input Ground Motion	Intensity*
10	0.5 0.8 1.4 2.0	50,000 100,000 150,000 250,000 500,000 Elastic	0.021 0.042 0.063 0.106 0.211	El Centro, 1940, E-W Taft, 1952, S69E Holiday Inn, 1971, E-W Pacoima Dam, 1971, S16E	1.5
20	0.8 1.4 2.0 2.4	250,000 500,000 750,000 1,000,000 1,500,000 Elastic	0.321 0.641 0.962 1.283 1.924	El Centro, 1940, E-W El Centro, 1940, N-S Taft, 1952, S69E Holiday Inn, 1971, E-W Pacoima Dam, 1971, S16E	1.5 1.0 .75
30	1.4 2.0 2.4 3.0	500,000 1,000,000 1,500,000 2,000,000 3,000,000 Elastic	0.286 0.572 0.859 1.145 1.717	El Centro, 1940, E-W Taft, 1952, S69E Holiday Inn, 1971, E-W	1.5
40	1.4 2.0 2.4 3.0	525,000 1,050,000 1,575,000 2,100,000 3,150,000 4,200,000 Elastic	0.161 0.323 0.484 0.645 1.717 1.291	El Centro, 1940, E-W Holiday Inn, 1971, E-W Pacoima Dam, 1971, S16E	1.5

*Spectral intensity of ground motion, divided by spectral intensity of the El Centro, 1940, N-S motion.

ANALYTICAL RESULTS

For the determination of maximum base shear for design purposes, V_{max} values should preferably be related to the known M_y at the base. Thus, the maximum base shears are plotted against the corresponding yield moments at the base in Figs. 2a, b, c, and d for 10-, 20-, 30- and 40-story walls, respectively, subject to earthquake input motions normalized to 1.5 times the spectral intensity of the 1940 El Centro, N-S record. Spectral intensity is taken as the area under the 5%-damped relative velocity response spectrum corresponding to 10 seconds of ground motion, between periods of 0.1 sec. and 3.0 sec. The V_{max} vs. M_y plot for all walls analyzed under earthquakes normalized to 1.5 times the spectral intensity of the 1940 El Centro, N-S record is presented in Fig. 3a. Similar plots for 20-story walls subject to ground motions normalized to 1.0 and 0.75 times the spectral intensity of the same record are presented in Figs. 3b and c, respectively.

The slopes and intercepts along the shear axes of the plots presented in Figs. 2 and 3 are listed in Table 2. A zero intercept and a slope of 0.67H would correspond to fundamental mode behavior, with the base moment M_y (corresponding to the maximum base shear V_{max}) equal to the yield moment M at the base. Also listed in Table 2 are intercepts along the shear axes made by lines with a fixed slope of $M_y/V_{max} = 0.67H$ (dotted lines in Figs. 2, 3) that best fit the points representing the M_y, V_{max} combinations obtained from dynamic analyses. The intercepts from Table 2 corresponding to a 0.67H slope of the best-fit M_y-V_{max} line represent the higher mode effects, as well as the effects of strain-hardening along the post-yield branch of the moment-rotation diagram, which causes M_v values to be larger than the corresponding M_y values.

TABLE 2
Maximum base shear vs. yield moment at base from dynamic analysis

No. of Stories	Ground Motion Intensity	Slope $M_y/V_{max}H$	Intercept V_{max}/W	Intercept V_{max}/W for Slope = 0.67H	Reference Figure Number
10	1.5	0.809	0.142	0.109	2a
20	1.5	0.619	0.127	0.121	2b
30	1.5	0.518	0.106	0.143	2c
40	1.5	0.561	0.113	0.135	2d
All	1.5	0.710	0.137	0.127	3a
20	1.0	0.674	0.079	0.077	3b
20	0.75	0.698	0.065	0.060	3c

*Spectral intensity of ground motion, divided by spectral intensity of the El Centro, 1940, N-S motion.

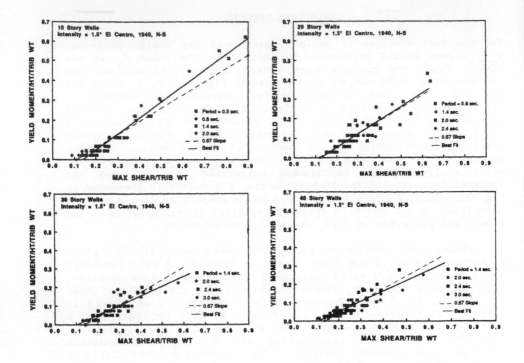

Figure 2. Maximum base shear vs. yield moment at bases of isolated
shearwalls; ground motion intensity normalized to 1.5
times the spectral intensity of El Centro, 1940, N-S
motion.

SYNTHESIS OF ANALYTICAL RESULTS

From Table 2 and Fig. 3, the maximum base shears in walls subject to
earthquakes of spectral intensities equal to 1.5, 1.0 and 0.75 times
that of the 1940 El Centro, N-S record are given by

$$V_{max} = 0.127W + M_y/0.67H \qquad (1a)$$

$$V_{max} = 0.077W + M_y/0.67H \qquad (1b)$$

and $$V_{max} = 0.060W + M_y/0.67H \qquad (1c)$$

respectively. If the spectral intensities of 1.5, 1.0 and 0.75
times that of the 1940 El Centro, N-S record are considered to be
approximately equivalent to peak ground accelerations of 0.5g, 0.33g
and 0.25g, respectively (the El Centro, 1940, N-S motion exhibited a
peak ground acceleration of 0.33g), Eqs. 1a through c may be
rewritten as

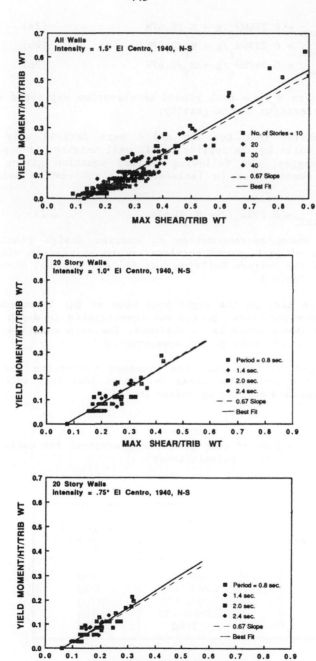

Figure 3.   Maximum base shear vs. yield moment at bases of isolated
shearwalls - varying ground motion intensities.

$$V_{max} = 0.254W\ddot{x}_g/g + M_y/0.67H \qquad (2a)$$
$$V_{max} = 0.233W\ddot{x}_g/g + M_y/0.67H \qquad (2b)$$
$$V_{max} = 0.240W\ddot{x}_g/g + M_y/0.67H \qquad (2c)$$

respectively, where $\dot{x}_g/g$ = peak ground acceleration expressed as a function of acceleration due to gravity.

If it is remembered that Eqs. 2b and c were derived only for 20-story walls, while Eq. 2a applies to all wall heights, it appears reasonable to suggest the following unified equation giving the maximum dynamic base shear in isolated walls subject to seismic excitation:

$$V_{max} = 0.25W\ddot{x}_g/g + M_y/0.67H \qquad (3)$$

To relate the above recommendation to current design practice, suggested values of peak ground acceleration to be used in various seismic zones of the Uniform Building Code [1] and the ANSI Standard [3] are given in Table 3.

Whether the first term on the right hand side of Eq. 3 depends on wall height and/or structural period was investigated in depth. No significant dependence could be established. The term was thus left dependent solely on the peak ground acceleration.

It is of interest to note that the proposed expression for the maximum value of dynamic base shear is almost identical with the expression recommended earlier by Aoyama in [4].

TABLE 3
Suggested design values of peak ground accelerations for various seismic zones

UBC/ANSI Seismic Zone	Range of Peak Ground Acceleration	Suggested Design Values
4	≥ 0.4g	0.4g
3	0.2g to 0.4g	0.3g
2	0.1g to 0.2g	0.15g
1	0.05g to 0.1g	0.075g
0	≤ 0.05g	0.025g

CONCLUSION

The significant conclusion emerging from this study is that the maximum shear force generated in an isolated shearwall subjected to earthquake-induced base motion is equal to the yield moment capacity of the base section, divided by two-thirds of the wall height (as would be expected on the basis of fundamental mode response) plus an additional shear force that is a function solely of the intensity of the ground motion (in terms of peak ground acceleration or some other measure).

It would have been of interest to record, with each maximum shear value, the corresponding state of deformation of the wall in terms of a ductility ratio or drift ratio. The absence of such deformation related information is a limitation of this study. Another arguable limitation is that the walls were assumed to behave elastically in shear. The effects of possible yielding of the walls in shear were not considered.

Work aimed at generalizing the above conclusion of this study for coupled-wall and frame-wall structures is currently in progress. The stated limitations will be overcome in this subsequent research.

ACKNOWLEDGEMENT

Partial financial support for this research, made available by the Reinforced Concrete Research Council, is gratefully acknowledged.

REFERENCES

1. International Conference of Building Officials, Uniform Building Code, Whittier, Calif., 1988.

2. Markevicius, V. P., and Ghosh, S. K., "Required Shear Strength of Earthquake Resistant Shearwalls," Research Report 87-1, Department of C.E.M.M., University of Illinois at Chicago, Chicago Ill., 1987.

3. American Society of Civil Engineers, Minimum Design Loads for Buildings and Other Structures (ANSI A58.1-1988, ASCE 7-88), New York, N.Y., 1989.

4. Aoyama, H., "Earthquake Resistant Design of Reinforced Concrete Frame Buildings with 'Flexural' Walls," Preprints, Second U.S.-Japan Workshop on the Improvement of Building Seismic Design and Construction Practices, San Francisco, Calif., 1986, pp. 5-1 - 5-29.

CONSTRUCTION, DESIGN AND RESEARCH
OF RC SHEAR WALL STRUCTURES IN CHINA

EHUA   FANG
Dept. of Civil Engineering
Tsinghua University, Beijing, China

ABSTRACT

High-rise RC shear wall buildings have been constructed widely in China in recent years. For the economy and functional requirement, different structural systems were evolved with a great amount of research work. This paper describes four shear wall structural systems, and tells briefly about the main research projects involving them and their primary design criteria.

HIGH-RISE RC SHEAR WALL STRUCTURES IN CHINA

In the 70's, a broad-scale construction of RC high-rise shear wall structures was started in China. Among them, more than 90% are made of cast-in-place concrete by using large panel forms or slip forms. Only those structures will be introduced in this paper.

The first 10-story shear wall apartment building was completed in Beijing in 1975 as a trial. After that, thousands of similar and taller buildings were built up. In the 80's, many other big and medium-size cities had such RC high-rise buildings constructed, most of which are apartments and hotels, and some are multi-use buildings. Having many merits, this kind of RC shear wall structure developed rapidly. It was found that:

1. By using large panel forms or slip forms, the load-bearing shear walls can be constructed speedily with less labour and fewer machines. Flat floor slabs directly supported by transverse walls can use precast or cast-in-place concrete. A variety of precast lightweight concrete panels can be used for exterior walls. Construction is easier and the construction time is reduced.

2. Because of no beams and columns projecting out, the smooth walls and ceilings are well accepted by the occupants and architects. Shear wall structures are their first choice for residential buildings.

3. Previous earthquakes demonstrated that most shear wall structures have less damage in both structural and nonstructural elements due to their great stiffness and high strength.

However, the defects and limitations of utilization in such structures are also obvious. Undoubtedly, the space is limited by orthogonally arranged walls and, therefore, inflexible.

With economic development and improvement of the living standard, multi-use and flexible space in an apartment building became necessary. For example, sometimes stores and offices are needed in the lower stories of an apartment building, and so partition arrangement should be various and flexible to satisfy the needs of different occupants. Thus, some improved shear wall structural systems were developed in China during the 80's. Since the earthquake resistance of structures is an essential requirement in the vast area of China, many research projects for tall shear walls have been done and some others are still in progress.

STRUCTURE SYSTEMS

Four systems which were evolved one by one will be described briefly below.

Small Bay Shear Wall Structure With Precast Floor Slabs
A typical configuration is shown in Fig. 1. Usually, precast prestressed hollow slabs in 3-4ᵐ span and about 1ᵐ width are used and supported by transverse walls. The precast slabs are rather economic in Beijing, China, but are not so good to connect integrately with the thin walls. The connection shown in Fig. 1 was tested under lateral deformation. The detailing and its design criterion, as well as corresponding construction requirement, were recommended in specification. The qualification of construction is critical. Such slabs and their connection are limited in use, only allowed for buildings lower than 50ᵐ in height.

The concrete interior longitudinal and transverse walls are always cast in place, while the exterior walls can be cast in place or precast panels can be used.

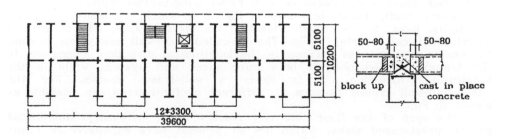

Figure 1.   The first system of shear wall structure

This kind of system (system 1) was the earliest type, which did not utilize fully the huge capacity of shear walls and was heavy in self-weight.

Medium Size Bay Shear Wall Structures with Flexible Space in the Lower Stories.
The typical configuration is shown in Fig. 2, in which a number of shear walls must go down to the ground (called G-wall hereinafter), while some walls are supported by columns (called C-wall hereinafter) in the first one or several stories (system 2).

The size of bays is enlarged to 5-6", so the cast-in-place concrete or composite flat slabs (for example, install prestressed thin slabs to the right position first, then cast the concrete on it in place to constitute a composite slab) have to be employed.

This kind of structure is of great advantage to the apartment buildings located by the roadside for setting up stores, or hotels and other buildings for commercial uses.

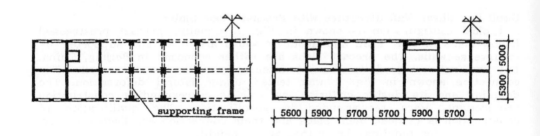

(a) plane of first floor   (b) plane of upper floors

Figure 2. The second system of shear wall structure

Large Bay Shear Wall Structures with Fewer Longitudinal Walls.
The typical configuration is shown in Fig. 3, in which the shafts composed by walls have to be used. Besides the shafts area, there are no longitudinal load-bearing walls. The so-called slab-wall subsystem beyond the shafts area acts as a frame in the longitudinal direction. Therefore, in the transverse direction, it is a pure shear wall system, while in the longitudinal direction, it is likely a wall-frame system. For the shaft-slab-wall system (system 3), different configurations are possible as shown in Fig. 3.

The span of the floor slab may reach 6-8". If necessary, unbounded partial prestressed slabs, which are at present more expensive in China than the cast-in-place RC slabs, can be used.

Architects and occupants can take advantage of such structure system, because the partitions can be arranged flexibly, and the rate of useful area increases.

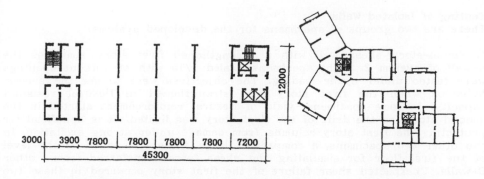

Figure 3.   The third system of shear wall structure

Shaft-Slab-Wall System with Partial Walls Supported by Beam-Column Frames (System 4)
The typical configuration is shown in Fig. 4. This system sums up the advantages of system 2 and 3. Because there are no longitudinal bearing walls except in the shafts, and due to the supporting beam-column frames, the flexible space exists in the lower stories. The multi-use of a shear wall structure building can be realized by using system 4.

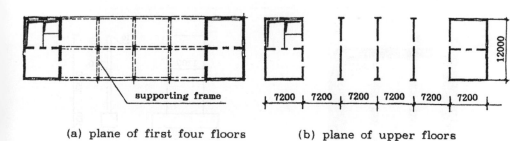

(a) plane of first four floors   (b) plane of upper floors

Figure 4.   The fourth system of shear wall structure

RESEARCH WORK

In order to investigate the seismic behavior and recommend design criteria for those developed systems (2, 3 and 4), a large number of research projects have been conducted since 1978. Research work was done cooperatively by engineers in construction companies and design institutes, as well as researchers in the China Academy of Building Research and Tsinghua University. The major projects and some useful results are described briefly in the following. Unless the name of the research institute is indicated, the projects were undertaken by the Dept. of Civil Engineering of Tsinghua University and were the responsibility of the author.

Testing of Isolated Walls.
There are two groups of specimens for the developed systems.

1. Seven-story shear wall with a strengthened first story simulates the G-wall of system 2. Eight specimens scaled 1/10 with or without openings were tested under cyclic loading. The objectives were to see the seismic behavior when the first story was strengthened in flexural resistant capacity. In some specimens, yield of flexural reinforcement started in the second story, so the drift of the first story was limited. It is beneficial for protecting the first story columns from damage under strong excitation. In the other two specimens, a comparable large force was added at the level of the first floor for simulating the shear force transferred from other C-walls. Unexpected shear failure of the first story occurred in these two specimens, because their shear-to-span ratio became 1.39 in this case. Fig. 5 is one of these specimens after failure.

2. Three cantilever walls were tested to investigate the effect of cross section and reinforcement type on the failure mode and seismic behavior. Two specimens have rectangular cross section: One of them (sw-1) has column-like reinforcement (flexural bars with closely spaced stirrups), another one (sw-2) has I shaped steels at both ends of it. The third wall (sw-3) has flanges, which are also reinforced with flexural bars and stirrups (see Fig. 6). All of them failed due to flexural-shear after the yield of flexural reinforcement. However, the SW-1 could not prevent the out-of-plane failure, and lost the strength suddenly with less plastic deformation.

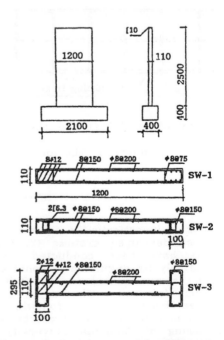

Figure 5. One example of G-wall specimen after testing

Figure 6. Specimens with different cross section shapes

Testing of Supporting Frame

1. Two specimens of 1/4 scaled shear wall with one-story supporting columns were tested at the Structural Laboratory of Research Division of the Beijing Design Institute in 1980. Fig. 7 indicates the loading pattern of the specimen. Test results show that the existence of G-wall (represented by a force acted horizontally in the opposite direction during tests) can effectively delay the damage of those columns.

2. Two specimens of 1/6 scaled 4-story supporting frame were tested in Tsinghua University recently to find the contribution of girders to such frames while strong column-weak beam criterion was still followed. Since the above wall was just like a strong girder, the plastic hinges and final damage occurred at the bottom of first story columns and the top of fourth story columns. It made the multi-story supporting frame similar to the single story supporting column. Fig. 8 shows the view of the four-story specimen.

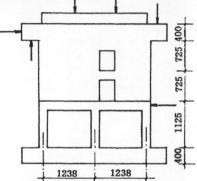

Figure 7. Test view of shear wall with supporting columns

Figure 8. Specimen of 4-story supporting frame

Shaking Table Tests of Structures

Three scaled structures were excited on the shaking table by inputting the earthquake waves which were also scaled. The structures are

1. Nine-story wall structure (1/24) with some walls supported by columns in the first floor (system 2). Two shafts located at each end of the structure. A running through crack occurred along the second floor and finally destroyed the structure (all the first story walls were strengthened).

2. Eleven-story wall structure (1/24) with some walls supported by columns in the first floor (also system 2). The difference between this one and the first structure was the absence of any flange at the ends of all transverse walls, including the end walls. So, except for the first floor, no shaft was formed, i.e., only one interior wall in the longitudinal direction made the upper structure just like a fish-bone. This structure was shaken on a large shaking table (see Fig. 9). After exciting by El Centro and white noise waves, cracks distributed mainly in the longitudinal wall and two end walls, the G-walls damaged in the first story by flexural-shear. In addition, the torsion effect increased as more crackings appeared, although the structure had a symmetric layout. It seems that, for torsion resistance, to have shafts in total height in such a system is necessary.

3. A 16-story scaled 1/18 structure was shaken again to study the behavior of system 4. There are 4-story supporting frames between two shafts. Fig. 10 shows the structure before setting on the shaking table. The x and y directions were shaken in turn. Obviously, the stiffness in the longitudinal direction of such a structure was lower than that in the transverse direction. The structural performance in the longitudinal direction was satisfactory. On the two end shafts, particularly on the two end walls, cracks distributed evenly along the height. During the exciting of 0.547g (peak acceleration) in the transverse direction, three transverse walls among the four between two shafts developed some cracks or concrete crush at the edge of 5th floor walls. This is because of the rather thin rectangular cross section of those walls. When going up to 0.834g (p.a.) input in the longitudinal direction, the connection between girders and shafts and at the bottom of shafts developed serious cracking. However, no heavy cracks could be found in the supporting frames.

Figure 9. Model structure of system 2   Figure 10. Model structure
             on shaking table   of system 4

Pseudo Dynamic Tests for 1/6 Scaled Structures
Three 12-story structures scaled 1/6 were tested by using the pseudo dynamic test method. All the tests were conducted in the Large Scale Testing Laboratory of the China Academy of Building Research. All three structures belong to system 2. Another 13-story structure of system 4 scaled 1/8 is in preparation of testing now. The main objectives of the pseudo dynamic tests are to check the design criteria and its reliability under earthquake conditions. Fig. 11 shows one of the structures under testing. Satisfactory results were obtained.

Elasto-Plastic Analysis
By using programs ULARC and DRAIN-2D, the elasto-plastic static and dynamic (time history) analyses have been employed. Force redistribution is obvious and it will increase the forces in slab-wall frame of system 3, also in supporting beam-column frame of system 4. The coefficients for design were recommended from the analyses.

Figure 11. Pseudo dynamic test of system 2

PRIMARY DESIGN CRITERIA AND RECOMMENDATIONS

The following criteria and recommendations are currently in effect in China. Some of them have been involved in the Design Specification for High-rise RC Structures. Here only the primary points are introduced.

1. It is necessary to set up shafts, which are going straight down to the ground from the top in system 2, 3 and 4. They act as the principle lateral resistant components. In addition to shafts, other isolated walls, no matter whether they are supported by columns or not, should be flanged or reinforced with section steel at the ends of walls, mainly in the region of plastic hinges under earthquake conditions.

2. In order to decrease the drifts of the lower stories, where the supporting columns are setting up, the lower part of G-walls should not only be thickened to increase stiffness, but also be strengthened.

It is required that, firstly, the ratio of shear resistant stiffness of the whole story between the upper and lower stories be limited. That is

$$r = \frac{G_{i+1} A_{i+1}}{G_i A_i} \cdot \frac{h_i}{h_{i+1}} \le 2 \tag{1}$$

where i and i+1 represent the story number beneath and above the transferring floor, respectively. G, A and h are shear modulus, total area of shear walls and story height in the corresponding story, respectively. Secondly, the plastic deformation in the lower part of the structure should be limited by adding more reinforcement in the G-walls.

3. In any one direction, the number of G-walls should not be less than 50% of the total walls in this direction. Meanwhile, it is better to keep the shear-to-span ratio of the G-walls higher than 1.5. Otherwise, design them as a squat wall in detailing (limit to the zone where $M/V l_w < 1.5$).

4. Considering the force redistribution which developed after plastic deformation occurs in G-walls, to increase the design force of frames is necessary. For system 3 in longitudinal direction, the shear force and corresponding bending moment resisted by walls beyond shafts should be multiplied by coefficient 2.5. For system 2 and 4, the shear forces (corresponding bending moments as well) resisted by supporting columns should not be less than 20% of the total story shear force, or not be less than 2% of that per column when the number of columns is less than ten.

At the same time, the design forces of girders should also be amplified. Supporting frames should be designed as ductile moment resistant frames. Special notice should be given to the bottom and top (beneath the wall) of the columns.

5. If the unbounded curved prestressed bars are used in slab of system 3 or 4 in seismic zone, partial unprestressed bars are needed (at least 0.4% content) to improve the ductility of the slab.

On the other hand, because the slabs act as continuous girders in slab-wall frame, the positive and negative bending moment in the slab under lateral load must be resisted by double reinforcement.

CONCLUDING REMARKS

China is a developing country with a large population. We need numerous residential buildings, which should be both economic and satisfactory in use. Furthermore, construction must not only consider the needs in the short term, but also be rational in the long-term plan.

The various shear wall systems introduced in this paper are only part of all the kinds of structural systems. However, with merits in construction, economy, and satisfying the basic requirements of the occupants, they are the most common types of tall residential buildings in China today.

REFERENCES

1. "Design Recommendations for High-Rise Structures (Systems and Detailing)" Shanghai Science Technology Publishing House, 1985 (in Chinese).

2. Fang, Ehua and Bau, Shi-Hua "Structural Design for High-Rise Structures" second edition, Tsinghua Univ. Press, 1990 (in Chinese).

3. Fang, Ehua "Seismic Behavior of Shear Walls Strengthened in First Story" Journal of Building Structures, Feb. 1989 (in Chinese).

4. Fang, Ehua and Fu, P.J., Zhu, H.L. "Shaking Table Tests of the Shear Wall Structures with Large Space in the First Story" Proceedings of 9WCEE, Vol.4, 1989.

5. Fang, Ehua "Failure Modes of RC Tall Shear Walls" in This Proceedings.

ASEISMIC DESIGN METHOD FOR A BUILDING WITH ECCENTRICITY

TOSHIKAZU TAKEDA   TETSUO SUZUKI   HIROAKI ETO
Technical Research Institute, Obayashi Corporation
4-640, Shimokiyoto, Kiyose-shi, Tokyo 204, Japan

ABSTRACT

An aseismic design method for a building with eccentricity is proposed in this report. Firstly, stresses of the individual members are determined considering torsional deformation of the building as a whole through static elastic three-dimensional frame analysis, and the reinforcement design is decided based on these stresses. Next, ultimate strength is determined through static inelastic two-dimensional frame analysis. The reinforcement design is adjusted upon repeatedly performing inelastic analyses to obtain a structure of beam-yielding type and having ductility along with satisfying the necessary ultimate lateral shear. Lastly, dynamic inelastic response analysis at the time of a severe earthquake is performed to verify the safety of the building.

INTRODUCTION

Buildings with eccentricity are liable to suffer local damage during an earthquake, which will hasten overall failure[1]. In the Japanese aseismic code[2], therefore, it is made a requirement for the necessary ultimate lateral shears of such buildings to be increased compared with a normal structure, especially, in case walls are provided eccentrically.

However, there are many points which remain unclear, such as the method of increasing the strength of each individual frame, and the method of evaluating ultimate lateral shear when considering torsional deformation.

In this report, taking a 6-story, 2*3-span building having wall eccentrically provided as an example, a proposal is made concerning the method of aseismic design for a building with eccentricity, along with which the behavior of the designed building in a severe earthquake is quantitatively evaluated to discuss the appropriateness of the design methodology.

ASEISMIC CODE IN JAPAN

A new aseismic design method for buildings has been in force in Japan since June 1, 1981[2]. The outline of this method is given below.

Lateral Seismic Shear

The lateral seismic shear Q_i of the i-th story above ground level shall be determined in accordance with the following formulas:

$$Q_i = C_i * W_i \quad \cdots\cdots\cdots\cdots\cdots\cdots\cdots\cdots\cdots\cdots (1)$$
$$C_i = z * R_t * A_i * C_0 \quad \cdots\cdots\cdots\cdots\cdots\cdots\cdots\cdots (2)$$
$$A_i = 1 + (1 / \sqrt{\alpha_i} - \alpha_i) \{ 2 * T / (1 + 3 * T) \} \cdots\cdots\cdots (3)$$
$$\alpha_i = W_i / W_n \quad \cdots\cdots\cdots\cdots\cdots\cdots\cdots\cdots\cdots\cdots (4)$$
$$T = 0.02 * h \text{ (sec)} \quad \cdots\cdots\cdots\cdots\cdots\cdots\cdots\cdots\cdots (5)$$

where, C_i : lateral seismic shear coefficient of i-th story

W_i/W_n : weights above i-th story and ground level, respectively

z : seismic hazard zoning coefficient (0.7~1.0)

R_t : design spectral coefficient, determined by type of soil profile and T

A_i : lateral shear distribution factor, determined by T and α_i

C_0 : standard shear coefficient, not less than 0.2 and 1.0 for moderate and severe earthquake motions, respectively

T : fundamemtal natural period

h : total height of building in meters

Design Procedure

Aseismic design of a building with large eccentricity shall be based on root 3, with an obligation to confirm the ultimate lateral shear Q_u, if based on the Japanese code. The shear Q_u of each story shall not be less than the necessary ultimate lateral shear Q_{un} determined in accordance with Eq. (6).

$$Q_{un} = D_s * F_e * F_s * Q_{ud} \quad \cdots\cdots\cdots\cdots\cdots\cdots\cdots\cdots (6)$$

where, Q_{ud} : lateral seismic shear for severe earthquake motion ($C_0 = 1.0$) determined in Section 1.1

D_s : structural coefficient given by Table 1

F_e : shape factor by eccentricity of stiffness R_e and given in Table 2.1, where R_e is determined by Eq. (7)

F_s : shape factor by variation of lateral stiffness R_s and given in Table 2.2, where R_s is determined by Eq. (8)

$$R_e = e / r_e \quad \cdots\cdots\cdots\cdots\cdots\cdots\cdots\cdots\cdots\cdots (7)$$

where, e : eccentricity of center of stiffness from center of mass

r_e : elastic radius defined as square root of torsional stiffness divided by lateral stiffness

$$R_s = r / \overline{r} \quad \cdots\cdots\cdots\cdots\cdots\cdots\cdots\cdots\cdots\cdots (8)$$

where, r : lateral stiffness, defined as value of story height divided by story drift caused by lateral seismic shear for moderate earthquake motions ($C_0 = 0.2$)

$\overline{r}$: mean lateral stiffness, defined as arithmetic mean of r's above ground level

TABLE 1 Structural Coefficient D_s for Building of Reinforced Concrete

Behavior of Members	Type of Frame		
	(1)Ductile moment frame	(2)Frame other than listed in (1)&(3)	(3)Frame with shear walls or braces
(A)Excellent Ductility	0.30	0.35	0.40
(B)Good Ductility	0.35	0.40	0.45
(C)Fair Ductility	0.40	0.45	0.50
(D)Poor Ductility	0.45	0.50	0.55

TABLE 2-1 Shape Factor F_e by Eccentricity of Stiffness R_e

Re	Fe
less than 0.15	1.0
$0.15 \leq Re \leq 0.3$	linear interpolation
more than 0.3	1.5

TABLE 2-2 Shape Factor F_s by Variation of Lateral Stiffness R_s

Rs	Fs
more than 0.6	1.0
$0.3 \leq Rs \leq 0.6$	linear interpolation
less than 0.3	1.5

BUILDING DESIGNED AND DESIGN PRINCIPLE

The plan and sections of the designed building are shown in Fig. 1. This is an office building of reinforced concrete construction having 6 stories above ground level, with moment-resisting-frame construction of 2 spans in the y direction, and moment-resisting-frame and walled-frame construction (having a shear wall at frame Ⓒ, row ②-③) of 3 spans in the x direction. Consequently, this is a building with eccentricity in the x direction.
The design procedure is shown in Fig. 2. Firstly, as a design for moderate earthquake, allowable stress design is made for $C_o = 0.2$. Next, as a design for severe earthquake, ultimate strength design is made for $C_o = 1.0$.
Ordinary designing would end here, but inelastic three-dimensional frame analysis is performed to examine the response behaviors during a severe earthquake to verify the appropriateness of the design methodology.
The features of the individual steps are cited below.

Design for Moderate Earthquake (Allowable Stress Design for $C_o = 0.2$)
1) Stress analysis under vertical load is performed by the moment distribution method.
2) Stress analysis under lateral load is performed by elastic pseudo-three-dimensional frame analysis with external force of A_i, and stress considering torsional deformation due to eccentricity is determined.
3) Stiffness reduction of shear wall and redistribution of stresses are not done.

Design for Severe Earthquake (Ultimate Strength Design for $C_o = 1.0$)
1) The construction is to be for total yielding of beam-yielding type.
   That is, hinges are provided at beam ends at each story, and column bases and wall base of the first story, for a construction where no member shows shear failure and ductility is excellent.
2) The necessary ultimate lateral shear Q_{un} is to be $0.45 \cdot Q_{ud}$ in the x direction and $0.30 \cdot Q_{ud}$ in the y direction. That is, D_s and F_s are to be 0.3 and 1.0, respectively, in both directions, with F_e to be 1.5 in the x direction and 1.0 in the y direction.
3) Calculation of ultimate lateral shear Q_u is done by inelastic two-dimensional frame analysis. In analyses, members are to be made models with rigid-inelastic rotational springs at member ends and restoring

force characteristic a bilinear model, which are simple. When the maximum story drift of each frame becomes 1/100 is taken to be the time of yielding mechanism, with the story shear force at this time the ultimate lateral shear. The design is to be reviewed in case Q_u is less than the abovementioned Q_{un}.

4) The yield strength of a member is based on the equation proposed by the Architectural Institute of Japan.

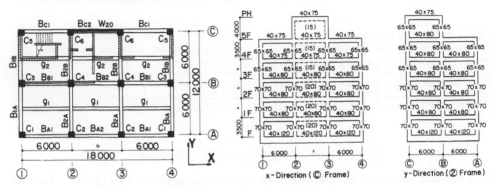

FIGURE 1 Designed Building

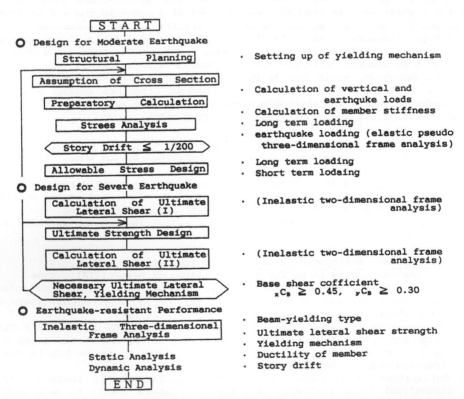

FIGURE 2 Design Procedure

Examination of Earthquake-resistant Performance

The ultimate lateral shear, yielding mechanism, ductility factors of stories and members are examined by inelastic three-dimensional frame analyses to confirm the appropriateness of this design methodology. The earthquake-resistant performance required is set as given below at this time.

1) Intensity of earthquake motion · · · · · maximum velocity 50 kine
2) Story drift · · · · · · not more than 1/100
3) Ductility factor of story · · · · · not more than 2
4) Ductility factor of member · · · · · not more than 4 for beam, not more than 2 for column and shear wall

DESIGN

Design for Moderate Earthquake

The materials used are given in Table 3 and the seismic forces for design in Table 4. Fig. 3 shows a part of the results of design obtained based on elastic pseudo-three-dimensional frame analysis. Only the end parts of the reinforcement design of beams are shown, but it may be seen that with regard to the x direction having eccentricity there is an increased amount of reinforcement in frame Ⓐ in which lateral deformation is large.

Table 5 gives story drift (δ/H), stiffness ratio (R_s), and eccentricity ratio (R_e). Although δ/H and R_s satisfy standards, regarding R_e, it becomes larger than 0.3 in the x direction for just about all stories, and it becomes necessary for the ultimate lateral shear to be increased in the design for severe earthquake.

TABLE 3 Material Properties

Concrete	Young's Modulus: $E_c = 2.1 \times 10^5 (kg/cm^2)$	
	Strength: $F_c = 210 (kg/cm^2)$	
Reinforcement	SD35 (D25mm)	Young's Modulus: $E_s = 2.1 \times 10^6 (kg/cm^2)$ Yielding Strength: $\sigma_y = 3500 (kg/cm^2)$
	SD30A (D10,13,16mm)	Young's Modulus: $E_s = 2.1 \times 10^6 (kg/cm^2)$ Yielding Strength: $\sigma_y = 3000 (kg/cm^2)$

SD35 : for main reinforcement of column and beam
SD30A : for wall reinforcement, stirrup and hoop

TABLE 4 Design Seismic Force

Story	High (cm)	Weight (t)	ΣW_i (t)	a_i	A_i	C_i	Q_i (t)
PH	401.2	61.5	61.5	0.04	2.79	0.56	34.3
5F	350.0	292.0	353.5	0.25	1.66	0.33	117.2
4F	351.3	257.2	610.6	0.43	1.41	0.28	172.1
3F	350.0	263.4	874.0	0.62	1.25	0.25	217.6
2F	350.0	267.9	1141.9	0.81	1.11	0.22	254.3
1F	370.0	267.9	1409.7	1.00	1.00	0.20	281.9
F	-	272.1	1681.8	-	-	-	-

$\Sigma H = 2170cm$, $\Sigma W = 1409.7t$, $T = 0.434sec.$, $Z = 1.00$, $R_t = 1.00$, $C_0 = 0.20$

TABLE 5   Stiffness and Eccentricity

Story	Story Drift		Shape factor of Lateral Stiffness				Shape factor of Eccentricity			
	X	Y	X		Y		X		Y	
	δ/H	δ/H	R_s	F_s	R_s	F_s	R_e	F_e	R_e	F_e
PH	1/1731	1/607	0.875	1.0	0.857	1.0	0.545	1.50	0.0	1.0
5F	1/1852	1/964	0.936	1.0	1.362	1.0	0.262	1.37	0.0	1.0
4F	1/1653	1/709	0.835	1.0	1.001	1.0	0.483	1.50	0.0	1.0
3F	1/1777	1/661	0.898	1.0	0.934	1.0	0.591	1.50	0.0	1.0
2F	1/1985	1/597	1.003	1.0	0.844	1.0	0.758	1.50	0.0	1.0
1F	1/2873	1/710	1.452	1.0	1.002	1.0	0.916	1.50	0.0	1.0

Design for Severe Eartquake

After completion of the design for moderate earthquake, the ultimate lateral shear(I) is sought by inelastic two-dimensional frame analysis. The ultimate lateral shear coefficients of the first story were 0.36 in the x direction and 0.34 in the y direction.

Compared with the necessary ultimae lateral shear, the value for the x direction is lower and that for the y direction higher. Therefore, ultimate strength designing is done based on the following principles.

1) Flexural Design of Beam
   Since the ultimate lateral shear is inadequate in the x direction, reinforcement design was altered so that the flexural strength of the beam woulud become (0.45/0.36) = 1.25 times the stress at the time of acting of the yielding mechanism. For the y direction, the design for moderate earthquake is left unchanged.

2) Flexural Design of Column
   For a column regarding which yielding is not set, reinforcement is designed so that flexural strength in the x direction of the cross section of design for moderate earthquake will be 1.88 times (= 1.25*1.5) and that in the y direction 1.5 times. For the first-story column base that will yield, the reinforcement design is made the same as for the first -story column head.

3) Flexural Design of Shear Wall
   Since the flexural strength $_wM_u$ of the wall is high ($_wM_u \geq 1.25*1.5_wM_m$), the reinforcement design for moderate earthquake is left unaltered, where the $_wM_m$ is the moment at yielding mechanism.

4) Shear Design
   Designing is done to secure the required shear strength.

The results of design are shown in Fig. 4 and Table 6. Fig. 4 shows the reinforcement design for only the first story, but compared with design for moderate earthquake, the feature is that the rate of increase in reinforcement arranged in frame Ⓐ where lateral deformation due to torsional deformation becomes large has been made higher.

Ultimate lateral shear(II) is determined after completion of ultimate strength design. The results are given in and Table 7. According to Table 7, the ultimate lateral shears in both directions of the individual stories exceed the necessary ultimate lateral shear set up in Section 2.2. It was ascertained that the yielding mechanism would be that of total yielding of beam-yielding type, with moreover, the wall yielding in flexure.

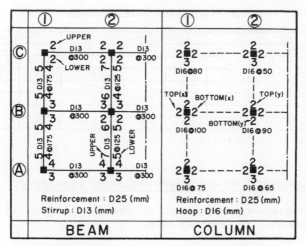

FIGURE 3 Design for Moderate Earthquake (Reinforcement Design of 1st Story)

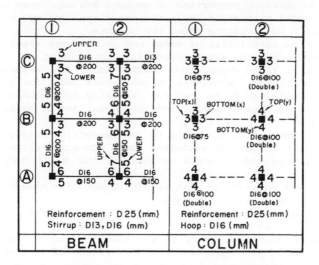

FIGURE 4 Design for Severe Earthquake (Reinforcement Design of 1st Story)

TABLE 6 Wall Reinforcement Design

Story	Thickness (cm)	Reinforcement Ratio (%)	Reinforcement
PH	15	0.40	D10 @ 225 double
5F	15	0.40	D10 @ 225 double
4F	15	0.40	D10 @ 225 double
3F	20	0.40	D10 @ 175 double
2F	20	0.41	D10 @ 150 double
1F	20	0.57	D10,13 @ 150 double

TABLE 7   Ultimate Lateral Shear

Story	Q_{ud} (t)	X Direction				$C_1 = \dfrac{Q_u}{Q_{ud}}$	Y Direction		
		Q_u(t)					Q_u(t)		
		A	B	C	Total		①, ④	②, ③	Total
5	583	61	62	140	263	0.451	38	53	182
4	856	102	91	205	398	0.465	63	78	282
3	1084	136	115	260	511	0.471	84	99	366
2	1267	163	134	304	602	0.475	101	116	434
1	1406	149	149	337	670	0.477	113	128	482

$$Q_{ud} = Z \cdot R_t \cdot A_i \cdot C_0 \cdot W \ , (\ C_0 = 1.0)$$

EXAMINATION OF EARTHQUAKE-RESISTANT PERFORMANCE

The earthquake-resistant performance in the x direction at the time of severe earthquakes will be examined here.

Inelastic Three-dimensional Frame Exact Analysis Method[3]
The various members are modeled as indicated below.
1) The flexural inelastic properties of members are expressed by a three-dimensional model with rigid-inelastic rotational springs at member ends for beams and a multi-spring model for columns.
2) The shearing characteristics of beams and columns are elastic.
3) Regarding the wall, inelastic properties for bending and shear are expressed using a three-dimensional member model for a perimeter column and an equivalent two-dimensional column member model for a wall panel.
4) The Takeda model[4] is to be used for restoring force characteristics of the rotational springs at member ends of beams, and of bending and shear of wall panels.
5) Multi-linear type models according to the respective material characteristics are employed for the restoring force characteristics of reinforcing steel springs and concrete springs of the multi-spring models of columns.
6) Deformation from pull-out of main reinforcement from beam-column joints is considered for beams and columns.
7) Beam-column joints are handled as rigid zones and the inelasticity of the member is judged by the location of the face.
   It is assumed that each story of the building has a rigid floor, and that the foundation is pin-supported in consideration of footing beams.

Static Analysis Results
In analysis, after initial analysis for long-term load, external forces of lateral force-distribution mode for design for moderate earthquake were incrementally applied to the centroids of the individual stories.
1) Story Shear Force Q_i and Story Deflection δ_i
   The relationship between Q_i and δ_i is shown in Fig. 5. It is when the story drift R_i is approximately 1/200 for the whole building that yielding as a story can be recognized with stiffness decreasing and shear force of each story becoming more or less level. When the maximum drift $R_{i,max}$ of frame B closest to the centroid is 1/100, the $R_{i,max}$ of frames A and C will be 1/78 and /150, respectively.
2) Ultimate Lateral Shear Q_u.
   When $R_{i,max}$ of frame B is 1/100 is taken to be the time of the yielding mechanism, and the story shear at that time is made the ultimate lateral shear. Ultimate lateral shear, deflection, and drift are all given in

Table 8.   In comparison of the ultimate shears with the results of inelastic two-dimensional frame analysis, the three-dimensional analysis result is approximately 18% larger for the entire building, and the three-
dimensional effect is recognizable.   By individual frame, frame Ⓐ has
roughly equal to two-dimensional analysis results, but with frame Ⓒ
which has a wall, the three-dimensional analysis provides a 37 to 47 %
higher value.   The ratio of shear force borne by a single shear wall in
the total x-direction shear force will be 52 to 58 %, which is sligthly
higher than in the design for moderate shear.

3) Yielding Mode
The hinge diagram and ductility factors at yielding mechanism are shown
in Fig. 6.   The maximum ductility factors μ of members are largest for
frame Ⓐ and 2.88 for beam and 1.68 for column, while μ for the frame
Ⓒ wall base is 1.79, and the earthquake-resistant performances set forth
originally are satisfied.

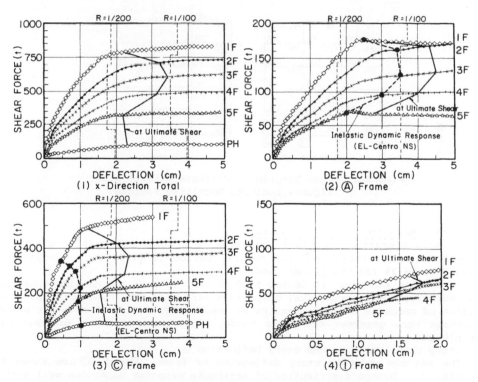

FIGURE 5   Story Shear~Story Deflection Relationship
            (Inelastic Three-dimensional Frame Analysis Result)

TABLE 8 Story Shear (Ultimate Lateral Shear) and Deflection

Story	Total (x-Direc.)			Ⓐ Frame			Ⓑ Frame			Ⓒ Frame			Ⓓ Frame (y-Direc.)		
	Q (t)	Disp (cm)	δ (cm)	Q (t)	Disp (cm)	δ (cm)	Q (t)	Disp (cm)	δ (cm)	Q (t)	Disp (cm)	δ (cm)	Q (t)	Disp (cm)	δ (cm)
5F	328 [1.25]	14.01	2.17 [161]	68 [1.11]	18.76	2.79 [125]	53 [0.85]	14.01	2.17 [161]	207 [1.48]	9.25	1.54 [227]	29	7.30	0.89 [393]
4F	481 [1.21]	11.84	3.11 [113]	99 [0.97]	15.97	4.01 [87]	102 [1.12]	11.84	3.11 [113]	281 [1.37]	7.71	2.20 [159]	41	6.41	1.35 [259]
3F	609 [1.17]	8.73	3.41 [103]	129 [0.95]	11.96	4.50 [78]	119 [1.03]	8.73	3.41 [103]	361 [1.39]	5.51	2.32 [151]	55	5.06	1.68 [208]
2F	711 [1.18]	5.32	3.17 [110]	168 [1.03]	7.46	4.32 [81]	126 [0.94]	5.32	3.17 [110]	417 [1.37]	3.19	2.03 [172]	61	3.38	1.80 [194]
1F	788 [1.18]	2.15	2.15 [172]	174 [0.95]	3.14	3.14 [118]	129 [0.87]	2.15	2.15 [172]	486 [1.44]	1.16	1.16 [319]	69	1.58	1.58 [234]

* Figures in [ ] at Q columns are ratios to ultimate lateral shear according to two-dimensional frame analysis in design for severe earthquake
** Values of frame B nearest to centroid are used as values in x direction displacements
*** Figures in [ ] at δ columns are inverse numbers of story drifts

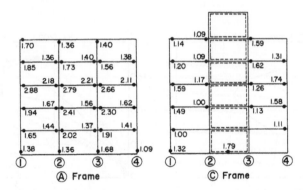

FIGURE 6 Hinge Diagram at Ultimate Lateral Shear
(Figures Indicate Ductility Factors)

Earthquake Response Analysis Results

The primary vibration mode in x direction is a torsional mode coupled with translation and rotation and the natural period is 0.354 sec. Damping is assumed to be 2% for the primary mode.

In dynamic analyses, EL-Centro 1940 NS,Taft 1952 EW and Hachinohe harbor 1968 NS(Tokachi-oki earthquake) components were used for input motions. The outline of response analysis results at input of EL-Centro NS wave for which responses were greatest are given below, where maximum input acceleration was 510.gal.

1) Story Shear Force Q_i and Story Deflection δ_i

The maximum response story deflection of frame Ⓐ and Ⓒ are shown in Fig. 5. Damage distribution of earthquke response coincides well with inelastic static analysis results, although the yielding mechanism is not yet formed and the torsional deformation is a little larger than the static result. The deformation of frame Ⓐ becomes larger due to torsional vibration, but the maximum story drift will be R ≤ 1/100 to satisfy the earthquake-resistant performance stipulated. Fig. 7 shows the response story shear and deflection characteristics of the first story of frame Ⓐ and Ⓒ during the first five seconds where maximum responses occured.

2) Hinge Diagram and Ductility Factor
   The hinge diagram and ductility factors result from maximum response
   values are shown in Fig. 8.   Compared with the static analysis results,
   the degree of damage of frame Ⓒ is small.   The maximum ductility
   factors μ of members of frame Ⓐ are 2.31 for beam and 1.42 for column,
   which satisfy the earthquake resistant performances.

According to the foregoing, the ultimate lateral shear of this building
exceeds the necessary ultimate lateral shear, along with which the yielding
mechanism is of beam-yielding type, while moreover, it has been confirmed
that even during a severe earthquake the degree of damage to frame Ⓐ will
be within the range that ductility can be expected.

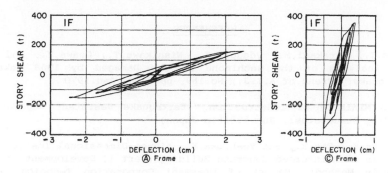

FIGURE 7   Story Shear〜Story Deflection Relationship at the First Story
            (Inelastic Three-dimensional Earthquake Response Analysis Result)

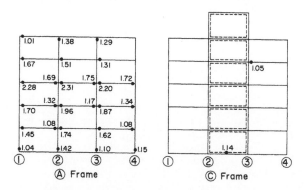

FIGURE 8   Hinge Diagram at Severe Earthquake Response (EL-Centro NS 50 kine)
            (Figures Indicate Ductility Factors)

CONCLUDING REMARKS

An aseismic design method for a building with eccentricty has been proposed and the appropriateness of the method has been confirmed. As the results, followings were concluded for structural design practice.

1) Reinforcement arrangement based on design forces considering torsional deformation of an entire structure is desirable.

2) The eccentric shape factor F_e specified in the Japanese aseismic code aiming less and even structural damage caused during the severe earthquake can be available in practice.

3) The comparison of static lateral loading capacity between two and three dimensional analyses provides larger value in three dimensional with considering orthogonal effect of a frame. Neverthless in the case of total yielding of beam-yielding type building, two dimensional analysis is sufficient to predict the capacity.

REFERENCES

[1] Okada,T., Murakami,M., Udagawa,K., Nishikawa,T., Osawa,Y., and Tanaka, H., :Analysis of the Hachinohe Library Damaged by 1968 Tokachi-oki Earthquake:, Trans.of A.I.J., No.167, pp47-58, Jan.1970

[2] Japan Ministry of Construction, :Earthquake Resistant Regulation for Building Structures:, Jun.1981(in Japan)

[3] Nagahara,K., Eto,H., and Yoshioka,K., :Three-Dimensional Inelastic Frame Analysis of Reinforced Concrete Building (Part 1) Development of Static Analysis Method:, Report of Obayashi Corporation Technical Research Institute, No.40, pp64-71, Feb.1990

[4] Takeda,T., M.A.Sozen and N.N.Nielsen, :Reinforced Concrete Response to Simulated Earthquakes:, Journal of the Structural Division, Proceeding of the A.S.C.E., vol.96, No.ST12, pp2557-2573, Dec.1970

CALCULATION OF DUCTILITY OF THE BRIDGE PIERS

TADAYOSHI ISHIBASHI
Master of The Construction Engeer Div.,
Tohoku Construction Office, East Japan Railway Company
1 - 1 - 1, Itsutsubashi, Aobaku,
Sendai City, Miyagi Pref., JPN

ABSTRACT

In order to find deformation capacity of roinforced concrete bridge piers under earthquake, cyclic horizontal loading tests were carried on the specimens of bridge piers for which span ratio, tensile reinforcement ratio, web reinforcement ratio and axial compressive stress have been modified, and the effect due to these factors against deformation capacity has been identified.

Furthermore, a study was made on the method to obtain ductilities for quantitative evaluation of deformation capacity of the reinforced concrete bridge piers, and a method capable of calculating thereof relatively accurately was proposed.

1. INTRODUCTION

In examination of safely of a reinforced concrete member during earthquake, the importance of increasing not only strength, but also the capability of the member to deform, or deformability, after yielding has come to be recognized. Civil structures such as railway bridge piers generally have specifications calling for tensile reinforcement ratio not more than 1 percent, shearing reinforcement ratio 0 to 0.4 percent, axial compressive stress intensity 5 to 20 kg/cm².

In order to experimentally study deformability of reinforced concrete railway bridge piers, horizontal alternating load experiments were conducted on rectangular bridge pier specimens having specifications in a range of shear span ratio of 1.5 to 4.0, tensile reinforcement ratio of 0.12 to 1.66 percent, and axial compressive stress intensity 0 to 40 kg/cm², and the ratio of shear strength and flexural strength of about 0.9 to 2.7, and the influences of these factors on deformability were investigated. Furthermore, a method of determining ductility (the quotient of maximum displacement when the envelope of the load-displacement curve is not less than yielding load (referred to as ultimate displacement) divided by displacement at time of yielding (referred to as yielding displacement)) of a reinforced concrete bridge pier from these experimental results was studied, and this will be reported since a method by which calculations can be made with comparatively good accuracy was found.

2. HORIZONTAL FORCE ALTERNATING LOAD EXPERIMENTS OF BRIDGE PIER SPECIMENS

(1) Specimen Configuration and Specifications

The specimen configuration is shown in Fig. 1. Examples of reinforcing bar arrangements are shown in Fig. 2. The mix proportions of the concrete used in making the specimens are given in Table 1. The reinforcing bars used were SD 35.

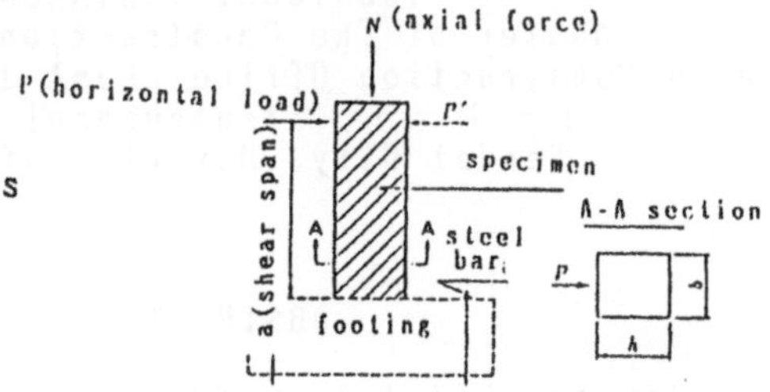

Fig.1 Specimen configuration

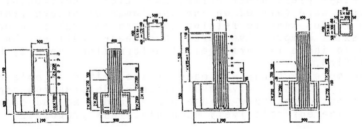

Fig.2 Examples of reinforcing bar arrangements

Axial reinforcing bars were extended lengths of 30 times bar diameter or more into footings from the footing tops, with the ends anchored by providing semicircular hooks. Shear reinforcing bars were made to surround axial reinforcement with ends anchored by hooking on to the axial reinforcement.

Table 1   Mix proportion of concrete

series	design strength kgf/cm²	kind of cement	biggest size of coarse aggregate (mm)	w/c (%)	cement	water	fine aggregate	coarse aggregate	admixture
I (IV)	270	normal Portland cement	25	49 (50)	367 (790)	170 (173)	755 (871)	1 011 (1 054)	0 868 (0.827)
II (III)	240	rapid hardening Portland cement	20	57.5	303 (709)	174 (164)	816 (818)	1 000 (1 031)	0 758 (0.733)
V	270	rapid hardening Portland cement	25	49	294	145	681	1 239	1.96

(2) Loading Method

The footing part of the specimen was fixed using prestressing steel bars, and static horizontal force alternate loading was done under a given axial force with the vicinity of the column head as the loading point. Application of horizontal load was done controlling load until the strain of tensile reinforcement at the cross section (bottom end of bridge pier) where maximum bending moment is produced had reached yielding strain, defining horizontal displacement of the loading point at the time of yielding as yielding displacement (hereinafter referred to as " δ_y "), and after yielding, alternating load was applied with δ_y as pulsating width, following which, alternating loads were applied controlling displacements with displacements of integral number of times δ_y as pulsating widths. The number of cycles at each displacement level was made 10 cycles as a rule, but the next level was moved on to when the load-displacement curve after a minimum of 3 cycles could be considered to be more or less identical to the preceding cycle.

(3) Yielding Displacement and Ultimate Displacement Obtained from Experiments
Ultimate displacement was considered as the maximum displacement when strength was not less than yielding load at the envelope of the load-displacement curve (Fig. 2). The valeus of yielding displacement (δ_y) and ultimate displacement (δ_u) obtained by experiments are given in Table 2.

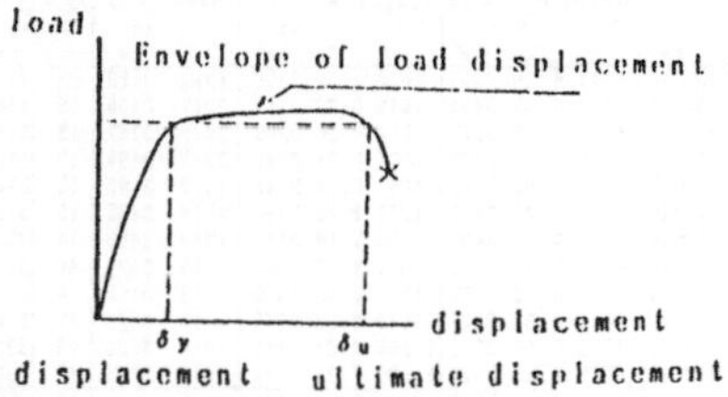

Fig.3 Envelope of load-displacement curve

3. METHODS OF CALCULATING DISPLACEMENT AND DUCTILITY OF REINFORCED CONCRETE BRIDGE PIER

The displacement of a specimen obtained by experiments is the aggregate of rotational displacement of the bridge pier from pull-out of axial reinforcement from the footing and the elasto-plastic displacement of the main part of the bridge pier. In case of a specimen which is of small cross section, the influence of pull-out of axial reinforcement is great, while with an actual structure of large cross section, this influence becomes little. To apply the ductility evaluation obtained from experiments with specimens to actual structures, it is necessary for evaluations to be made separating displacement. The methods of calculating the various displacements and method of calculating ductility of a reinforced concrete bridge pier are described below.

3.1 Rotational Displacement due to Pull-out of Axial Reinforcement

a) Amount of Axial Reinforcement Pull-out
When tensile force acts on axial reinforcement due to horizontal load application, a phenomenon of axail reinforcement being pulled out from the footing occurs. When axial reinforcement has been pulled out, rigid body rotation of the bridge pier in accordance with the amount of pull-out occurs,

and displacement of the free end is added to this.

The amount of pull-out of axial reinforcement from the foot
ing may be obtained by the procedure below.

The amount of pull-out at the top of the footing is
hypothesized, and the bond force from the top of the footing
toward the interior, the reduction in stress intensity, and the
reinforcing bar strain are calculated. The sums of the amounts
of strain are deducted from the amount of pull-out previously
hypothesized, and since there is a hook at the end of the bar,
the hypothesis of the amount of pull-out at the top is
repeatedly modified until slippage at the end becomes zero, and
by this, it is possible to obtain the amount of pull-out
approximately.

Table 2　Specimen specifications and experimental results

series	specimen number	section a/d	section b×d cm	tensile reinforcement number P	tensile reinforcement σsy kg/cm²	hoop reinforcing bar number P'w	hoop reinforcing bar σsy kg/cm²	side reinforcing bar number σsy	side reinforcing bar kg/cm²	axial stress σo kg/cm²	axial stress σc kg/cm²	Pyl ton	Pul ton	Py2 ton	Pu2 ton	Pu2/Pul	δy cm	δu cm	δyl cm	δul cm	δyoc cm	δuo cm	μo cm
I	1	4.0	40×40	4D13 0.36	3883	1D10 0.36	3366	2D13	3883	15	330	8.3	10.2	10.8	13.2	1.29	0.69	5.52	0.207	0.273	0.389	5.25	13.50
	2	4.0	40×40	4D19 0.82	3490	1D10 0.36	3366	2D19	3490	15	334	13.2	16.8	18.3	20.7	1.23	1.10	4.40	0.312	0.389	0.483	4.01	8.30
	3	4.0	40×40	4D25 1.45	3227	1D10 0.36	3366	2D25	3227	15	334	19.2	24.6	24.2	28.2	1.15	1.30	5.85	0.376	0.447	0.518	5.40	10.42
	4	4.0	40×40	4D19 0.82	3490	1.5D6 0.24	3740	2D19	3490	15	331	13.1	16.8	18.6	23.8	1.42	1.10	5.50	0.312	0.392	0.498	5.11	10.26
	5	4.0	40×40	4D19 0.82	3490	1D6 0.16	3740	2D19	3490	15	334	13.2	16.8	18.5	21.2	1.26	1.10	4.40	0.312	0.389	0.519	4.01	7.73
	6	4.0	40×40	4D19 0.82	3490	1D10 0.29	3366	2D19	3490	15	331	13.1	16.8	17.6	21.8	1.26	1.05	5.25	0.312	0.392	0.494	4.86	9.84
	7	4.0	40×40	4D19 0.82	3490	1.5D6 0.18	3740	2D19	3490	10	331	12.5	16.0	16.6	23.4	1.46	1.04	5.20	0.306	0.384	0.496	4.82	9.72
	8	4.0	40×40	4D19 0.82	3490	1D10 0.36	3366	2D19	3490	40	331	16.7	20.6	21.0	25.7	1.25	1.05	5.25	0.349	0.438	0.545	4.81	8.83
	9	4.0	40×40	4D19 0.82	3490	1D10 0.36	3366	2D19	3490	0	330	10.9	14.4	12.3	14.8	1.03	1.00	6.00	0.291	0.365	0.411	5.64	13.72
	10	4.0	40×40	4D16 0.57	3525	1D10 0.36	3366	2D16	3525	15	334	10.3	13.0	13.8	16.7	1.28	0.90	4.50	0.251	0.324	0.428	4.18	9.77
	11	4.0	40×40	4D16 0.57	3525	1.5D6 0.24	3740	2D16	3525	15	330	10.3	13.0	13.0	16.0	1.23	0.91	5.40	0.252	0.329	0.455	5.07	11.14
	12	4.0	40×40	4D16 0.57	3525	1.5D6 0.24	3740	2D16	3525	15	330	10.3	13.0	15.4	18.4	1.42	0.90	5.40	0.252	0.329	0.480	5.07	10.56
II	1	4.0	60×40	9D13 0.54	3676	2D6 0.11	3800	0		10	298	12.2	13.4	14.0	14.5	1.09	0.50	2.00	0.221	0.287	0.269	—	—
	2	4.0	60×40	9D13 0.54	3670	0		0		10	298	12.2	13.4	14.1	15.1	1.13	0.55	2.20	0.221	0.287	0.300	—	—
	3	3.0	60×40	6D13 0.36	3670	2D6 0.11	3800	0		10	300	12.2	13.6	16.2	17.8	1.31	0.37	1.48	0.157	0.230	0.155	—	—
	4	3.0	60×40	6D13 0.36	3670	0		0		10	298	12.2	13.6	14.2	15.0	1.10	0.44	1.76	0.157	0.230	0.183	—	—
III	1	2.5	40×50	3D13 0.21	3880	0		0		5	244	7.3	8.2	10.2	10.2	1.07	0.33	—	0.100	0.123	0.171	—	
	2	2.5	40×50	7D13 0.48	3880	0		0		5	244	14.3	15.7	15.3	15.7	1.00	0.47	0.96	0.132	0.160	0.234	0.80	3.42
	3	2.5	40×50	7D13 0.48	3880	1D6 0.14	4500	0		5	240	14.3	15.6	15.4	16.6	1.06	0.56	3.41	0.132	0.160	0.206	—	
	4	2.07	40×50	4D10 0.16	4010	0		0		5	244	7.7	8.3	9.7	10.3	1.24	0.18	1.09	0.06	0.078	0.087	—	
	5	2.07	40×50	6D13 0.41	3880	0		0		5	240	15.3	16.7	17.5	17.7	1.06	0.36	0.90	0.098	0.116	0.140	0.78	5.57
	6	2.07	40×50	6D13 0.41	3880	1D6 0.18	4500	0		5	240	15.3	16.7	17.8	18.1	1.08	0.28	1.68	0.098	0.116	0.116	—	
	7	1.5	40×50	3D10 0.12	4010	0		0		5	244	8.4	9.4	12.2	12.2	1.30	0.06	0.37	0.034	0.045	0.020	—	
	8	1.5	40×50	4D13 0.28	3880	0		0		5	244	15.0	16.5	18.0	19.0	1.15	0.28	1.41	0.056	0.070	0.062	—	
	9	1.5	40×50	4D13 0.28	3880	1D6 0.18	4500	0		5	240	14.9	16.5	17.3	18.2	1.10	0.19	0.80	0.056	0.070	0.043	—	
IV	1	2.5	40×40	7D16 0.99	3530	1D10 0.40	3370	0		10	344	20.9	23.5	23.6	23.9	1.02	0.51	2.62	0.190	0.238	0.187	2.38	12.73
	2	2.5	40×40	7D16 0.99	3530	1D10 0.45	3370	2D16	3530	10	344	23.3	28.2	25.9	29.5	1.05	0.46	2.71	0.179	0.225	0.223	2.49	11.17
	3	2.5	40×40	7D16 0.99	3530	1D10 0.51	3370	5D16	3530	10	344	26.7	34.6	33.8	38.6	1.12	0.53	2.65	0.185	0.233	0.235	2.42	10.30
	4	4.0	40×40	6D19 1.23	3490	1D6 0.11	3740	0		10	420	15.7	17.3	17.9	22.7	1.31	0.81	3.28	0.304	0.356	0.506	2.92	5.77
	5	4.0	40×40	6D19 1.23	3490	1D10 0.36	3370	4D19	3490	10	420	19.2	25.3	22.8	30.7	1.21	0.91	5.46	0.312	0.382	0.525	5.08	9.77
V	1	4.0	40×50	6D16 0.65	4000	0		0		5	357	12.0	13.2	12.7	17.2	1.30	0.55	—	0.239	0.291	0.545	—	
	2	2.5	40×50	6D13 0.41	4040	0		0		5	346	13.2	14.5	14.0	14.8	0.96	0.27	1.07	0.106	0.133	0.190	0.94	4.95
	3	2.0	40×50	6D13 0.41	4040	0		0		5	330	11.8	13.3	12.5	12.5	0.94	0.20	0.79	0.071	0.088	0.104	0.70	6.73
	4	1.5	40×50	6D13 0.41	4040	0		0		5	276	21.9	24.0	—*18.5	0.77			—					
	5	1.5	40×50	5D16 0.54	4000	0		0		5	299	27.4	29.9	—*29.8	1.00			—					
	6	2.5	40×50	6D16 0.65	4000	1D6 0.09	3660	0		5	346	19.2	21.0	17.9	19.7	0.94	0.31	1.25	0.138	0.166	0.200	1.09	5.40
	7	2.0	40×50	5D16 0.54	4000	1D6 0.10	3660	0		5	346	20.5	22.5	22.1	22.1	0.98	0.29	1.17	0.096	0.113	0.147	1.06	7.21
	8	1.5	40×50	5D16 0.54	4000	1D6 0.16	3660	0		5	330	27.4	29.9	26.1	31.0	1.04	0.26	2.08	0.116	0.141	0.108	—	
VI	1	4.0	40×40	6D22 1.66	3950	1D10 0.27	3940	0		10	359	21.7	23.1	18.9	22.8	0.99	0.77	3.83	0.474	0.576	0.535	3.25	6.07
	2	4.0	40×40	6D22 1.66	3950	1D10 0.42	3940	0		10	359	21.7	23.1	22.9	8.99	0.83	5.81	0.474	0.576	0.512	5.23	10.21
	3	4.0	40×40	6D22 1.66	3950	1D10 0.38	3940	4D22	3950	10	359	26.9	34.3	23.7	33.6	0.98	0.92	5.53	0.511	0.621	0.584	4.91	8.41
	4	4.0	40×40	6D22 1.66	3950	1D13 0.58	4010	4D22	3950	10	359	26.9	34.3	26.0	33.6	0.98	0.86	6.19	0.511	0.621	0.562	5.57	9.9

*1:failure　*2:shear breaking

The amount of pull-out at yielding calculated and the
amount of pull-out obtained from experiments are compared in
Table 3,　while strain distributions inside footings are
compared in Fig. 3. It may be said that the two are in good
agreement.

After $2\delta_y$, reinforcing bar strain at the bottom of the
bridge pier becomes extremely large at 10,000 to　15,000 μ.
A comparison of the strain obtained by calculations with
ultimate strain as 15,000 μ and the measured value is shown in
Fig. 4, and there is good agreement.

Table 3 Reinforcing bar
pull-out amounts (unit: mm)

		test	calculation
III	1	0.339	0.363
	2	0.466	0.445
	3	0.501	0.445
	4	0.316	0.299
	5	0.435	0.445
	6	0.453	0.445
	7	0.265	0.252
	8	0.359	0.387
	9	0.396	0.398
IV	1	0.462	0.445
	2	0.394	0.445
	3	0.493	0.439
	4	0.439	0.445
	5	0.505	0.445

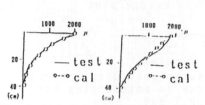

Fig.4 Axial reinforcement
strain distribution
in footing (at yielding)

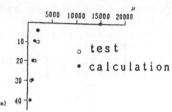

Fig.5 Axial reinforcement
strain distribution
in footing (at ultmate state)

b) Rotational Displacement of Bridge Pier due to Pull-out of
Axial Reinforcement

The strain of reinforcing steel at the bottom end of the
bridge pier in the ultimate state becomes 5,000 to 15,000 μ at
2 δ_y , and subsequently, there is a slight amount of increase,
but there is little change. When displacement is made to
increase, the increase in displacement of the main part of the
bridge pier becomes larger than the rotational displacement due
to pull-out of axial reinforcement, and the propotion of the
total displacement made up by rotaional displacement becomes
small. Although the ultimate strains of axial reinforcement at
the bottom end of bridge piers differ according to the
individual specimens, the ratio to total displacement of the
difference between the rotational displacement due to pull-out
of reinforcement obtained with strain of axial reinforcement at
the bottom end of the bridge pier as 10,000 μ and the
rotational displacement obtained similarly with axial
reinforcement strain as 20,000 μ is 2 to 3 percent and small
(see Fig. 5). The amount of pull-out in the ultimate state is to
be obtained assuming reinforcing bar strain at the bottom end of
the bridge pier as 20,000 μ giving consideration to the
measured value of reinforcing bar strain near the ultimate
state and the fact that the degree of influence of pull-out on
total displacement in case of different reinforcing bar strains
is small.

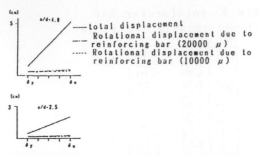

Fig.6 Rotational displacement due to reinforcing
bar pull-out and total displacement

When the amounts of
pull-out of axial
reinforcement at yielding and
at the ultimate state have
respectively been obtained, it
will be possible for
rotational displacement to be
calculated based on Fig. 6.

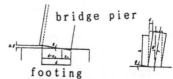

Fig.7 Rotarional displacement
due to reinforcing bar
pull-out

In Fig. 7, with

$\Delta \ell$: amount of axial reinforcement pull-out
d : effective height of member
x_n : distance from compression fiber to neutral axis
(equilibrium of force, according to law of
maintenance of plane)

the rotational angle θ of the bridge pier will be

$$\theta = \Delta \ell / (d - x_n) \qquad \cdots \cdots \cdots (1)$$

Consequently, rotational displacement δ_1 due to pull-out of
axial reinforcement will be obtained as

$$\delta_1 = h \cdot \theta = h \cdot \Delta \ell / (d - x_n) \qquad \cdots (2)$$

The relationships of the amounts of pull-out at yielding
and in the ultimate state in the experiment obtained by
calculation (respectively put as $\Delta \ell_y$ and $\Delta \ell_u$) with the
ratio D / ϕ between center-to-center spacing D of reinforcing
bars and reinfocing bar diameter ϕ will be as shown Figs. 8
and 9. As a consequence, when anchorage is adequate with
anchorage length of 30 times or more of reinforcing bar diameter
secured, with ends of reinforcing bars provided with hooks, and
with concrete strength in a normal range of 240 to 350 kg/cm²,
the relationship can be obtained using the equations below.

$$\Delta \ell_y = 0.070 - 0.0054 (D / \phi) + 0.00017 (D / \phi)^2 \qquad \cdots (3)$$

$$\Delta \ell_u = 0.083 - 0.0054 \, (D/\phi) + 0.00015 \, (D/\phi)^2$$
$$\cdots (4)$$

provided that $3 \leq D/\phi \leq 16$ (unit: cm).

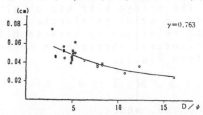

Fig.8 Relationship of reinforcing
bar pull-out amount
and D/ϕ (at yielding)

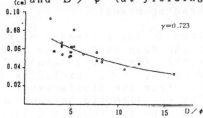

Fig.9 Relationship of reinforcing
bar pull-out amount
and D/ϕ (at ultimate state)

3.2 DISPLACEMENT OF MAIN PART OF BRIDGE PIER AT YIELDING

For rigidity of a
bridge pier in
determining the
displacement of the main
part of the bridge pier
at yielding, the part
without occurrence of
cracking was considered
to be effective over the
entire cross section,
while at the bottom of
the pier, since
reinforcing bars had
failed, concrete on the
tension side was ignored,
and the rigidity of the
intermediate part was
considered to vary
linearly. Meanwhile, an
example of the strain
distribution of axial
reinforcement in the

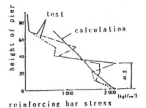

Fig.10 Example of reinforcing
bar stress intensity
of main part of bridge pier

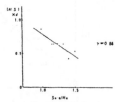

Fig.11 M.S. and $S_u \cdot a / M_u$

bridge pier proper at yielding (making measurements on
two axial reinforcing bars) is shown in Fig. 10 , and

increase in reinforcing bar strain due to the influence of diagonal cracking may be seen. The distance (hereinafter referred to as "M.S.") of the position above the bottom end of the bridge pier where reinforcing bar strain had reached yield strain from the bottom of the bridge pier was obtained based on the results of measurements on axial reinforcing bar strain inside the pier proper. The relationship between M.S. and the ratio of shear strength to flexural strength is shown in Fig. 11. Eq. (5) may be estimated from Fig.11.

$$M.S. = (1.5 - 0.7 \cdot S_u \cdot a / M_u) \cdot d \qquad \cdots \cdots (5)$$

provided that $0 \leqq M.S. \leqq d$, and, where,
$\quad$ S_u : shear strength, according to Eqs. (6), (7), (8)
$\quad$ a : shear span,
$\quad$ M_u : ultimate bending moment

$\quad$ With consideration given to the influence of M.S., the displacement (δ_{yocal}) of the main part of the bridge pier at yielding was obtained (Fig.12). The displacement (δ_{yotest}) of the main part was obtained by $\delta_y - \delta_{yl}$ excluding the rotational displacement δ_{yl} from the displacement yielding δ_y obtained by experiments. The relatiomship between δ_{yocal} and δ_{yotest} is shown in Fig. 13. The two roughly coincide with each other.

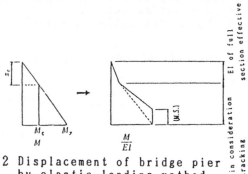

Fig.12 Displacement of bridge pier by elastic loading method

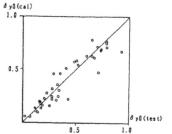

Fig.13 Displacement of main part of bridge pier at yielding (cm)

3.3 EXAMINATION OF DUCTILITY OF MAIN PART OF REINFORCED CONCRETE BRIDGE PIER

The value remaining after deducting the rotational displacement δ_{ui} of bridge pier due to pull-out of axial reinfocement from δ_u obtained from experiments is considered as the ultimate displacement δ_{uo} of the main part of the bridge pier. The displacement δ_{yo} of the main part of the bridge pier at yielding is determined lowering rigidity giving consideration to cracking.

The ratio δ_{uo}/δ_{yo} of δ_{uo} and δ_{yo} is to be referred to as the ductility μ_o of the main part of the bridge pier only. In general, it is said that the ductility of a reinforced concrete member is increased when shear strength becomes large compared with flexural strength, and the relationship between μ_o and the ratio of shear strength and flexural strength will be examined here.

Shear strength S_u is obtained by $S_c + S_v$, where, shear strength S_c due to other than hoop reinforcement will be according to the following:

Case of $1.5 \leqq a/d \leqq 2.5$

$$S_c = 3.5 \cdot (a/d)^{-1.166} \cdot f'c^{1/3} \cdot \beta_p \cdot \beta_d \cdot \beta_n \cdot b \cdot d$$
$$\cdots \cdots \cdots (6)$$

Case of $2.5 < a/d$

$$S_c = 0.94 \cdot (0.75 + 1.4d/a \cdot f'c^{1/3} \cdot \beta_p \cdot \beta_d \cdot \beta_n \cdot b \cdot d)$$
$$\cdots \cdots \cdots (7)$$

Shear strength due to hoop reinforcement

$$S_v = A_w \cdot \sigma_{sy} \cdot z \cdot (\sin\theta + \cos\theta)/s \qquad \cdots \cdots (8)$$

where,

$\beta_p = (100\,P_t)^{1/3}$, $\qquad \beta_d = (100/d)^{1/4}$, $\qquad \beta_n = 1 + 2\,M_o/M_u$
$f'c$: compressive strength of member
d : effective height of member
P_t : tensile reinforcement ratio
M_u : ultimate bending moment
M_o : limit bending moment at which tensile stress occurs in member cross section
A_w : cross-sectional area of one set of hoop reinforcement at section s
σ_{sy}: yield point stress intensity of hoop reinforcement
θ : angle of hoop reinforcement with member axis
$\quad z = d/1.15$

From data of specimens in which hoop reinforcing bars are not arranged, Eq. (9) is obtained from the relationship between μ_o and strength ratio ($S_u \cdot a/M_u$).

$$\mu_o = -1.9 + 6.6 \cdot S_u \cdot a/M_u \quad (y = 0.892) \qquad \cdots \cdots (9)$$

From data of specimens in which hoop reinforcing bars are arranged and not arranged, Eq. (10) is obtained.

$$\mu_o = -1.9 + 6.6 \cdot S_u \cdot a/M_u + (13P_w - 1.6) \cdot P_w \qquad \cdots \cdots \cdots (10)$$

provided that $0.9 \leqq S_u \cdot a/M_u \leqq 2.7$
$(13P_w - 1.6) \cdot P_w \geqq 0$
$P_w = 100 \cdot A_s /b \cdot s \leqq 0.6\%$

Eq. (10) expresses the ductility of the main part of the bridge pier.

The comparison of μ_{oc} obtained by Eq. (10) and μ_{ot} is shown in Fig. 14.

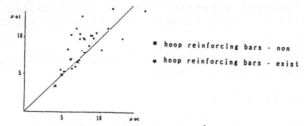

Fig.14 Comparison of μ_{oi} and μ_{oc}

3.4 CALCULATION OF DUCTILITY

The ductility μ of a reinforced concrete bridge pier can be estimated by using the abovementioned results. That is, μ may be obtained by Eq. (11) calculating deformation δ_{yo} of the main part of the bridge pier at yielding, the rotational displacements, and δ_{yi} and δ_{ui}, respectively, due to pull-out of axial reinforcing bars at yielding and in the ultimate state, and ductility μ_o of the main part of the bridge pier.

$$\mu = \frac{\delta_u}{\delta_y} = \frac{\delta_{uo} + \delta_{ui}}{\delta_{yo} + \delta_{yi}} = \frac{\mu_o \, \delta_{yo} + \delta_{ui}}{\delta_{yo} + \delta_{yi}} \qquad \cdots \cdots (11)$$

$$\left(\because \quad \mu_o = \frac{\delta_{uo}}{\delta_{yo}} = f\left(\frac{S_u \cdot a}{M_u}, P_w\right) \right)$$

On the other hand, when deformability (ductility) required of a member is determined by a means such as earthquake response analysis, it will be possible for the shear reinforcement required to be obtained from Eqs. (10) and (11).

THEME V:

RESEARCH AND DEVELOPMENT NEEDS

481

THEME V: RESEARCH AND DEVELOPMENT NEEDS

Wednesday Morning Session, January 16, 1991

Chairman: T. Paulay

Reporter: J. A. Ramirez

Summary of Remarks by T. Paulay

Workshop participants are exhorted to contribute to the last session for a fruitful outcome of the workshop. Areas of possible contribution:

- improvements to the various models for membrane shear.

- improvements in regard to the practical applications.

- establish a set of research priorities to be considered by the National Science Foundation.

The reporters for each of the previous sessions were asked to present a summary, including highlights, discussions, and research needs. Each reporter's summary would be followed by individual oral discussions within a one-half hour allotted time. Due to time limitations, participants were encouraged to send written comments to the organizers of the workshop. The outcome of this session will be included in the final report of this workshop.

Summary for Monday Morning Session

by P. Balaguru, Reporter

Four state-of-the-art reports and three papers were presented during the Wednesday morning session. The discussions revolved primarily around two issues. The first issue dealt with the shear ductility. The questions raised were: Does shear ductility exist? Can it be used? and Can it be increased? The general feeling was:

- Shear ductility does exist and it can be predicted.

- Even though shear ductility can be used to dissipate energy under earthquake loading, more research is needed to evaluate the load-deformations relationship at later stages of cyclic loading.

- More rational design methods need to be developed for utilizing the shear ductility.

The second issue dealt with the size effect in shear. Three causes were mentioned, namely the crack size, the aggregate interlock in cracks, and the fracture mechanics. The aggregate interlock is effected by a number of factors, such as aggregate size, shape, and strength as compared to the matrix strength. High-strength concrete and lightweight concrete are expected to have less interlock resistance.

The future research areas as related to this session were identified below:

- The load-deformation relationships of shear walls in various configurations and load histories, and the interaction of frame and shearwalls.

- Evaluation of high-strength concrete for shear.

- Evaluation of other composites, such as fiber-reinforced concrete, for improving shear ductility.

- Reliability analysis is required for strength and ductility in shear.

- The load-deformation response of shear elements with various bar orientations, load histories, and confinements.

Discussion No. 1 — H. Kupfer, Technical University of Munich, Germany:

As we have learned from this interesting meeting, the horizontal reinforcement of a shear wall — acting like the stirrups of a beam — should not yield before the vertical reinforcement of the tension chord is yielding. Yesterday, I showed a truss model made of wooden slats and rubber bands and you have seen that, in case of high longitudinal strains, the aggregate interlock forces between diagonal struts are acting in unfavorable direction, so that the diagonal compression forces produced by shear will become more steep relatively to the beam axis and the tension forces in the stirrups will increase. I think, we should take into account this behavior in design of the horizontal shear reinforcement of shear walls subjected to earthquake forces, to avoid yielding of shear reinforcement.

Another point is to develop the I-shaped cross-section and the whole structure of a shear wall in such a way that the shear cracks are already existing when the bending cracks come into being, because in this case the shear crack direction and the shear bearing behavior is more favorable. It should be even possible to avoid the two fans of shear struts near the abutment and also the alternating bending crack opening along the whole construction joint between abutment and shear wall by the use of special design ideas.

Discussion No. 2 — J. C. Walraven, Delft University of Technology, The Netherlands:

We have discussed quite a number of aspects concerning the bearing capacity of reinforced and prestressed concrete members loaded in shear. One of the questions that came up was how high strength concrete should be treated. In all buildings codes, the shear capacity is formulated as a function of number of parameters, such as the strength of the concrete, the longitudinal reinforcement ratio, the shape and the size of the cross-section, and the shear reinforcement. I think that for high strength concrete a further parameter may be highly influential, and that is the ratio particle strength to cement-matrix strength or, in other words, the percentage of particles which break through when cracks in the concrete are formed. Fracture of the particles results in relatively smooth cracks: the lack of aggregate interlock may limit the redistribution of forces in the member, and may even lead to dangerous situations in cases that axial tensile forces occur. Axial tensile forces may lead to cracking of the concrete over the whole depth of the cross-section and may, as, such, drastically reduce the shear capacity.

A further task for the near future is the formulation of rational design models for shear, with a wide applicability and logical transitions. Logical transitions are required from small to large members from short to slender members, from statically determinate to statically indeterminate members, and from regions with a steady flow of forces to regions with discontinuities. The various truss models, as discussed here, provide a good basis. An interesting concept is certainly the new Japanese Code, with its superposition of strut-tie models and truss models.

Further research is also necessary with regard to the interaction between the ultimate limit state and serviceability limit state. Requirements with regard to the serviceability limit state may impose limits on the applicability of some plasticity models.

Finally I would like to suggest that we make an inventarisation of shear projects which are being carried out now and which are planned for the near future, so that efficient exchange of information is possible.

Discussion No. 3 — P. G. Gambarova, Polytechnic University of Milan, Italy:

With reference to the papers presented by Prof. Walraven (on Aggregate Interlock) and Prof. Kupfer (on Transverse Reinforcement in beams subjected to shear), I would like to emphasize that one of our main concerns should be to try to extend what has already been ascertained for normal concrete, to special or new materials, such as high-strength, lightweight and fiber-reinforced concretes, since many aspects of these materials are quite new.

Since we cannot carry out as many tests as we have done so far on normal concrete, a careful balance between future experimental research work and mathematical modelling will help in assessing to what extent the constitutive and behavioral laws of normal concrete still apply to new materials.

With regard to this, some relatively simple models, such as those based on limit analysis and truss analogy, may be instrumental in assessing and clarifying the role of the various parameters that characterize the new materials.

To be practical, I would like to mention two problems, which are under investigation in the R/C Laboratory of our Department of Structural Engineering: the role of fibers (bridge effects) on shear behavior of R/C and P/C thin-webbed cracked beams, and the splitting of high-strength concrete in anchorages and splices (bond degradation).

Discussion No. 4 — D. H. Kim, Sunkyong Ltd., Korea:

I didn't know when I should speak, because I might say something other than this theme. Even though I have designed a few hundred structures, both buildings and bridges, and military structures, I think I am a kind of outsider for this concrete subject because, recently, I have been working on composite materials which are used mostly by aerospace people. But, just now, I heard two gentlemen, and I thought I should give my opinion here. Sometimes the outsider's opinions are important for the recommendations purposes.

First of all, to me, concrete is a composite made up of both particulates and fibers. By fiber, including steel bars, I mean the length-to-diameter aspect ratio of infinity, not only short ones.

When we analyze the composites in general, we do use the generalized coordinates, and both forces and displacements, rotational and linear, which means all six components for each. So, to me, analysis and design should always be in three dimensions. This means that the moment and the shear must be considered at the same time, so that analysis becomes much simpler.

Another thing is the word "ductility" of concrete which is quite new to us. There is, in concrete shear walls, a behavior which we may call "ductility." Whatever matrix is used, a polymer, ceramic, metal or cement composite, when acted on by impact or vibration, micro-cracking starts and propagates. This micro-cracking absorbs the dynamic force. If the force is not completely absorbed by the crackings, then, there becomes the pull-out and eventually cutting-off of the fibers, one after the other. These will absorb additional energies. This means that this structure becomes tough against the impact and cyclic loading. Toughness means the capability to absorb energy. If you see from outside of the structure, the external energy from earthquake is, at least partially, dissipated. This mechanism of energy absorption may be called "ductility," and the concrete shear wall certainly has it. There are certain ways of formulating and characterizing this kind of absorption and dissipation.

This ductility may be increased by proper design.

The first method is to use reinforcements with as small diameters as practical and make the bond between the reinforcement and the matrix "weaker" or "flexible." Some ceramic composites are reinforced by "weak" fibers and the interface is made weaker to increase toughness.

Another method is to increase the toughness of the matrix (cement mortar in this case) itself by adding other matrices like polymers. In order to minimize the micro-cracking, we add some material which increases the toughness or tries to alter the material properties. Toughness may be increased at the sacrifice of strength. However, the required strength can be obtained by adding proper reinforcements.

As an example, the standard specification for fiber reinforced thermosetting polymer pipes, ASTM D2992-A/B, requires cyclic loading of at least 150 million cycles and 100,000 hours of continuous loading, and if any single drop of water leaks, this structure is supposed to have failed. Any pipe which survives such test under certain pressure is rated as this pressure grade. Because, we believe, any structure, with whatever ductile material we used, will have micro-cracking, progressing to large cracking, we may apply the similar method used for composite pipes to the concrete walls. We solved such problem of pipes by adding either water-stopping layers, or some material to increase ductility, eventually increasing the fatigue strength. As an example, when a thermosetting polymer, such as epoxies, polyesters, some other noble matrices, is used, we add either some kind of thermoplastics or elastomers to increase the ductility at the sacrifice of tensile strength. We must trade off some properties to achieve our purposes. I think the same thing can be done with reinforced concrete.

To summarize, we can increase the ductility — which we often call toughness — of concrete walls by

1) addition of short fibers. Fibers could be polymer, carbon, or steel.

2) addition of additives, including polymers.

3) addition of long fibers, which have length-to-diameter aspect ratio of infinity.

For the third concept, proper processing method must be studied. However, this method could be one of the best schemes to have enhanced toughness and ductility when the structure is exposed to the dynamic forces, such as from earthquake.

In design and analysis, the reinforced concrete may be treated as fibrous composites. Then we may be able to establish the mechanisms of

1) Energy absorption by ductile matrix.

2) Fiber pull-out and cutting which will gradually absorb and dissipate the external force, such as earthquake.

3) Bond between the matrix (concrete) and particulate (aggregate) which will contribute to toughness.

There was a paper on analyzing the concrete wall with lamination theory. My opinion on this is that

1) It is an advanced concept treating the panel much closer to the reality.

2) It could be much better if the bond-stretching coupling action and the interlaminar stress concepts were discussed.

My final recommendation is that we pay some attention to a kind of interdisciplinary cooperation. This may improve our progress much further.

Summary for Monday Afternoon Session

by R. G. Oesterle, Reporter

Eight papers presented in this session dealt with experimental studies of shear walls and nuclear containment structures, as well as the two state-of-the-art reports for theoretical studies of membrane shear behavior. The issues raised for discussions included:

- What is the difference between infilled framed shear walls and flexural cantilever shear walls?

- Is the current design of nuclear structures using elastic assumptions adequate?

- Is the practical application of design procedures more important than the accuracy?

Further researches recommended are:

- Prediction of shear deformation.

- Effect of foundation deformation as it relates to tests being conducted on shear walls with very stiff foundations.

- Effect of variable M/V during cyclic loading.

- Variability in the softening coefficient.

- Inelastic behavior of nuclear containment structures.

- Use of diagonal reinforcement to improve shear ductility.

- Is shear ductility a useful concept? How should it be defined?

Discussion No. 1 — J. C. Walraven, Delft University of Technology, The Netherlands:

I have a remark with regard to the use of diagonal reinforcement. In the formulation of building codes, a question which comes up regularly is the reduction of shear capacity by axial tensile forces. Interesting tests have been carried out by Regan, at the Imperial College in London, in about '69. The testing principle is shown in the figure.

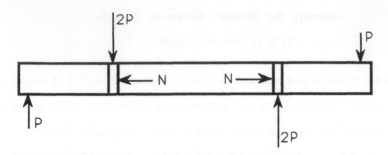

If the tensile force N is applied earlier than the transverse forces P and 2P, the inner part of the beam is cracked in a direction perpendicular to the beam axis. At the moment inflexion point, the crack is open over the full depth. Nevertheless the influence of tension on the shear capacity is normally small. This is predominantly a result of the interlocking effect of the crack faces. In high strength concrete, however, the cracks intersect the aggregate particles, so that the crack faces are relatively smooth and the shear capacity of the crack is small. In such a case, a much severer reduction of the overall shear capacity has to be feared. It is remarkable that shear reinforcement, perpendicular to the beam axis, offers no remedy. Inclined shear reinforcement would, however, function well.

Discussion No. 2 — J. O. Jirsa, University of Texas, U.S.A.:

As I listened to the talks over the last couple of days, one of the things that struck me was that I think that there are two groups of people looking for some common ground. It is evident in this list of topics presented. Many of the sessions were called "Membrane Shear" and I look at the membrane shear tests and I see that in those tests you're doing almost everything you can to avoid boundary conditions. The boundaries are intended to represent a section out of a large element that is subjected to almost identical conditions throughout. I think the nuclear containment vessel certainly fits into that area, as does the massive offshore structure. Then you've got the other condition where the boundary conditions dominate. I think that's true of all or *almost* every structure other than the large panel systems. I have trouble looking at a structural wall in a building and saying that what we know about the panel tests applies to a shear wall in a shear-wall-frame system. I can't remember all the details, but there hasn't been a whole lot done in cyclic loading in the membrane tests — mostly monotonic tests to predict the load-deformation response under a whole variety of different reinforcing conditions or material changes. The difference between these applications is that in one case you've got shear and nothing else to rely on. In the other area, in buildings and the other kinds of structures, you need a system where flexural ductility is present, and you want to make sure that shear doesn't become a problem. Those two philosophies don't really go together. They're two different approaches which are not interchangeable. I think the recommendations will have to address those two different issues. Professor Paulay already mentioned that he is worried about the details. When you're testing panels, you're not worried about the details. You do everything you can *not* to worry about the anchorage at the ends and the hooks pulling out. So, what bothers me about trying to come up with something that is uniform for all shear applications is that there is a very different design philosophy and performance requirement for these two types of elements or structural conditions.

Discussion No. 3 — R. N. White, Cornell University, U.S.A.:

I want to suggest that there may be a strong link between items 7 and 9 listed on the board. Perhaps shear ductility can be guaranteed only if we provide diagonal reinforcement. This may be a provocative statement, but it seems to me that if you have only orthogonal reinforcement with extensive orthogonal cracks, you may be in a dangerous situation in many types of structures. Some diagonal steel may be required to guarantee what we call shear ductility, and, again, I am groping as to what the latter really is. Orthogonal steel alone may not provide hysteresis loops that enclose adequate areas when the structure is subjected to severe cyclic loading.

The second point is that we need practical methods for predicting shear deformations, that is, for predicting effective shearing modulus for concrete as a function of normal stress. As consulting engineers are trying to do these analyses, they are struggling to define the value of shear modulus to be used for a structure that is undergoing cracking. Finite element analysis approaches are highly dependent upon adequate material properties definition.

The last issue is the effective tensile strength of concrete in real structures, where you have drying effects, shrinkage effects, and restraint effects that are producing cracking. Simple laboratory specimens usually don't have these effects, but a real structure may be badly cracked before it sees any applied loadings.

Discussion No. 4 — H. Aoyama, University of Tokyo, Japan:

I would like to just comment on shear ductility. When I made a presentation the day before yesterday about the restoring-force characteristics of containment structures, I did not make any further discussion about the shear ductility. But the way we are doing with the containment structures is simply to idealize the nonlinear behavior into certain kind of mathematical model and it is subjected directly to the time-history analysis for earthquake excitation. So, in that kind of method of design or analysis, we are already saying that shear ductility exists and it can be used in the design. However, if the question is raised in the form of whether we can express shear ductility in the form of, say, Newmark ductility, in other words, if we can divide the elastic response by, say, a square root of twice μ minus one, where μ is shear ductility, of course, the answer would be that it is very questionable. We may not be able to use shear ductility in that way. So, it is all the matter of definition: What do you mean by shear ductility?

Discussion No. 5 — G. Corley, Construction Technology Labs, U.S.A.:

I would like to follow up on something Professor White brought up. This workshop focuses primarily on shear in the plane of the structure. However, there is a related issue where shear strength perpendicular to the plane of the structure is greatly reduced by induced tensions due to shrinkage, movement of the structure, or loads applied in-plane. Under repeated applications of a moderate out-of-plane load, shear failures may occur under working loads. I have been unable to find much in the literature on this subject. This is a subject that we should *get* some information on in the future. None of the theories suggested in this workshop touch that subject. Data on out-of-plane shear strength under repeated loads in combination with in-plane tension are needed.

Discussion No. 6 — B. Thürlimann, ETH Zürich, Switzerland:

I think the term ductility has two aspects. If you have a statically loaded structure subjected to a foundation settlement, it is a one-time occurrence. If you have cyclic loading in an earthquake, that is quite different. We should distinguish between these two cases. I think a certain detail can have sufficient shear ductility to accommodate a foundation settlement, to redistribute unknown residual forces. However, in the case of cyclic loading, the same detail may exhibit a completely insufficient behavior. Hence we should distinguish the two cases.

The large majority of structures are not subjected to earthquakes. In many cases, they are, however, subjected to settlements and/or unknown residual forces. We should also develop structural details leading to an adequate ductility for these less-demanding cases.

Summary of Tuesday Morning Session
by C. French, Reporter

This session covered fourteen papers in the theoretical studies of membrane shear behavior, including six in elements and members, three in shear walls, and five in the application of finite element methods. The following topics were raised and discussed:

Elements and Members:

- Rotating crack model vs. fixed crack model.
- Shear transfer across cracks.
- Design of shear reinforcement for thin-webbed I-section and box beams.

- Effectiveness of aggregate interlock, particularly in cyclic loading, high-strength concrete, and lightweight concrete.
- New material properties to enhance cyclic load results.

Shear walls:

- Interactive computer program for design.
- Modelling under reversed cyclic loading for pull out, shear slip, and deteriorating compression struts.
- Shear contribution is not very effective due to severe pinching of the hysteretic loop.
- Relative contribution of shear and bending to deformation.

Application of Finite Element Methods:

- Shear failure due to instability.
- Analytical investigation using crack strain distribution.
- Modelling the three-dimensional effects of the U.S.-Japan frame-wall structures.
- Using micro- and macro-models to track the load-deformation behavior.
- Prediction of energy dissipation by FEM.
- Reduction of tensile strength of concrete, f_t, to reflect shrinkage stress.

Discussion No. 1 — J. Kollegger, University of Kassel, Germany:

I would like to make two comments concerning, first, the significance of panel tests and, second, the results of panel tests. Today we have very powerful computers. We have good finite element programs. What we are lacking are material models to realistically analyze reinforced concrete structures. You can apply material models which come from panel tests in all types of surface structures. Professor Collins showed us possible applications in his lecture yesterday (e.g., offshore structures, containment vessels, shear walls, and others). If you want to analyze detailing problems, you would not take the results coming from panel tests; you would take the concrete results from triaxial tests, you would use results from bond tests for the bond modelling and you would include every single reinforcement bar in your model. Results from panel tests can only be applied if you analyze structures of larger scale.

My second point is concerning the results of these panel tests. In North America and Japan, it seems as if everybody is following the original Vecchio-Collins experimental results, which showed a strong degradation of the concrete strength under static loading. In France, there have been experiments published in 1968 and 1972 by Demorieux and Robinson. There have been experiments by Schlaich and Schaefer at Stuttgart published in the beginning of the '80's which have been carried out at the same time as the Vecchio-Collins' experiments. Experiments on panels were done by Eibl in Karlsruhe. And we have done some tests at Kassel. In all of the mentioned experiments, the large reduction of the compressive strength of concrete was not measured. And, so, although it might seem an easy problem — just pulling and pushing concrete, and subjecting it to shear, in a panel experiment — I feel there is still not a real answer found.

Discussion No. 2 — H. Noguchi, Chiba University, Japan:

I would like to make discussion on the conclusions of the state of the art on the theoretical studies on membrane shear behavior in Japan. I had introduced two models, macroscopic models and microscopic models, in my presentation. First of all, I would like to consider the role of macroscopic models. The main role is a powerful design tool. For that object, the following five items are necessary in the future research. The first item is the compressive deterioration of cracked concrete and bond behavior. This is manifested by a reduction factor of the concrete compressive strength in the theory of plasticity. The constant value is usually used, for example, 0.7. But is it a constant value?

The second item is the verification by investigating the internal stress flows by FEM. There are many assumptions in the macroscopic models. Figure 12 of my presentation showed the principal stress distribution of a shear wall obtained from FEM. This principal stress distribution is divided into the stresses in two mechanisms: truss mechanism and the arch mechanism. The stress in the truss mechanism is not so dominant in this proportion of the shear wall, but stress in the arch mechanism is rather dominant, especially along the diagonal concrete strut. From the comparisons between this macroscopic models and the stress in the arch mechanism, you can find good agreement between these two. And, so, FEM will be a good tool for the verification of the macroscopic models.

The third item is an approach from the upper bound theory of limit analysis. Unfortunately, there are not so many approaches from the upper bound theory of limit analysis. In Japan, Professor Minami is trying to apply this theory to shear walls, beams, and columns. In the upper bound theory, the condition of compatibility on deformation is established. In the lower bound theory, the equilibrium condition of force is established, but there is no compatibility on deformation. If the compatibility on deformation is established, we can analyze the deformation. This item is connected to the fourth item: the analysis of deformation. This item will be connected to the evaluation of ductility, because we must be very careful about the failure mode of a joint shear failure after the beam flexural yielding under large deflection in the ductile design.

And the final item for the macroscopic models is the verification and improvement of accuracy in wide range for influencing factors, including high-strength concrete. The accuracy in wide range is very important, because this will be connected to the universality of the model. Professor Paulay advised me in this workshop that the application to beam-column joints from the knowledge of membrane shear is very important in the design. From the macroscopic models, we can contribute to the application to beam-column joints like this from the knowledge of membrane shear. In fact, in Japan, Dr. Ichinose, Professor Watanabe, and Mr. Fujii are trying to carry out these applications. I hope they can analyze beam-column joints from macroscopic aspects successfully.

Another model is the microscopic model. In yesterday's session, Professor Okamura mentioned the roles of FEM. Of course, it is difficult to consider the roles of FEM in the future, but now we are considering mainly two roles. The first is in place of experimental approach. This approach will deepen understanding of the shear-resisting mechanisms. And this approach is more effective than experimental works, because it saves time and human expenses. We are trying to apply FEM to the development of high-rise building with ultra-high-strength materials in Japan. The projects must be done in five years. We wish to select necessary specimens for the experimental works. The second role is a design tool. In Japan, JCI Guideline for Application of FEM to Concrete Structures was published, but they are limited mainly to the elastic analysis. But near future, we can apply nonlinear FEM to concrete structures, especially large-scale structures. We have six items as future research subjects for the microscopic model. The first item is the generalization of compressive deterioration and compressive failure of cracked concrete, including high-strength concrete under monotonic and reversed cyclic loading. In the new project for high-strength R/C structures, various basic tests have been done. For example, in Chiba University and Nihon University and Hazama Corporation, the following test results were observed. In the case of normal strength under 40 MPa, the reduction factor for the concrete compressive strength decreased from 1.0 to about 0.6 as the average tensile strain across cracks increased. But for high-strength concrete, there was remarkable reduction for the small tensile strain region. And the reduction factor is getting down to about 0.3 or 0.4 for 100 MPa high-strength concrete.

The second item is the evaluation of shear deterioration of cracked concrete subjected to reversed cyclic loading. As for this item, many researchers have discussed and so now let me skip to the third item that is the evaluation of bond-slip behavior based on the bond mechanism of deformed bars. In the bond mechanism of deformed bars, the effect of wedge action causes the splitting crack along a longitudinal bar. The fourth item is a generalization of the judgment of failure modes. There are many failure modes: flexural yielding, flexural compressive failure. These failure modes are rather easy to predict, but it is rather difficult to define shear failure in FEM analysis. As for the shear tension failure in the analysis under load control or displacement control, the strength decay has not been observed. But in the test, load decreased after the peak. And so, when shear tension failure occurs, it is difficult to predict in the analysis. We must define the failure point probably from the strain aspect. As for shear compression failure in the analysis under displacement control, we can obtain a descending branch. But, is it true, as compared with the test results? The fifth is the invention of investigative methods of the FEM analytical results. Output is automatically given from FEM, but we must check it in order to grasp the shear resisting mechanisms and verify the macroscopic models for developing shear design equations. The last item is an application of nonlinear FEM to the design of RC structures. But, this must be skipped due to the lack of time. But in the future, we will make a revised guideline especially about reliability of code, nonlinear finite element analysis, and how to use nonlinear FEM for the design of RC structures.

Discussion No. 3 — T. Hsu, University of Houston, U.S.A.:

I will try to answer some of the problems raised by Dr. Kollegger and Professor Noguchi regarding the softening of the concrete compression struts. The softening coefficients obtained in Paris, Kassel and Stuttgart were in the range of 0.6 to 0.8, while those obtained in Toronto could be as low as 0.25. In the last few years, our research at the University of Houston was directed towards bridging the difference between the European work and the Canadian work. To date, we have found that the softened coefficient is affected by four variables. The differences in these four variables could explain the differences between the European work and the Canadian work.

The first variable is the principal tensile strain in the direction perpendicular to the concrete compression struts. This principal tensile strain at failure has a strong effect on the softening of the concrete struts. The larger the tensile strain, the lower the softening coefficient. On the one hand, the tensile strains of steel in the European panels were often limited within the yield strain, say 0.002. On the other hand, the steel in the Canadian panels were allowed to be stretched way beyond the yield strain, say up to 0.02.

The second variable is the angle of reinforcement with respect to the cracks. In most of the European tests, the rebars cross the cracks perpendicularly, i.e., 90°. In most of the Canadian tests, however, the rebars cross the cracks at 45°. Panels with 45° rebars will give a softened coefficient lower than that for panels with 90° rebars. The difference can be explained physically as follows: when a rebar crosses a crack at 90°, the straight centerline of the bar remains straight after cracking; when a rebar crosses a crack at 45°, however, a "kink" will result in the centerline of the bar at the crack, the crack width being the length of the kink. This kink will destroy the bond and crush the concrete in the vicinity of the crack, resulting in lower softening coefficient. This is why Professor White suggested the use of 90° rebars in the nuclear containment structures.

The third variable is the load sequence. Our tests show that sequential loading (first tension, followed by compression) produced higher softening coefficients than proportional loading. This was because the slight release of tension in sequential loading prior to failure reduced the stress concentration at the crack tips, thus requiring a higher compressive stress to propagate the cracks. Most of the European panels were subjected to sequential loading, while almost all of the Canadian tests utilized proportional loading.

The fourth variable is the difference between the fixed angle and the rotating angle. The fixed angle is the angle between the longitudinal steel and the principal compressive stress defined by the applied membrane stresses. This fixed angle determines the direction of the cracks. The rotating angle is the angle between the longitudinal steel and the principal compressive stress in the concrete. When the percentages of steel are equal in the two orthogonal directions, the fixed angle is the same as the rotating angle. Therefore, no shear stress will develop in the cracks between the concrete struts. When the percentages of steel are not equal in the two orthogonal directions, the fixed angle will not be the same as the rotating angle, causing shear stress between the concrete struts. This shear stress is almost proportional to the difference between the fixed and rotating angles. By weakening the concrete struts, the shear stress will lower the softening coefficient. In the Canadian tests the proposed lower bound for softening coefficients was determined by panels with different percentages of steel in the two orthogonal directions.

Professor Walraven mentioned that shear resistance due to aggregate interlock will be relatively lower in high strength concrete than in normal strength concrete. This is because aggregates were observed to split open at the failure surfaces in high strength concrete. We are currently studying the fifth variable, namely, concrete strength. If Professor Walraven's thinking is correct, the softening coefficient of high strength concrete is expected to be lower than that of normal strength concrete for panels where the rotating angle does not coincide with the fixed angle.

In summary, research in the constitutive laws of reinforced concrete membrane elements is still very new. We are still struggling to identify the variables that govern the stress-strain relationships. These relationships are quite complex, because they are two-dimensional. That means the stresses and strains in one direction are strongly affected by the stresses and strains in the orthogonal direction.

Discussion No. 4 — H. Okamura, University of Tokyo, Japan:

Actually, I do not agree with Dr. Noguchi's comments *(laughter)*. I would like to say that I have confidence to predict the behavior of this kind of reinforced concrete members not only under monotonic but also under reversed cyclic loadings. This type of structures such as shear walls are very easy to be analyzed, because the relative displacements between steel and concrete in the wall is limited at least at the boundary of members, and the bond problem in this region is easy to be treated. For this type of structures, the smeared crack model is capable. This is entirely different from a beam-column joint. We made a program which is available now to the public and displacements, stresses, crack patterns are to be calculated. The verification is important for nonlinear analysis, we have to be very careful for applicable items, which are limited for each models. Usually any models are based on some experiments of different levels. For example, Professor

Collins' model is based on reinforced concrete panel tests. And our models are based on one-dimensional reinforced concrete members' tests. The model for cracked concrete is based on the tensile test of reinforced concrete. The length of specimens is about 3 meters long, as we wanted for the specimen to have many cracks. The compression model is based on the simple test, and the shear transfer model is also based on simple tests. As the tests are very limited especially in boundary conditions, we have to verify the applicability of the models constructed from this kind of simple tests for the two-dimensional reinforced concrete test results. Sometimes a model could not be applied because of the restriction of the test method. Especially for bond problems, we could not have the generalized test method. Therefore, the finite element analysis of smeared crack approach is difficult to be applied for the part where the concrete and steel do not show the same behavior and the relative displacements occur between the two materials. This relative displacement is depending on various factors, especially on the boundary conditions. Therefore, I think we cannot use a usual finite element method to generalized field where the effect of bond is important. For example, the beam-column joint is one of the most difficult portions. However, I would like to say again that this type of structure, such as reinforced concrete shear walls is not necessary to be tested anymore, because we already have the very good analytical model. I discussed with Professor Thürlimann last night that for this type of structures, a macroscopic model such as the truss model is to be used for design, and the nonlinear analysis can be used for checking the true behaviors or more detailed safety, as the calculation of the envelope requires only one hour by a personal computer of 16 bit. If we use a much better computer, it can be done more quickly.

Summary of Tuesday Afternoon Session

by S. Wood, Reporter

This session includes four papers on the effect of shear on energy dissipation and seven papers on the design for earthquake resistance. Two observations were given: First, the application of panel testing results to shear walls in building may be questionable for walls two-stories high or more, and second, the effect of decreasing shear strength with increasing deformation would demand further research.

Possible topics for discussion were:

- Determination of shear design forces in earthquake.
- Determination of shear ductility by performance demand?
- Effects of torsion in buildings.
- Behavior of walls with multiple openings under monotonic and cyclic loading.
- Repair and strengthening of structural frames by infilled walls.
- The importance of understanding the behavior of real structures under earthquake loading.

Discussion No. 1 — T. Hsu, University of Houston, U.S.A.:

When I was a student in China in the 1950's, earthquake design of a building was simply to fill in the structural frame with wall panels. This is what we call today the "framed shear wall concept." Research in earthquake engineering at that time was simply to find the best functional pattern for installing the wall panels. The patterns may be vertical or horizontal or checkerboard. As I understand, at that time, the Chinese learned this frame shear wall concept from the Japanese.

When I became a young engineer in the United States in the 1960's, I learned that the way to design a structure to resist earthquake was to make a structure ductile and that the desired ductility could be derived only from the bending deformations of the structural members. When a shearwall was required, we should build a cantilever shear wall and then attach the building frame to it. The forces in the wall and the frame could be analyzed by taking the compatibility at all story levels between the wall and the frame. The shearwall was supposed to resist the lateral loads by bending and to develop a flexural plastic hinge at the base. This new concept of cantilever shearwall sounded clear and correct to me. But I kept wondering what was wrong with the old framed shear wall concept. The overall perception at the time was that the wall panels in the framed shearwalls, which are subjected to shear, could not develop shear ductility.

Two major advances were achieved in the 1970's and 1980's to revolutionize our understanding of the membrane shear action in reinforced concrete. The first advance was the development of the various truss model theories, which gave us a clear concept of how the steel and concrete jointly resist the applied shear. The second was the extensive experimental research conducted around the world, leading to the understanding of the constitutive laws of reinforced concrete in shear. The combination of these two advances allowed us to predict the post-cracking shear deformations. Based on these theoretical and experimental research, I became convinced that wall panels subjected to shear could develop shear ductility if correctly designed, and that the old framed shearwall concept could be valid.

In the past few years, we, at the University of Houston, had made some parametric analyses of framed shearwalls. We found that the shear ductility of the wall panels decreased with increasing constraint of the surrounding frame, especially at the critical location adjacent to a large foundation. This is what Professor Jirsa is concerned about. In view of these findings, my enthusiasm for framed shearwalls had became more cautious. I could not understand how the wall and the frame interact. I am not sure whether sufficient shear ductility could be developed to make the framed shearwall a cost-effective structure. I could not agree with Professor Okamura that we don't need more tests on framed shearwalls.

My enthusiasm for framed shearwall is rekindled by Mark Fintel's presentation at this workshop. From 30 years of observations of building failures in earthquake he concluded that ".... not a single concrete building containing shear walls has ever collapsed. While there were cases of cracking of various degrees of severity; no lives were lost in those buildings. Of the hundreds of concrete buildings that collapsed, most suffered excessive inter-story distortions that in turn caused shear failure in the columns. Even where collapse of frame structures did not occur and no lives were lost, the large inter-story distortions of frames caused significant property damages." It appears that framed shearwalls are just as effective as cantilever shearwalls in resisting earthquake. However, we are far from understanding how the framed shearwall works. Since the framed shearwall concept came originally from Japan and many structures had been built based on this concept, I would like to learn from the Japanese delegation their experience with such structures.

Another point. Professor Okamura said that his methodology of modelling the behavior of framed shearwalls was to first model the small elements of steel and concrete, and then build them up to a composite structure by the finite element method. I agree that this is a valid method. In fact, in our workshop paper we have used Professor Okamura's micro-model of treating steel stiffened by concrete. Using his micro-model, we formulate the macro-model of constitutive laws for reinforced concrete as a whole. In my opinion, there is no inconsistency between micro-model and macro-model.

Discussion No. 2 — T. Tanabe, Nagoya University, Japan:

Relating to the energy absorption which is calculated from the load and displacement hysteretic curves, we are not successful enough to pursue this curve until the last end of the load. Almost all the calculations of experiment stop somewhere around the maximum point. On the other hand, my calculations showed that there is a phenomenon of snap back to occur. If this is the case, a final point of loading — I mean the zero load after the failure — looks like to come back almost to the original point. Then, no energy absorption. So I think we should be, if we want to be accurate, able to pursue numerically the curve to the zero load. Otherwise, I think we are not so much accurate enough. My calculations also showed that there exists some kind of out-of-plane mode of failure. I found out in the paper of Mrs. Fang from Tsinghua University that out-of-plane failure mode of the shear walls was observed. I think we are not so much informed yet really to predict the ultimate condition of shear-failing-type members of reinforced concrete. So, I am partly opposing what Professor Okamura said.

Comment by T. Paulay, University of Canterbury, New Zealand:

May I intervene for one minute to express my regret that Professor Fang could not attend our deliberations. Most of you may not know it, but she is a lady who has been involved in research relevant to shear wall structures for many years and has accumulated an enormous quantity of experimental evidence, largely using micro-concrete models, sometimes up to twenty stories. I just wish to report to you that she has made important contributions to this area. It is very sad that we couldn't welcome her here.

Discussion No. 3 — P. G. Gambarova, Polytechnic University of Milan, Italy:

Talking about shear behavior of plane, thin elements subjected to shear with or without torsion, may you comment on possible interaction between in-plane shear and torsion?

Though a certain level of torsion has to be expected in plane, thin elements subjected predominantly to shear, it seems to me that scanty test results are available: I remember the tests carried out by Prof. Leonhardt (rectangular sections with a side ratio of 5-6) and I can recall some tests carried out by Prof. Iori in our Department (with a side ratio of 10). I performed also a few tests on P/C beams with open-type section (channel beams). In the former cases, a relatively complex spatial truss forms; while in the latter case, the initial behavior is in good agreement with the plane-section assumption, and the subsequent behavior is more complex, since, after skin cracking, warping torsion becomes predominant.

On shear-torsion interaction, Prof. Thürlimann and Dr. Maier may be probably more precise, since they have carried out many well-known tests on shear-walls.

Discussion No. 4 — B. Thürlimann, ETH Zürich, Switzerland:

Statically a building is nothing else than a cantilever turned upside through 90 degrees. If you have a cantilever beam, the dead load is acting perpendicular to its axis. Now you turn the beam up, you have a building subjected to horizontal forces. In addition, some axial load is acting on it.

In the case of a building with a box section subjected to torsion, the shear flow in the four sides can be determined using the theory of circulatory torsion. By superposition of the shear due to transverse load and the shear flow due to torsion each plate can be analyzed and designed separately.

The influence of a secondary effect due to warping of the flat sides into hyperbolic paraboloids needs consideration. The imposed curvature reduces the strength of the concrete compression field. The effective concrete strength should be reduced accordingly.

About some 20 years ago, we conducted tests on box beams to study their behavior and strength. These tests should apply to buildings with a box section as well as indicated above.

References:

(1) Paul Lüchinger and Bruno Thürlimann, "Versuche an Stahlbetonbalken unter Torsion, Biegung und Querkraft (Tests on R. C. Beams subjected to Torsion, Bending and Shear)," Report No. 6506-5, Inst. of Structural Engineering, Swiss Federal Institute of Technology, Zurich, July 1973

(2) Paul Lüchinger, "Bruchwiderstand von Kastenträgern ans Stahlbeton unter Torsion, Biegung und Querkraft (Strength of R. C. Box Beams subjected to Torsion, Bending and Shear)", Dissertation No. 5886, Inst. of Structural Engineering, Swiss Federal Institute of Technology, Zurich, 1977

Discussion No. 5 — D. H. Kim, Sunkyong Ltd., Korea:

I would like to make some brief comments on what we discussed in this session.

First, any panel, whether it is a wall or a slab, it may be under in-plane shear, out-of-plane bending and shear, and any kind of boundary forces and displacements. So, instead of considering the in-plane shear problem separate from the out-of-plane bending, we may treat the panel as a three-dimensional problem. Instead of handling it by either "truss model" or "arch model" which can yield only approximate in-plane stresses, we may approach the problem by the laminate theory. Complete solution of the three-dimensional problem may be possible by the use of laminate theory.

The reinforcement (type, distribution, sizes, and orientation) plays important role for the panel strength. We may be able to design the optimum structure by utilizing laminate theory.

The problem of a panel with openings can be solved. There are many theories available.

In fact, any two-dimensional problems, including vibration and buckling problems, with any boundary conditions, including mixed boundaries, can be solved.

The problem is the panels in a building interact with the frame. The panel, a wall or slab, is with unknown boundary conditions which must be given by the enclosing frame. These boundary conditions of the walls or the slabs can be solved only if we can solve the whole structure, including the frames and panels which are interacting together. We may have to stress to put more effort to develop the method of analyzing such three-dimensional problems of whole structure. Even if we get such solution, the boundary condition of the panels will be a mixed one. Solving panel problems, including in-plane shear and out-of-plane bending, with boundary conditions of displacements and forces mixed, may turn out to be exciting to the engineers.

Discussion No. 6 — J. O. Jirsa, University of Texas, U.S.A.:

The interaction between frames and panels comes up very strongly in an area that hasn't been mentioned much here. When we are looking for problems in the earthquake area that are likely to get increasing attention, I think we have to look at the area of repair and strengthening. If we take existing structures, one of the things we always identify are the weaknesses, and very often those involve shear — in all kinds of different ways. It may be shear in joints, shear in elements with inadequate transverse reinforcement, or whatever. And as soon as you start talking about adding some sort of new strength to that structure, it almost always boils down to transferring shear — either between interfaces or from one material to another. For example, you might take a frame system and change it to a wall system. There is a great need for information to evaluate shear strength of existing systems. We need to have some sort of program to estimate what is the state of the current structure. I am very impressed with Dr. Okamura's programs and what he's able to do from his experience. As long as we hire him, I feel confident that we'll have good results. But the ability to evaluate using a finite element analyses or whatever approach, provides a way of getting a handle on the problem. Then there are all these other practical and design problems involving modifying the structure and satisfying the shear requirements that make the structure work.

Discussion No. 7 — Irina Scherbakova, Interpreter for T. Zhunusov, KazNIISSA, U.S.S.R.:

On behalf of Professor Zhunusov, of the Soviet Union Kazakhistan, I would like to say a few words. He was highly impressed by all the reports made here and especially he liked the equipment and devices which you showed on the slides, which induce force effects. But, as his recommendation, he would like to pay attention to the behavior of structures in case of real earthquakes. He is well-acquainted with the consequences of such disastrous earthquakes, which have taken place lately. For example, Armenia (1988), California (1989), Northern Iran (1990), and the last earthquake in his republic, just a few months ago, in eastern Kazakhistan (1990). And the analysis of all these earthquakes proves the important role of the diaphragms, or you call it shear walls. However, the application of these diaphragms, connected with the columns of the frame, may lead to brittle failure. That's why they study composite diaphragms. And toward this topic, his report was devoted.

In conclusion, he would like to stress the importance of carrying out research that deals with seismic effects, and we must take into account these factors, as well as the constructive models of buildings.

Discussion No. 8 — T. Paulay, University of Canterbury, New Zealand:

As chairman may I have the privilege of making a simple contribution myself to this fascinating, interesting issue which designers face? What shear force should we design for? Once we have it, half the problem is solved, but what is it? You describe very clearly two types of structures for which design and performance criteria are really quite different. When you compare a nuclear power facility with a thirteen- or fourteen-story building in Los Angeles, of course, the design lateral forces are very much smaller for the building. We accept these with the full knowledge that we will have to pay for it by having to accept much larger deformations. Therefore the design is based on ductility. Once you do have a ductile structure, it is really not that difficult to find out what the shear force is. Dr. Ghosh gave you an example of a simple cantilever. You know what is the flexural strength and you can work out what is the maximum force which is applied to a certain level to mobilize, to develop this flexural strength and then you can design for that shear.

I would like to suggest a research topic. A great deal has been said about ductility and I am not going through the details of ductility from shear but would emphasize ductility in general for structures in earthquake regions, particularly buildings which are designed for a relatively low lateral-force resistance. We in New Zealand share your concern for uncertainty in predicting earthquake-induced actions. The code tells us what to do, but is a code to be believed? A severe earthquake does not read the code, so it may impose something else than what the code envisaged. Yet, we would like to have a good structure. I believe that if you can provide large ductility capacity, then you provide a tolerant structure. Within this discussion I cannot explore all the means by which all shear dominant structures can increase their ductility capacity. One simple design approach attempts to develop a structural system with a ductility capacity which is in excess of what the maximum demand might be. In this case, we may be less concerned about the dynamic response, about the future intensity of the earthquake, because we know that we can accommodate it. The trouble starts when the ductility demand may exceed the structure ductility capacity and thus we may end up with a brittle failure. I feel that there is much promise from the point of view of the designer to explore by research the means by which stable ductility, to improve hysteretic damping, can be increased, hopefully, to a magnitude which is larger than what we have budgeted for.

OPEN SESSION DISCUSSION

by J. C. Walraven, Delft University of Technology, The Netherlands

We are planning a number of research programs on shear. The first project concerns the development of rational models for lightweight concrete members. The application of lightweight concrete is again actual in our country, after a period of years in which it was not competitive on the building market. The new interest in lightweight concrete can partially be explained by the fact that we are running out of our traditional gravel aggregate. Moreover, lightweight aggregate particles can be made from fly-ash, which is a waste product. Tests are carried out on large flanged thin-webbed beams. It is investigated if rational models, as developed recently by Collins, Kupfer/Bulicek and Gambarova, can be extended to lightweight concrete members. A major aspect being studied is the capacity of cracks to transmit shear forces.

The second subject is the behavior of statically indeterminate deep beams. A few years ago, some articles were published by MacGregor and Rogowski, dealing with plastic truss models and their application. Our research can be seen as a continuation of their research. Key words are "superposition of strut/tie and truss systems," "shear ductility," "sensitivity to differential settlements."

The third subject is high strength concrete. We will carry out tests on aggregate interlock to study the reduction of the interlocking capacity as a result of a particle fracturing. Another aspect is the softening behavior of HSC in tension and its significance for the explanation of size effects. Furthermore, we intend to study the crushing behavior of HSC-webs. I would like to encourage the conduction of panel tests on high strength concrete. The equipment available here in Houston offers excellent facilities for such tests!

WRITTEN DISCUSSION

by A. H. Mattock, University of Washington, U.S.A.

The various relations between maximum concrete compressive stress and principal tensile strain, ε_1, shown in Fig. 8 of Professor Collins' report, all indicate a maximum value of concrete compressive stress equal to f_c'. This is justified by the results of the panel tests. This maximum value of concrete compressive stress is not solely a property of the concrete. If it were, the maximum value would be 0.85 f_c', the uniaxial compressive strength of the concrete. The higher values obtained in some of the panel tests with low values of ε_1 must result from lateral restraint of the concrete by the reinforcement when the concrete dilates rapidly just before failure. Some of the scatter in the test data in this region is probably due to variation in the amount of reinforcement in the panels. Unless there is some appropriate minimum panel reinforcement requirement, it would appear appropriate to use an upper limit of 0.85 rather than 1.0 for the expression relating f_{2max} and ε_1. The upper limit of 0.9 used by Hsu and his associates is more reasonable.

Post-Workshop Submission

**A Supplement to Herbert Kupfer and Hans Bulicek's Paper,
entitled "A Consistent Model for the Design of Shear Reinforcement in
Slender Beams with I- or Box-Shaped Cross-Section"**

DESIGN PROPOSAL

based on a uniaxial compressive stress field with the inclination ϑ leading to a required shear reinforcement ratio of

$$\omega = \nu \cdot \tan \vartheta$$

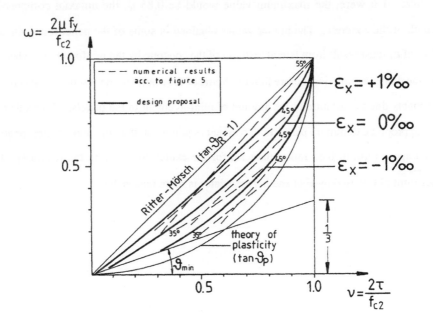

The following formula for calculating ϑ results from a linear interpolation of $\tan \vartheta$ between $\tan \vartheta_R = 1$ and $\tan \vartheta_P$ in terms of ϵ_x by taking into account unfavourable crack inclinations φ :

$$\tan \vartheta = 1 - \frac{2 - 10^3 \epsilon_x}{4} (1 - \tan \vartheta_P)$$

$$\not< \tan \vartheta_P = (1 - \sqrt{1 - \nu^2})/\nu$$

$$\not< 1/3$$

safety factors have to be added according to the relevant codes

INDEX OF CONTRIBUTORS

Printed and bound by CPI Group (UK) Ltd, Croydon, CR0 4YY
01/11/2024
01782615-0018